Fundamentals of Automatic Control

Fundamentals of Automatic Control

Robert C. Weyrick
Community and Technical College
The University of Akron

McGraw-Hill Book Company

New York
St. Louis
Dallas
San Francisco
Düsseldorf
Johannesburg
Kuala Lumpur
London
Mexico
Montreal
New Delhi
Panama
Paris
São Paulo
Singapore
Sydney
Tokyo
Toronto

Library of Congress Cataloging in Publication Data

Weyrick, Robert C
Fundamentals of automatic control.

1. Automatic control. I. Title.
TJ213.W43 629.8 74-10896
ISBN 0-07-069493-1

FUNDAMENTALS OF AUTOMATIC CONTROL

101112 KGPKGP 89

The editors for this book were Alan W. Lowe and Alice V. Manning, the designer was Tracy A. Glasner, and its production was supervised by Valerie Klima and Laurence Charnow. It was set in Medallion by York Graphic Services, Inc. It was printed and bound by Kingsport Press, Inc.

To my mother

Contents

Preface

Automatic control is an essential part of modern technology. A machine tool, a chemical process, a space vehicle—virtually all segments of industry rely on automatic control systems to meet high performance requirements.

Today's control systems received much of their impetus from the military demands of World War II, and a well-established body of linear control theory evolved during the 1940s and 1950s. Automatic control methods based on this theory have proven themselves in practice and have been widely applied. Although recent developments have centered on the control of complex nonlinear systems, the classical linear methods remain the foundation of automatic control, and it is the purpose of this text to present the fundamentals of these methods of analysis and design.

Engineering technicians are increasingly involved with control systems, not only from the standpoint of testing, calibration, and maintenance but also in the more standardized aspects of design. For this reason many engineering technology programs include an introduction to automatic control. Since the mathematics content of these programs generally includes an introduction to calculus, the subject of automatic control can be presented to technicians at a level that combines a mathematical basis with intuitive reasoning.

The various books on automatic control that have been published number well into the hundreds. Therefore, the question may be raised, "Why another book on this subject?" A cursory examination will reveal that most available texts are written at an engineering level. They emphasize the theoretical aspects of control systems and assume a thorough working knowledge of calculus and differential equations. Only a limited number of control books have been directed toward the needs of the technician, and these books tend to be descriptive in nature and to make minimal use of mathematics. This book attempts to bridge the gap between the elementary works and advanced works.

Although the theory and methods in the book are intended for a general controls application, the servomechanism is used throughout as a representative control system. Not only are servomechanisms extensively employed in industry but they are also well suited to the laboratory experience which is highly recommended in the study of automatic control.

In regard to prerequisites, a basic knowledge of physics, including mechanics and electricity and magnetism, is assumed. Some background in electronics would be helpful in understanding amplifier topics; however, this is not vital to the overall subject. The book does not require an extensive mathematics background but does suppose a sound knowledge of algebra, trigonometry, and complex numbers. An introductory knowledge of calculus will be helpful in studying control theory from a mathematical point of view; however, this need not be extensive. A familiarity with the concepts of differentiation and integration should prove adequate. Differential equations are used in the text; however, they are introduced and explained as required for an understanding of the subject matter.

The first chapter of the book introduces the basic concept of negative feedback in automatic control and the block-diagram representation of control systems. Chapter 2 is concerned with the representation of physical systems by means of mathematical models. Chapter 3 describes a number of the devices used in automatic control categorized as transducers, power actuators, and amplifiers.

Chapter 4 describes the transient performance of simple control systems with a minimum of mathematical development. This discussion is continued in Chap. 5, which deals with the widely used Laplace transform method of time-domain analysis. This method is not covered rigorously but is presented as a working tool with emphasis on its application to control systems.

The frequency-response method of analysis and design is extensively covered in Chaps. 6 and 7. This method is largely graphical in nature and therefore should be readily comprehended and applied by the engineering technology student.

Chapter 8 discusses the testing of control systems from a hands-on point of view.

Nonlinearities and several other topics of an advanced nature are briefly introduced in the concluding chapter. Although a thorough treatment of nonlinearities is beyond the scope of this book, the technician should be aware of their presence and the limitations that they impose on linear analysis.

The book is appropriate for a one-term course in automatic control. Depending on the level of the students and the scope of the course, it would not be necessary to include all chapters. For a less analytical approach, Chaps. 2 and 5 may be largely deleted without loss of continuity. Chapter 3 could be covered selectively, omitting the sections covering devices not common to servomechanisms.

Although the book is specifically intended for the engineering technology student, it should also be useful to the practicing engineer or technician who wishes an introduction to automatic control.

The uniqueness of a basic text is not in the contents as such but in the presentation in a manner appropriate to a certain group of readers. For the subject matter of this book, the author is generally indebted to the many individuals who have contributed to the field of automatic control. In particular I would like to single out my former associates at the Goodyear Aerospace

Corporation with whom I worked on numerous control projects ranging from missile guidance to industrial process control.

I must also express sincere appreciation to my aunt, Miss Helen Hassler, for her indispensable help in converting the manuscript to typed form. Finally I am most grateful to my wife, Helen, for her continuous encouragement and understanding during the seemingly endless task of preparing the manuscript.

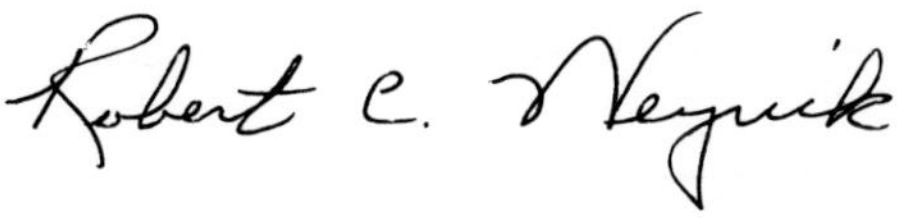

Fundamentals of Automatic Control

Chapter One

introduction

1-1 GENERAL INTRODUCTION

Automatic control is playing an ever-increasing role in almost all phases of our modern way of life. From familiar household uses such as the automatic bread toaster and thermostatically controlled furnace to sophisticated control systems for power generation and space exploration, automatic control influences our daily living.

In a sense, the advent of automatic control represents a second industrial revolution. The industrial revolution of the nineteenth century made greater quantities of power available for man's use. Wind power and water power were largely displaced by the introduction of steam power. Fuels including coal, gas, and petroleum and its byproducts became the primary sources of energy for manufacturing and transportation. The form of energy that has played the greatest part in industrial development is electricity, and this is likely to continue with the conversion of nuclear energy to electrical form.

To utilize power effectively, it has been necessary to learn to control and regulate it. The industrial revolution of the twentieth century has been marked by man acquiring this ability. With power and the means of control of it, much of the physical and mental work can be done by machines with a higher degree of performance than would otherwise be possible.

Many persons engaged in industrial activities need to work with control systems, if only to use them. Consequently, to the engineer and technician a knowledge of automatic control is useful for its own sake. However, several other benefits are also to be acquired. Automatic control systems are physical systems which have dynamic behavior. Hence the study of control theory provides a basis for the general understanding of dynamic systems. Such understanding is vital since a system must be understood in order for it to be controlled effectively.

Control systems frequently employ components of many types, including mechanical, electrical, electronic, hydraulic, pneumatic, and combinations thereof. Hence, one who works in the control field must be familiar with the principles and characteristics of a broad range of devices.

Automatic control is an interdisciplinary subject since it draws on knowledge from many differing fields. Its study can be beneficial in bringing together a number of subjects that have been considered separately and applying them to a common problem.

The philosophy of automatic control also has other far-reaching implications. Although we will be entirely concerned with engineering applications of automatic control, its concept is being extended to other areas. Economic, business, and political systems are being increasingly regarded in terms associated with automatic control. Although the variabilities of human behavior diffuse the analysis of such systems, as the understanding of their dynamics increases, the ability to control them will increase also.

In addition the human body itself includes numerous complex control systems among its functions. The control of body temperature, chemical composition, and physical actions such as driving an automobile are examples of such body systems.

1-2 HISTORY OF AUTOMATIC CONTROL

Although earlier applications may have existed, the first control system not involving a human being is considered to be the flyball governor developed by James Watt in 1788. Watt concluded that a man opening and closing a steam valve was not the best way of keeping the speed of a steam engine constant. His governor, shown in Fig. 1-1,

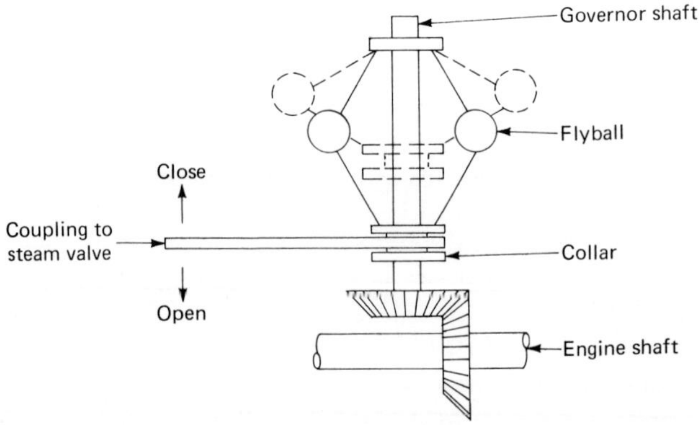

Figure 1-1 Flyball governor

is a completely mechanical device that connects to the shaft of a steam engine. As the speed of the governor shaft increases, the balls rotate faster and move outward. The coupling to the steam control valve then moves up the shaft, closing the valve and slowing the engine. A decrease in speed results in the opposite motion of the flyballs and a corresponding increase in engine speed.

It soon became apparent that an unfortunate byproduct of attempts to improve the accuracy of automatic control is instability. This is characterized by a tendency for the system to oscillate or hunt about a specified operating condition. In the case of governor control of the steam engine, the speed would tend to rise and fall about the desired value. Early efforts to eliminate this problem were largely on an intuitive or trial-and-error basis.

The first theoretical analysis of automatic control was a study of the stability of the flyball governor by James Clerk Maxwell in 1868.[1] Shortly thereafter other significant contributions to the theory of automatic control were made by Hurwitz in 1875, Routh in 1884, and Liapounov in 1892. Further applications of governors to other engines soon appeared; and, in the early 1900s, their application in industrial control began. About the same time, automatic control devices were studied for such uses as ship steering.

Modern interest in control is generally regarded as beginning in 1934 when Hazen,[2] drawing on the work of earlier originators, published the first precise theory of automatic control systems. About this time work by Nyquist,[3] Black,[4] and Bode[5] in the design of stable amplifiers for telephone systems gave added impetus to the development of feedback theory.

The demands of World War II greatly accelerated work in the field of automatic control. In particular the need for precise and rapid control of ships, aircraft, and radar antenna systems led to significant advances in theory and practice. With the lifting of wartime security restrictions in 1945, rapid progress has taken place in the control field. Since that time many books and innumerable articles and papers have been written and the application of control systems in the industrial and military fields has been almost without limit.

1 *J. C. Maxwell, "On Governors," Proc. R. Soc. (Lond.) vol. 16, 1868, p. 270.*
2 *H. L. Hazen, "Theory of Servomechanisms," J. Franklin Inst., vol. 218, September 1934, p. 279.*
3 *H. Nyquist, "Regeneration Theory," Bell Syst. Tech. J., vol. 11, January 1932, pp. 126–147.*
4 *H. S. Black, "Stabilized Feedback Amplifiers," Bell Syst. Tech. J., vol. 13, January 1934, pp. 1–18.*
5 *H. W. Bode, "Amplifiers," U.S. Patent 2,123 178 (1938).*

During the 1940s and 1950s two distinct yet interrelated approaches to the analysis and design of control systems evolved. Development in the United States centered on frequency-response methods, while much significant work in the U.S.S.R. used transient analysis or time-response techniques. In recent years these two approaches have become melded, and much contemporary analysis is based on simultaneous application of both methods.

Much in the way of modern automatic control has been made possible by two parallel developments: semiconductor devices and electronic computers. The successive use of electron tubes, transistors, and integrated circuits has provided control equipment and measuring instruments offering precision, reliability, and economy. The advent of the computer age has had a marked influence on control systems. High-speed digital computers have extended the realm of control to large and complex systems involving a number of interrelated quantities. Digital computers are finding extensive application in industrial process control systems in which many variables are measured and controlled simultaneously to achieve optimum performance.

1-3 OPEN-LOOP AND CLOSED-LOOP CONTROL

If we examine the word *control,* we find several meanings given in the dictionary, e.g., command, direct, govern, and regulate. Thus, a control system may be regarded as a group of physical components arranged to direct the flow of energy to a machine or process in such a manner as to achieve the desired performance.

The word *automatic* means self-moving or self-acting; thus an automatic control system is a *self-acting* control system.

An important distinction applied to control systems, whether automatic or otherwise, is that between *open-loop* and *closed-loop* operation. Automatic control, including this distinction, can perhaps be best introduced by means of a simple example.

Assume that the oven shown in Fig. 1-2 is heated by an electric heater controlled by a switch that provides several levels of current to the heating element. The setting of the switch represents the *input quantity* since it activates the system to produce the output. The temperature of the oven is the *output* or *controlled quantity.* Since the current to the heater is the quantity being altered, we can think of it as the *manipulated quantity.*

For a given setting of the control switch the oven temperature will reach a value related to the heater current and heat losses through the oven walls. If the temperature is unsatisfactory, this fact by itself can

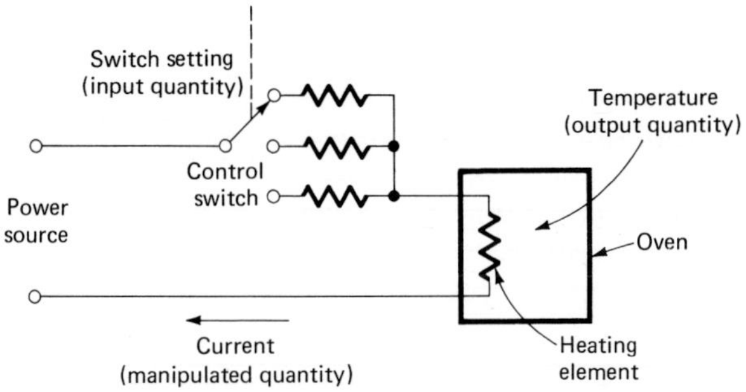

Figure 1-2 Oven-temperature control, open-loop

in no way alter the input to the oven control. Thus it can be said that the output quantity has no effect on the input quantity. The control in this case is identified as an *open-loop control system*. If some condition such as the ambient temperature surrounding the oven should change, the oven temperature will also change. Thus the open-loop control system cannot correct for changes that disturb the controlled quantity.

A human being can be added to the system for the purpose of maintaining the oven temperature at a desired value. This is shown in Fig. 1-3. By observing a thermometer within the oven, the person could alter the position of the control switch to more nearly achieve the required temperature. It is important to note that the addition of the human operator has provided a means by which the output is *fed back* and *compared* with the desired value. Actual control of the heating element depends on the *error* or *difference* between the desired and actual temperatures. Any necessary change is made in the direction

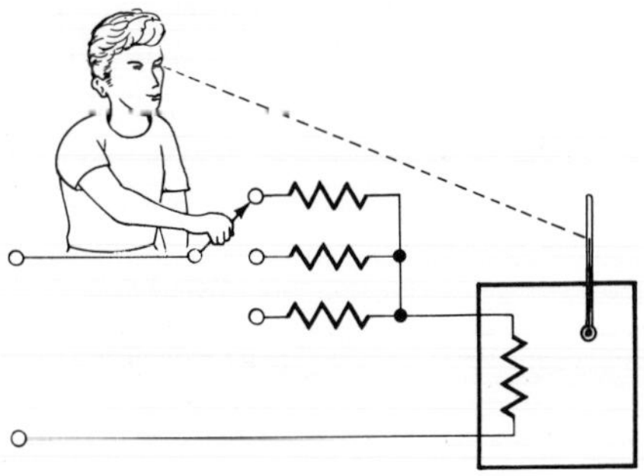

Figure 1-3 Oven-temperature control, manual closed-loop

of reducing this error. For example, if the temperature were low, the heater current would be increased to provide more heat.

Systems in which the output has an effect upon the input are called *closed-loop control systems* or, more commonly, *feedback control systems*. The feedback nature of the system in Fig. 1-3 may be seen by tracing the closed loop. The function of comparing the actual temperature with the desired value is performed in the mind of the operator, and the command is executed by the response of his muscles. Even though closed-loop control of the oven is established by the operator, the results may be less than satisfactory. The temperature will cycle above and below the desired value as the operator adjusts the switch even though the excursions would tend to decrease with operator experience. A change in the level of the power source or in the ambient temperature surrounding the oven would adversely affect the system's output and cause the operator to "relearn" a switching pattern. In addition, watching a thermometer and operating a switch would be a dull and time-consuming task. This arrangement could be described as a *manual closed-loop control system*.

To further improve performance and obtain more precise control, the human may be replaced by a mechanical, electrical, or other form of comparison and control unit. Figure 1-4 shows the oven control system with closed-loop control implemented on an automatic basis, that is, without human intervention. The temperature is now measured by a thermocouple, a device that generates an electrical voltage proportional to temperature (see Sec. 3-10). This voltage is fed back and compared with a reference voltage that represents the desired temperature. The difference between the two voltages is amplified electronically and controls the current to the heating element. Since the voltage feedback is subtracted from the reference voltage, it is known as a *negative-feedback signal*.

The example has illustrated how automatic closed-loop control is

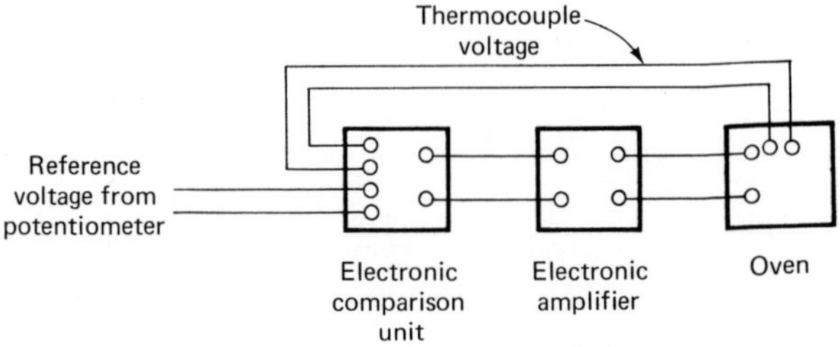

Figure 1-4 Oven-temperature control, automatic closed-loop

employed to perform a task more rapidly and consistently than can be accomplished by a human being. It also frees an individual from a rather menial task. The use of negative feedback in a closed-loop control system tends to maintain a desired value of a quantity or condition by measuring the existing value, comparing it to the desired value, and using the difference as a means of initiating action to reduce the difference. *The negative-feedback concept has become the foundation for automatic-control-system design.*

1-4 CONTINUOUS AND DISCONTINUOUS CONTROL

Within the distinction between open-loop and closed-loop control, a further classification may be made. This is based on the manner in which the actuating signal is used to control the transfer of energy from source to load. Two categories of control that are possible are *discontinuous* or *digital* control and *continuous* or *analog* control.

A discontinuous or digital signal varies in a discrete manner and may take only certain discrete values between its limits. The simplest form of discontinuous control is *on-off control* in which the function of the input signal is to turn the power flow off or on. Once this switching has been carried out, no further control is possible other than reversing the operation. Since only two discrete states are available, the operation may be regarded as discontinuous. The lamp circuit shown in Fig. 1-5*a* is an example of on-off control.

Another form of discontinuous control is *step control* in which energy transfer is possible at one of several possible rates. For example, a transformer with a tapped secondary and a multiposition switch may be connected to the lamp as shown in Fig. 1-5*b*. In this case a number of voltages are available which provide discrete increments of power to the lamp. In addition the oven-temperature control shown in Figs. 1-2 and 1-3 represents a step control system.

With analog control the signals throughout the system vary in a continuous manner, and any value between the available maximum and minimum can be obtained. In Fig. 1-5*c* a variable resistor is placed

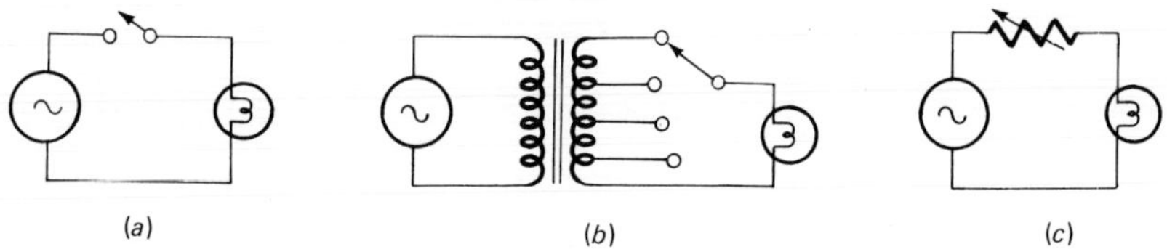

Figure 1-5 (a) On-off, (b) step, and (c) continuous control of a lamp

in series with the lamp. A change in resistance changes the rate of energy transfer to the lamp in a continuous manner. The oven control in Fig. 1-4 is also an example of continuous control.

In this text we shall consider only continuous control systems for several reasons. Continuous control is employed much more widely than discontinuous control, especially in basic control applications. In addition continuous control readily lends itself to an understanding of feedback control theory using relatively uncomplicated mathematics. The same is not true of the study of discontinuous control. As a consequence discontinuous control is looked upon as an advanced field of study not generally within the scope of a basic controls text.

1-5 BLOCK DIAGRAMS

Control systems are composed of a number of interrelated components. Before proceeding further, we should become familiar with the representation of a system for purposes of analysis and design. To simplify the visualization of a system, it is often illustrated by a number of interconnected boxes or blocks. Such a representation is known as a *block diagram*. The blocks do not necessarily represent a complete device but sometimes identify a function that occurs in the system. A block may be considered individually only when its output is essentially unaffected by connection to the input of another block. When interconnected by directed line segments, the blocks show the functional operation of the system. Each connecting line represents some quantity or variable in the system, and the arrow shows the unidirectional flow of information. Thus the functional information presented in a block diagram tends to stress *what* is happening in the system rather than showing the means by which it is accomplished.

The ratio of the output signal from a block to the corresponding input indicates the manner in which the signal is transferred through the block. Hence this ratio is defined as the *transfer function* of the block. Since systems invariably have a number of interconnected blocks, the

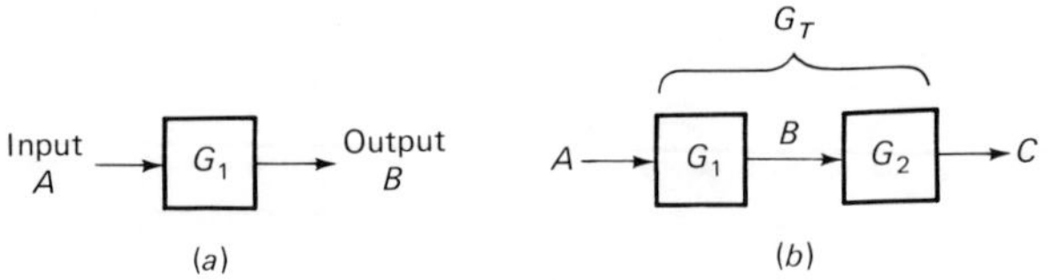

Figure 1-6 (a) Individual block and (b) cascaded blocks

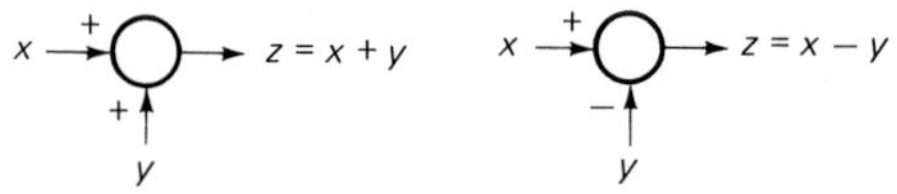

Figure 1-7 Summing-point representation

analysis is more readily carried out if a shorthand notation for the transfer function is introduced. A common method of presentation uses the letter G with a suitable subscript. Figure 1-6*a* shows a block that exhibits a simple proportional or gain function having magnitude G_1. For this block

$$\text{Transfer function} = \frac{\text{output}}{\text{input}}$$

or

$$G_1 = \frac{B}{A} \tag{1-1}$$

where A and B are the input and output quantities, respectively.

Before proceeding further, it should be pointed out that relatively few system elements possess a transfer function wherein the output responds immediately to a change in input. Typically a time delay or lag is present in the output response. Therefore the transfer function should have provisions for characterizing the time-dependent or dynamic behavior of the block. Such transfer functions will be developed in later chapters. For the immediate purpose of presenting block-diagram representation, we will limit transfer functions to those having a gain characteristic only.

When blocks are connected in cascade—that is, the output of one block is coupled to the input of another block—their transfer functions may be combined. The overall transfer function is the product of the individual transfer functions of the blocks. This applies to any number of blocks in cascade that have linear transfer functions (linearity will be discussed in Sec. 2-2). As an example, the overall transfer function G_T of the cascaded blocks in Fig. 1-6*b* is

$$G_T = \frac{C}{A} = \frac{B}{A} \times \frac{C}{B} = G_1 \times G_2 \tag{1-2}$$

A point for summing information is indicated by a circle on the block diagram. Two or more variables may be added or subtracted at a *summing point*. Figure 1-7 shows two variables being combined. A variable having a plus sign at a junction goes through the junction without change in sign; a variable having a minus sign goes through with a sign reversal.

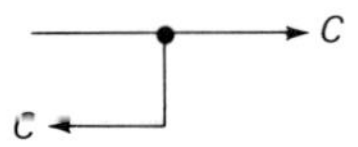

Figure 1-8 Takeoff-point representation

The quantities being combined at a summing point must be in the same units; for example, two voltages may be combined but a voltage and a current cannot be brought together. Any number of variables may enter a summing point.

A *takeoff point* is used when the output of a block is applied to two or more blocks. A takeoff point is simply represented by a dot as shown in Fig. 1-8.

To illustrate the block-diagram representation of a system, let us again examine the oven-temperature control described in the preceding section. This system is shown in block-diagram form in Fig. 1-9. In this particular diagram the various blocks may be identified with elements of the system. In addition the units associated with the transfer function of each block are shown in parentheses.

The block diagram tends to clarify the physical understanding of the system and provides a convenient basis for system analysis. We shall also see that block diagrams are useful in pointing out the similarity between apparently unrelated systems. It should be emphasized that the lines connecting one block with another represent the flow of control information within the system. The main sources of energy for the system need not be included in the block diagram.

1-6 SIMPLIFICATION OF BLOCK DIAGRAMS

The block diagram that is initially drawn for a system may contain a large number of blocks and signal paths and be more complicated than is desirable. In such cases a simplification may be performed to reduce the block diagram to a form with fewer blocks. Rearranging

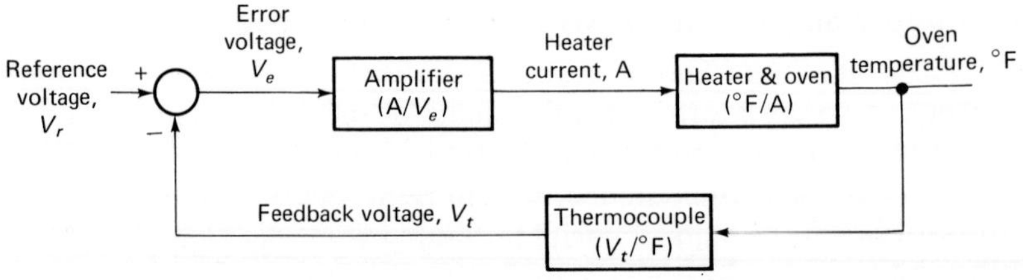

Figure 1-9 Block diagram of oven-temperature control

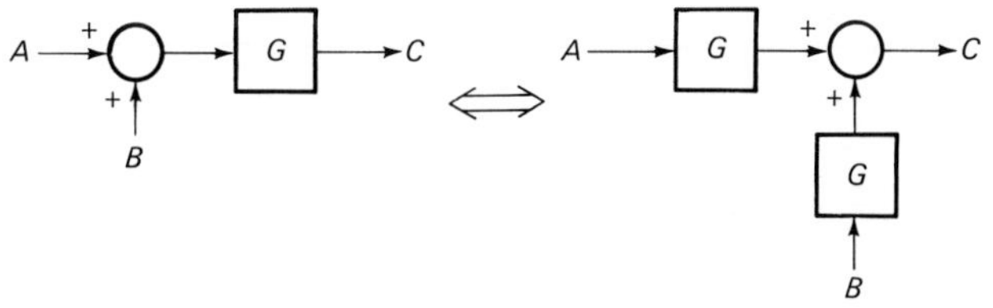

Figure 1-10 Moving summing point around a block

a system diagram to effect simplification is termed *block-diagram algebra* since it is analogous to simplifying algebraic equations. However, reducing a block diagram has the advantage of providing a better understanding of the interrelationships of the various elements in the system as compared to simplifying the system equations.

The combining of cascaded blocks into a single block, as described in Sec. 1-5, is obviously one step toward simplification. As another possibility, it is sometimes useful to move a summing point around a block as indicated in Fig. 1-10. Note that the inputs A and B are introduced through blocks incorporating the function G around which the summing point was moved. Applying the distributive property of algebra, we see that $C = (A + B)G = AG + BG$. Thus the two diagrams are equivalent.

Table 1-1 gives a number of transformations useful in simplifying block diagrams. These transformations can be verified by showing that the outputs from the two equivalent diagrams are the same. Note that the original and equivalent identities can be used interchangeably.

1-7 CLOSED-LOOP TRANSFER FUNCTION

Since certain functions and types of variables are commonly associated with feedback control systems, a generalized block diagram may be formulated and the associated closed-loop transfer function derived. Some standardization of the symbols and terminology relating to feedback control systems has been achieved and is used in this case.

Figure 1-11 is a general block diagram of a closed-loop control system. It is important that the terms used in this diagram be clearly understood.

System Variables

The *desired value* V is an external signal applied to the system to command a specific output from the process. The desired value is sometimes called the system *command* or *set point*.

TABLE 1-1 BLOCK-DIAGRAM IDENTITIES

Transformation	*Original*	*Equivalent*
1. Combining blocks in cascade	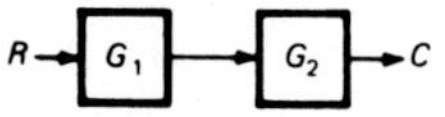	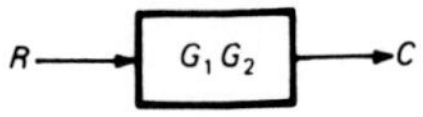
2. Combining blocks in parallel	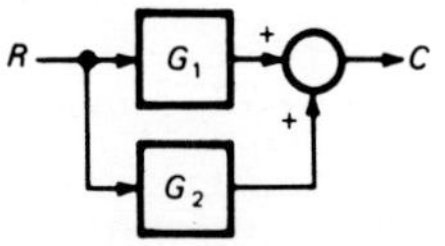	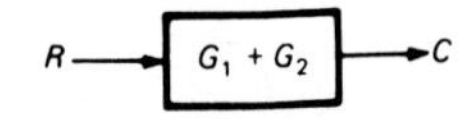
3. Moving summing point beyond a block	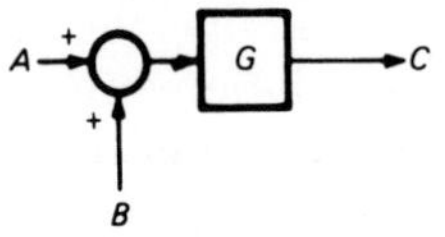	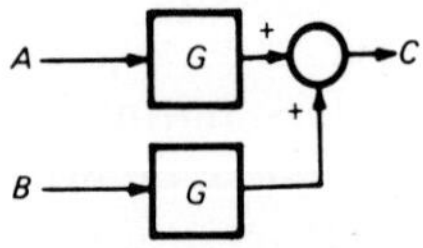
4. Moving summing point ahead of a block	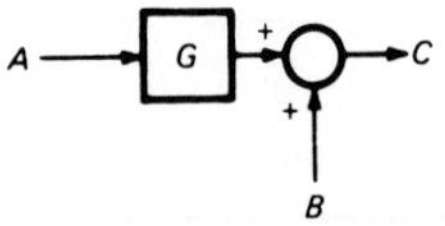	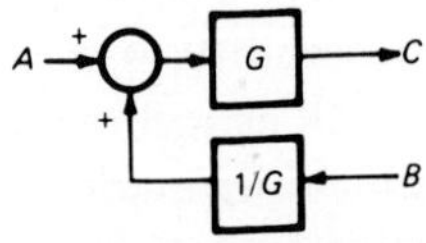
5. Moving takeoff point beyond a block	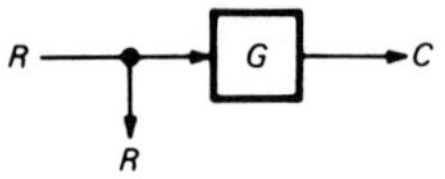	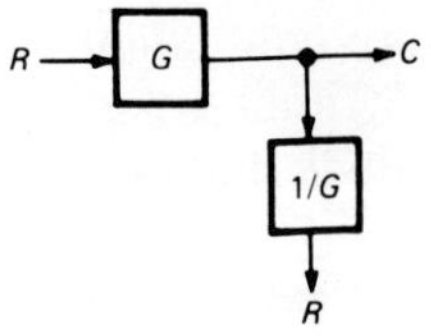
6. Moving takeoff point ahead of a block	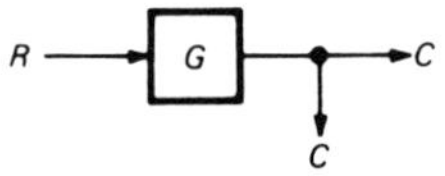	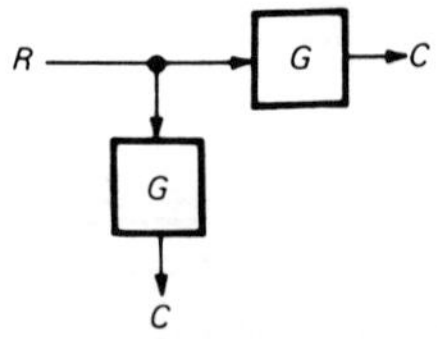
7. Simplifying closed-loop system	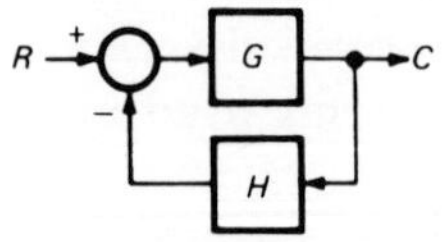	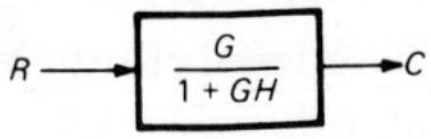

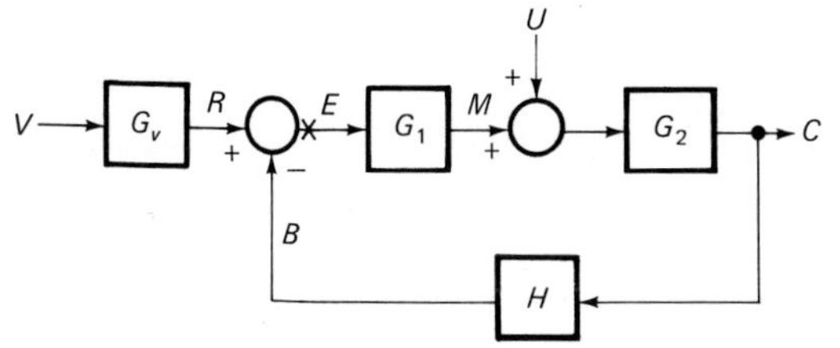

G_v Reference input elements
G_1 Control elements
G_2 System elements or process
H Feedback elements
V Desired value
R Reference input
C Controlled output
M Manipulated variable
U Disturbance input
B Feedback signal
E Error or actuating signal

Figure 1-11 General block diagram of control system

The *reference input R* is derived from the desired value and is a signal external to the control loop. It serves as the reference of comparison for the feedback signal.

The *controlled output C* is the process quantity being controlled.

The *manipulated variable M* is the control signal which the control elements apply to the process.

The *feedback signal B* is a signal which is a function of the controlled output and which is summed with the reference input.

The *error* or *actuating* signal E is the algebraic difference between the reference input and feedback signals and provides the signal applied to the control elements.

The *disturbance input U* is an unwanted input signal to the system that tends to cause the controlled output to differ from the value commanded by the reference input. Disturbance inputs are due to changes in the load on the system. For example, a change in the ambient temperature surrounding the oven described in Sec. 1-3 is a disturbance since it changes the heat input required by the oven. Obviously the response of the system to a disturbance input should be minimal.

System Elements

The *reference input elements* G_v convert the desired value to a reference input signal.

The *control elements* G_1, sometimes called the controller, are the components that act on the error signal and generate a control signal to the process.

The *feedback elements H* are the components required to measure or sense the value of the controlled output and convert it to a usable feedback signal.

The *system elements* or *process* G_2 is the body, process, or machine for which a particular quantity is to be controlled.

The control system shown in Fig. 1 11 is made up of two basic parts. The blocks labeled G_1 and G_2 make up the transmission path between the actuating signal and the controlled output and are known as the *forward-gain path* of the system. Block G_1 includes the control elements, and block G_2 is the process or system being controlled. For simplicity the product $G_1 \times G_2 = G$ will be used to represent the forward gain in analyzing Fig. 1-12. The *feedback path* through block H is the transmission path from the controlled output to the feedback signal. The feedback and reference input signals are compared at the summing junction, and their difference represents the error signal. Thus, at the summing junction

$$E = R - B = R - CH \tag{1-3}$$

Since the transfer function of the forward-gain block G is C/E, $E = C/G$. Substituting for E in Eq. (1-3) gives

$$\frac{C}{G} = R - CH \tag{1-4}$$

Solving Eq. (1-4) for the closed-loop transfer function C/R yields

$$\frac{C}{R} = \frac{G}{1 + GH} \tag{1-5}$$

Note that this relationship is shown as Identity 7 in Table 1-1. Equation (1-5) reveals an important fact about closed-loop systems. If the forward gain G is large so that the product $GH \gg 1$, the closed-loop gain reduces to

$$\frac{C}{R} \cong \frac{1}{H} \tag{1-6}$$

Thus we see that closed-loop behavior depends on the characteristic of the feedback path and is essentially independent of the characteristic of the forward gain, including the process itself. Should the forward

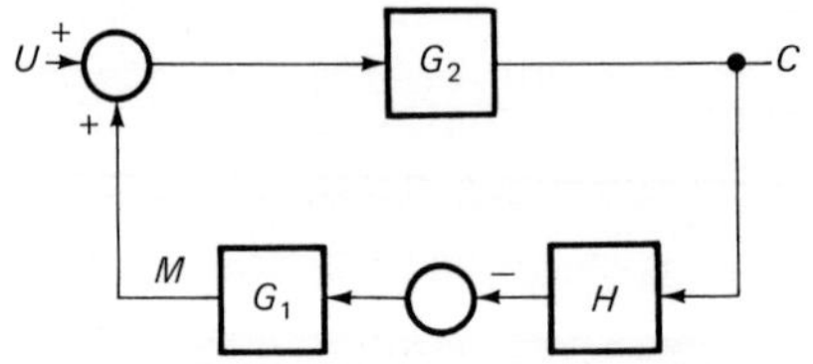

Figure 1-12 System with disturbance input

gain change for some reason, the feedback in the system will obviate the effect of the change on the output. Thus it is unnecessary to calibrate the system except for the feedback element (e.g., the thermocouple in Fig. 1-4).

If the closed loop is interrupted at some point (say point X in Fig. 1-11), we note that the gain around the loop is GH. This gain is referred to as the *open-loop transfer function* and is often taken as the ratio of the feedback signal to the error signal. Thus

$$\text{Open-loop transfer function} = \frac{B}{E} = \frac{CH}{E} = \frac{EGH}{E} = GH \tag{1-7}$$

If G is taken as the forward gain, Eq. (1-5) can be interpreted as

$$\text{Closed-loop transfer function} = \frac{\text{forward gain}}{1 + \text{open-loop transfer function}}$$

A special case of Fig. 1-11 that is often encountered in practice involves a direct comparison of the output and reference input. Since the feedback block now has a value of unity, Eq. (1-5) becomes

$$\frac{C}{R} = \frac{G}{1 + G} \tag{1-8}$$

Such a system is known as a *unity-feedback system* and occurs where the controlled output is to be a replicate of the reference input.

The effect of a disturbance input U may be analyzed by assuming that the reference input has a zero value.[6] In this case, the system may be viewed as shown in Fig. 1-12.

In a manner similar to the foregoing analysis, we have

$$C = G_2(U + M) = G_2U - G_1G_2HC \tag{1-9}$$

Solving Eq. (1-9) for the response C/U gives

$$\frac{C}{U} = \frac{G_2}{1 + G_1G_2H} \cong \frac{1}{G_1H} \tag{1-10}$$

for $G_1G_2H \gg 1$. From Eq. (1-10), we note that the effect of the disturbance input is reduced as the magnitude of the open-loop transfer function is increased. Thus, we see that a high controller gain G_1 can provide a precise controlled output that is largely insensitive to system disturbances.

The following example illustrates the use of the block diagram identities of Table 1-1 and shows the simplification of the overall transfer function of a closed-loop system.

6 *The principle of superposition, to be described in Sec. 2-2, is applicable in analyzing the disturbance input.*

EXAMPLE 1-1

Simplify the block diagram shown in Fig. 1-13, and determine the transfer function C/R.

Solution

We note that there are two feedback paths in the system. If these two paths were in parallel, a single feedback function could be obtained using Identity 2 in Table 1-1. Hence we shall try to rearrange the system elements to accomplish this. Using Identity 4, the inner summing point can be moved to the left as shown in Fig. 1-14*a*. In addition, the inner takeoff point can be moved to the right using Identity 5. The series blocks and the parallel feedback blocks can then be combined to give the reduced diagram shown in Fig. 1-14*b*. Using Identity 7, the closed-loop transfer function is

$$\frac{C}{R} = \frac{G_1G_2G_3}{1 + G_1G_2G_3[(H_1/G_1) + (H_2/G_3)]} = \frac{G_1G_2G_3}{1 + G_2(G_3H_1 + G_1H_2)}$$

1-8 ACCURACY AND STABILITY

A prime requirement of control systems is that the controlled variable be maintained accurately within specified limits as varying conditions disturb the system. Closed-loop control tends to make for accurate performance since the negative feedback endeavors to continually reduce any error to an acceptable value. However, under certain conditions, the corrective action can produce unstable operation. In a *stable system*, the response to an input will reach and maintain a useful value within a reasonable interval of time. On the other hand an *unstable system* is ineffective in maintaining the controlled variable at the desired value. The output may drive to an extreme limiting value,

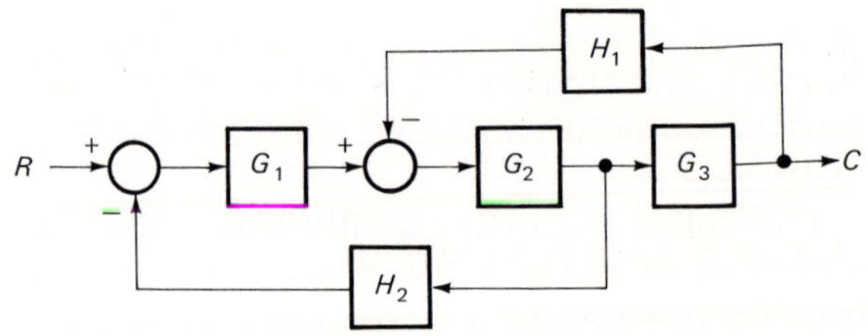

Figure 1-13 Block diagram for Example 1-1

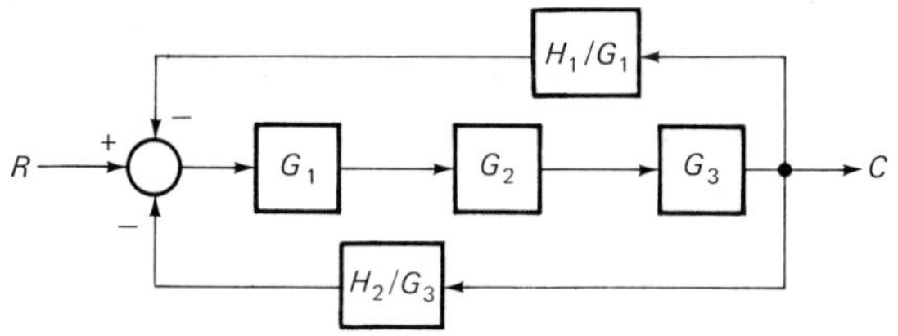

(a)

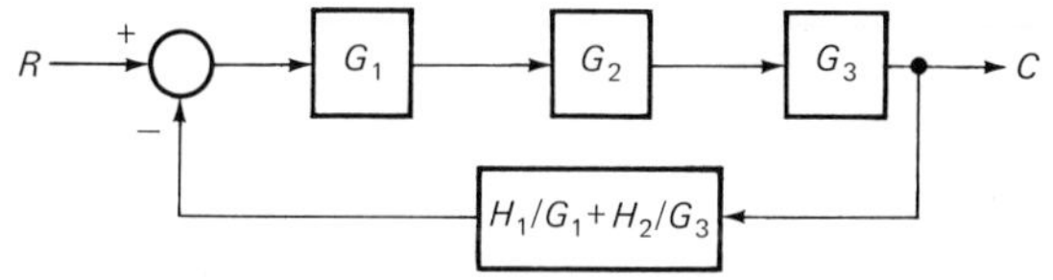

(b)

Figure 1-14 Reduction of block diagram of Example 1-1

or violent sustained oscillations may take place rendering the control useless or even destructive. If stable closed-loop performance is likened to a manual system with a skilled operator, an unstable system may be compared to manual control with an untrained and irresponsible operator.

The time delays or lags in the response of the various control elements and/or the controlled system are the cause of instability. To explain unstable behavior, let us assume that the corrective action that occurred in a system is no longer needed; that is, the error signal has become 0. However, inherent time delays in the system may prevent the stopping of control action in time to prevent an overshoot by the controlled variable. The overshoot will then result in corrective action in the opposite direction. This process of continuous corrective action can build up to sustained and possibly dangerous variations in the output.

From another point of view, the time delays in the system may cause the feedback to reinforce rather than oppose the input signal. The feedback in this case is now positive rather than negative and can result in a continuous oscillation. It should be noted that positive feedback is sometimes desirable, as in an oscillator circuit that generates a desired sinusoidal output.

It is important to note that the requirements of accuracy and stability are mutually incompatible. The closed-loop analysis in Sec. 1-7 shows that system accuracy is improved as loop gain is increased. However,

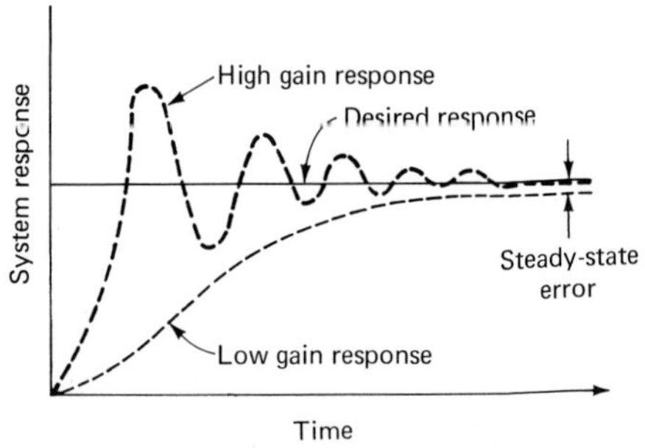

Figure 1-15 System response to step-input command

an increase in gain also tends to make the system unstable since a given error signal causes greater corrective action. This characteristic is indicated in Fig. 1-15, which shows the response of a system to a sudden change in input command. In the case of a system with low gain, the transient response tends to be slow and the final or steady-state error may be large. On the other hand, a high-gain system exhibits a significant overshoot and oscillations. However, when the system finally stabilizes, the steady-state error may be quite small.

In general the control system designer is faced with meeting requirements of accuracy, stability, and speed of response. Inevitably compromises must be made in order to achieve the best possible system.

1-9 APPLICATIONS OF AUTOMATIC CONTROL

This chapter has served to introduce the subject of automatic control, which involves the use of negative feedback in closed-loop systems. Succeeding chapters will be concerned with the mathematical representation of control elements and systems, descriptions of various control system components, and methods of system analysis, design, and testing. To give added perspective to the subject of closed-loop operation, we will conclude this chapter by briefly describing several types of control applications. Although the scope of automatic control is virtually unlimited, we will limit this discussion to examples which are commonplace in modern industry.

1-10 SERVOMECHANISMS

Although a servomechanism is not a control application per se, this device is commonplace in automatic control. A *servomechanism*, or "servo" for short, is a closed-loop control system in which the con-

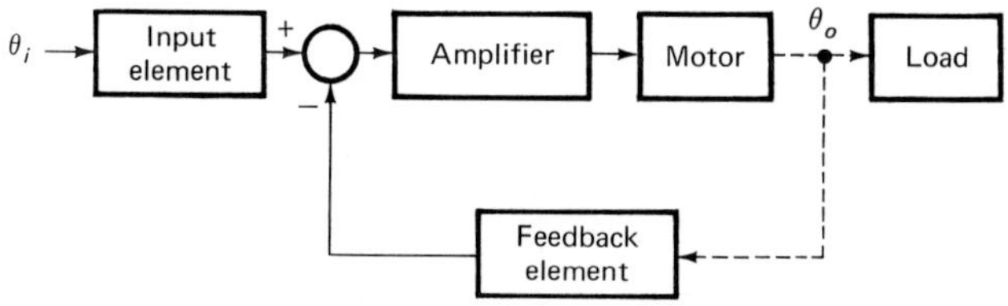

Figure 1-16 Block diagram of positional servomechanism

trolled variable is mechanical position or motion. It is designed so that the output will quickly and precisely respond to a change in the input command. Thus we may think of a servomechanism as a following device, and it is from this that the name is derived from the words *servus* (Latin for servant or slave) and *mechanism*.

Figure 1-16 shows a block diagram of a typical servomechanism incorporating an electric motor. The controlled variable is the mechanical shaft angle θ_o indicated by the dotted line. For this reason, the device is known as a *positional servomechanism*. The input command must specify the desired angle and is identified as θ_i. The feedback element provides a measure of θ_o in a form that can be compared with the input command. The signal representing the difference between θ_o and θ_i is applied to an amplifier which in turn drives the motor to reduce the error to 0.

Another form of servomechanism in which the rate of change or velocity of the output is controlled is known as a *rate* or *velocity servomechanism*. The diagram of Fig. 1-16 is applicable where the feedback element provides a measure of the velocity of θ_o.

1-11 PROCESS CONTROL

Process control is a term applied to the control of variables in a manufacturing process. Chemical plants, oil refineries, food-processing plants, blast furnaces, and textile mills are examples of production processes to which automatic control is applied. Process control is concerned with maintaining at a desired value such process variables as temperature, pressure, flow rate, liquid level, viscosity, density, and composition.

The liquid-level control system illustrated in Fig. 1-17 is an example of the closed-loop control of a process variable using conventional analog equipment. The level in the tank will remain constant if the inlet and outlet flow rates are equal. If the rates differ, the level will rise or fall. The feedback element is a float mechanism that provides a signal proportional to the level. This value is compared to the desired

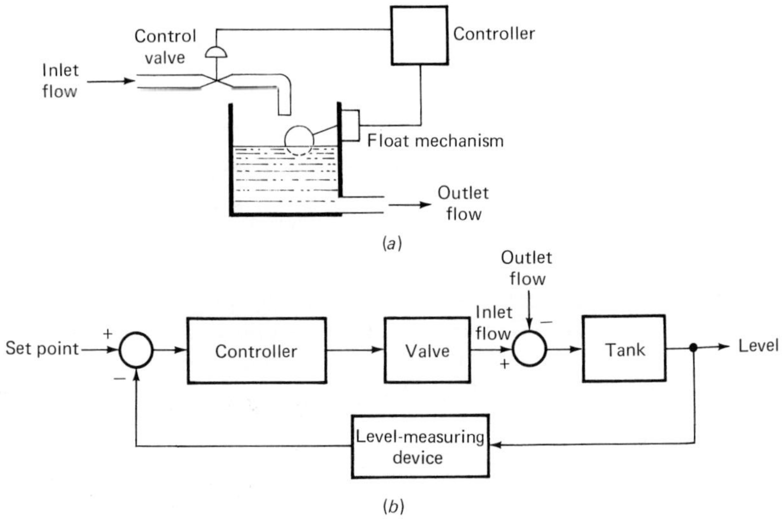

Figure 1-17 Liquid-level control system

value or set point to generate an error signal. The controller then adjusts a valve on the inflow to the tank to restore the level to the desired value.

In a typical process there may be anywhere from a few of these control loops to upwards of 100 or more. In recent years, process control has been extended to bring all these individual control functions under integrated control using a digital computer. The computer performs a supervisory function by analyzing the process conditions against desired performance criteria and determining the changes in process variables to achieve optimum operation.

Computer control of a process is illustrated by the diagram in Fig. 1-18. The block labeled Digital computer contains the elements found in a computer: memory, input and output sections, and arithmetic unit. The speed of the computer enables it to handle a large number of process variables in a time-sharing arrangement. The controlled variables are the outputs of the process that relate to the product quality and production rate. The system samples the controlled variables one at a time in a repetitive cycle. The computer calculates the desired value for each manipulated variable in the process that affects one or more of the controlled variables. This value is introduced as a set point for the analog controllers on the manipulated variables. The analog-to-digital (A/D) converter converts the measured analog signals into digital form for the computer. The digital-to-analog (D/A) converter converts the digital computer output into analog set-point signals for the individual control loops.

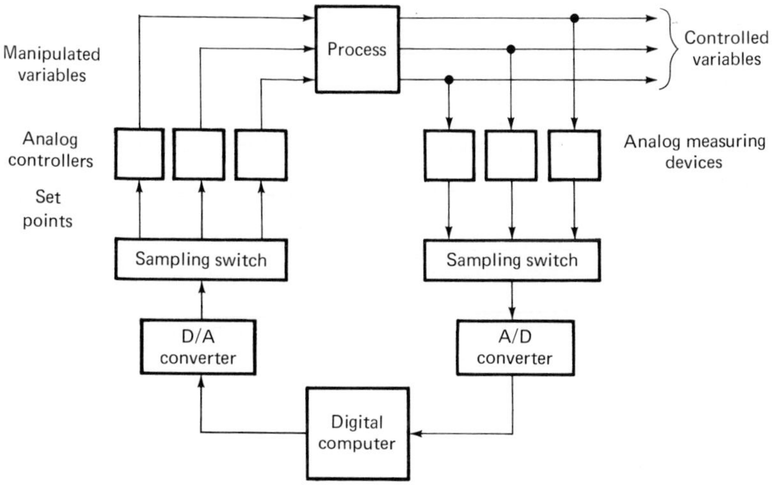

Figure 1-18 Block diagram of computer control of manufacturing process

Much current work in process control involves extending the use of the digital computer to provide direct digital control (DDC) of the process variables. In direct digital control the computer calculates the values of the manipulated variables directly from the values of the set points and the measurements of the process variables. The decisions of the computer are applied directly to digital actuators in the process. Since the computer duplicates the analog controller action, these conventional controllers are no longer needed.

1-12 POWER GENERATION

The electric power industry is primarily concerned with energy conversion and distribution. Large modern power plants which may exceed several hundred megawatts of generation require complex control systems to account for the interrelationship of the many variables and provide optimum power production. Control of power generation may be generally regarded as an application of process control, and it is common to have as many as 100 manipulated variables under computer control.

The advent of nuclear reactors is a significant advance in power generation, and the use of these reactors is expected to increase significantly during the next few decades. For proper operation the power level of the reactor must be maintained at a desired value or vary in a prescribed manner. This must be accomplished automatically and safely with minimal human supervision. The position of graphite regulator rods within a reactor determines the rate of nuclear fission and therefore total power generation. If the power output differs from the desired value, the difference generates a control signal that moves the regulator rods in the proper direction to achieve the desired power level in the reactor.

Automatic control has also been extensively applied to the distribution of electric power. Power systems are commonly made up of a number of generating plants. As load requirements fluctuate, the generation and transmission of power is controlled to achieve minimum cost of system operation. In addition, most large power systems are interconnected with each other, and the flow of power between systems is controlled.

1-13 NUMERICAL CONTROL

There are many manufacturing operations such as boring, drilling, milling, and welding, to name but a few, which must be performed with high precision on a repetitive basis. *Numerical control* (N/C) is a system that uses predetermined instructions called a *program* to control a sequence of such operations. The instructions to accomplish a desired operation are coded and stored on some medium such as punched paper tape, magnetic tape, or punched cards. These instructions are usually stored in the form of numbers—hence the name numerical control. The instructions identify what tool is to be used, in what way (e.g., cutting speed), and the path of the tool movement (position, direction, velocity, etc.).

The numerical control process may be described with the aid of the simplified block diagram shown in Fig. 1-19. The drawing defining the desired part is used to determine the program necessary to produce the part. The program sheet is written in an appropriate programming language such as APT (Automatically Programmed Tools). A digital

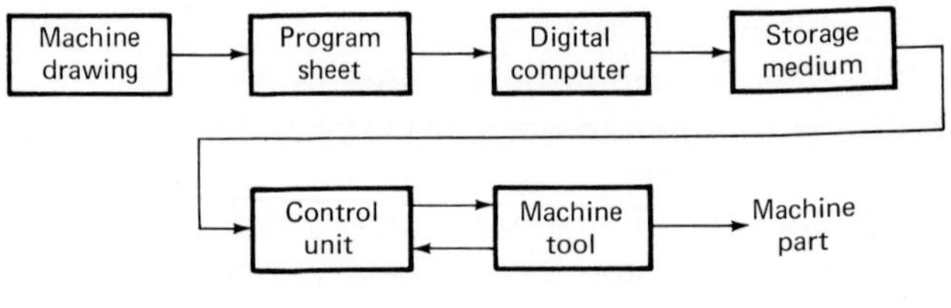

Figure 1-19 Block diagram of numerical control process

computer then performs the calculations to convert the program instructions to a storage medium for control of the manufacturing operations. The machine tool operating in conjunction with the control unit follows the instructions on the storage medium to produce the part. Information regarding the machine tool operation may be fed back to the control unit to obtain precision operation with errors reduced to a minimum.

1-14 TRANSPORTATION

To provide mass transportation systems for modern urban areas, large, complex systems are needed. Several automatic transportation systems now in operation have high-speed trains running at several-minute intervals. Automatic control is necessary to maintain a constant flow of trains and to provide comfortable acceleration and braking at station stops.

Aircraft flight control is another important application in the transportation field. The aircraft flight path is monitored by a device such as a gyrocompass, radio beacon, or radar system and compared with the desired flight path. The elevator, rudder, and other aircraft control surfaces are then actuated to correct any differences. This has proven to be one of the most complex control applications due to the wide range of system parameters and the interaction between controls. Aircraft control systems are frequently adaptive in nature; that is, the operation adapts itself to the surrounding conditions. For example, since the behavior of an aircraft may differ radically at low and high altitudes, the control system must be modified as a function of altitude.

Ship-steering and roll-stabilization controls are similar to flight control but generally require far higher powers and involve lower speeds of response.

PROBLEMS

1-1. List five examples of automatic control devices found in the home. Identify whether the operation of each is open-loop or closed-loop and discontinuous or continuous.

1-2. In a chemical process control system, the chemical composition of the product is controlled as shown below. A measurement of the composition is obtained by using an on-stream infrared analyzer. The valve on the additive stream may be driven by a small electric motor. Devise a means of automatically controlling the composition, and draw a block diagram of the system.

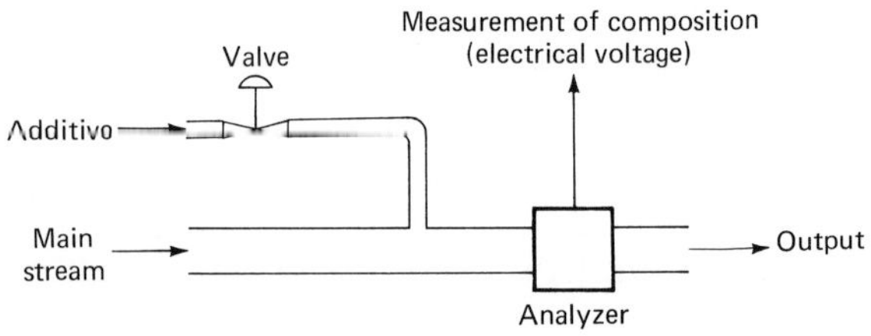

Prob. 1-2

1-3. The power level of a nuclear reactor is controlled by the position of graphite control rods (as the rods are inserted, the power level drops). The power level may be measured by an ionization chamber whose output current is proportional to power level. Devise a means of automatically controlling reactor power, and draw a block diagram of the system.

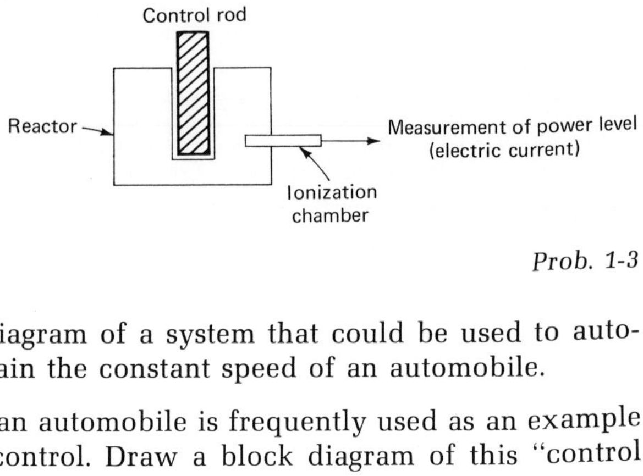

Prob. 1-3

1-4. Draw a block diagram of a system that could be used to automatically maintain the constant speed of an automobile.

1-5. The steering of an automobile is frequently used as an example of closed-loop control. Draw a block diagram of this "control system."

1-6. The teacher-student learning process may be regarded as a feedback process intended to reduce error to a minimum. The student may be considered the process, and the desired output is the knowledge acquired. Starting with the general block diagram of a control system, construct a block diagram of this learning process.

1-7. Windmills have tail pieces that keep the rotor facing the wind. A shift in wind direction causes a force to be exerted on one side of the tail swinging the rotor into the wind. What kind of control is this? What is the command? What is the controlled variable?

1-8. The feedback signal in a closed-loop control system is inadvertently connected so that it adds to rather than subtracts from the reference input. What effect might this have on the controlled output?

1-9. Simplify the following block diagrams, and determine the transfer function C/R.

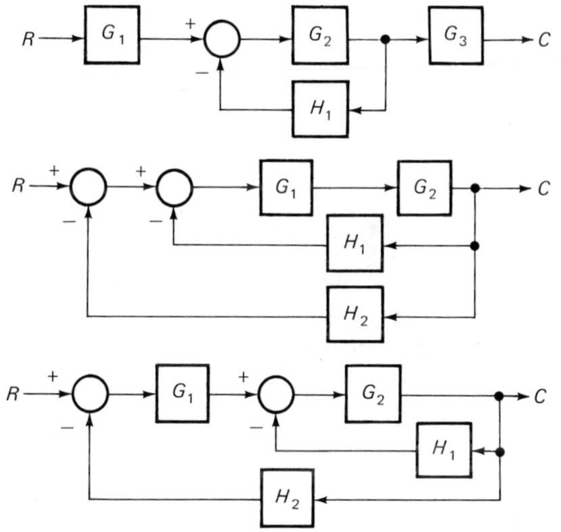

Prob. 1-9

1-10. Simplify the following block diagram, and determine the equation for C as a function of R and U.

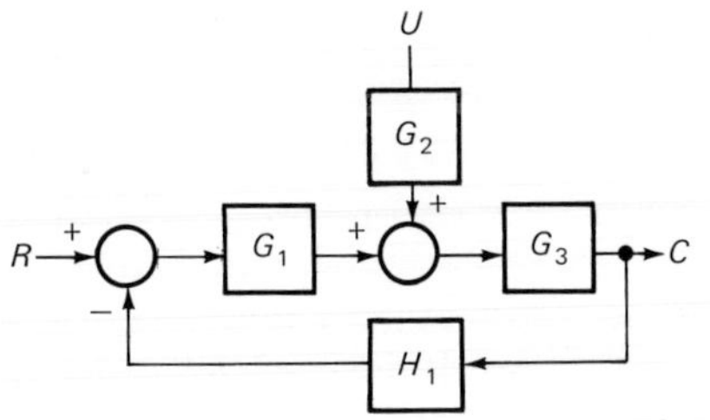

Prob. 1-10

1-11. Determine the value of the transfer function C/R for the block diagram below.

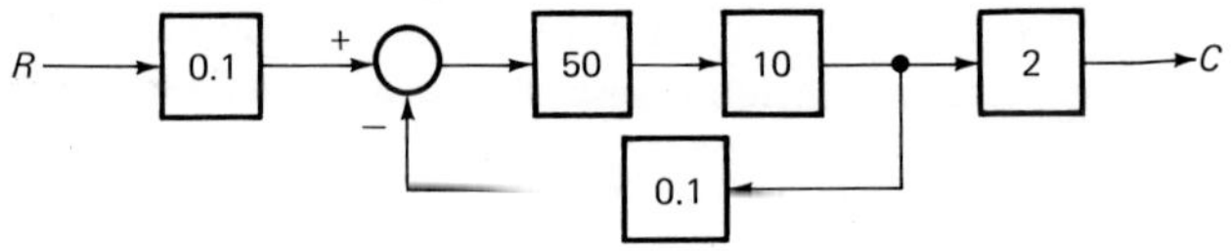

Prob. 1-11

1-12. Determine the value of G to meet the input-output requirements of the system below.

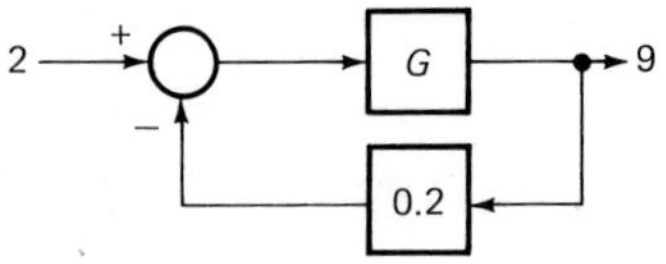

Prob. 1-12

1-13. Determine the value of H to meet the requirements of the system below.

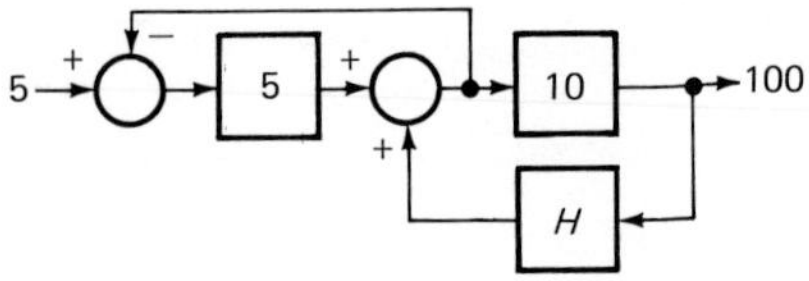

Prob. 1-13

1-14. Determine the value of the output C for the system below.

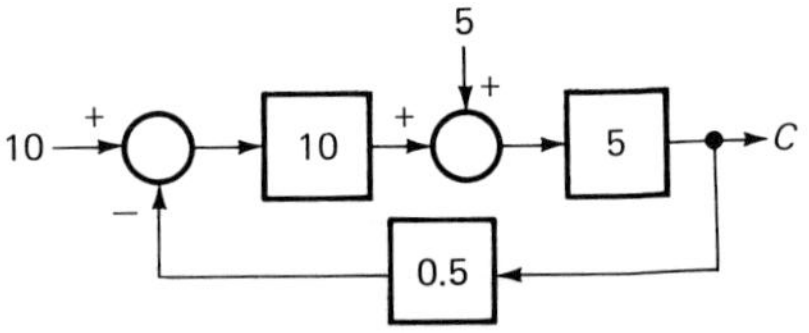

Prob. 1-14

1-15. Simplify the following block diagram.

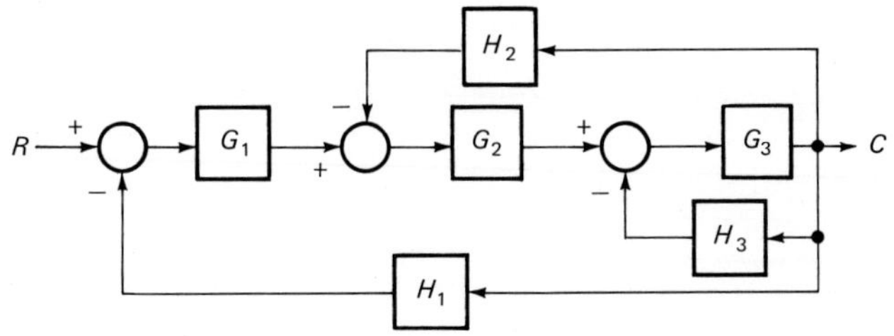

Prob. 1-15

Chapter Two

mathematical models

2-1 INTRODUCTION

In the preceding chapter automatic control systems were examined in a descriptive fashion with little use of mathematics. This discussion served to introduce the concept of negative feedback in closed-loop control systems. However, to analyze and design automatic control systems on other than a trial-and-error or an intuitive basis, some knowledge of the underlying mathematics is essential.

To understand the behavior of systems, whether simple or complex, one must obtain relationships between the variable quantities in the system. These relationships take the form of mathematical equations; these equations are known as a *mathematical model* of the system.

As indicated in Chap. 1, a control system consists of a combination of operating components arranged in such a way that a definite informational flow pattern is established. The characteristics of the various components may be expressed by transfer functions that relate the respective input and output variables. When combined together, these individual transfer functions yield an overall system model.

In our introductory consideration of control systems the equations defining their behavior were algebraic in nature. Values of the variable quantities in the system were established by the system parameters and, once established, did not change so long as the parameters remained constant. This condition, which is independent of time, represents *steady-state operation* of the system. The steady-state values of quantities are customarily denoted by capital letters.

However, a control system is not static in nature. The output must frequently respond to changes in input commands, and we are concerned with the manner of this response. The system is also subject to disturbances that vary with time. In addition, a major problem confronting the controls designer is instability or unwanted oscillatory variations in the output. All these considerations require that a control system be analyzed under conditions of *dynamic* or *transient operation*. In this case the system variables are not constant but are changing with time. We will use lowercase letters to represent time-varying quantities. Thus $x(t)$ would represent a quantity x whose value varies as a function of time t. For convenience $x(t)$ is generally written as simply x.

The equations that describe dynamic systems contain not only the variables themselves but also the rates of change or derivatives of these variables. Such equations are known as *differential equations*. The purpose of this chapter is to present methods of writing differential equations for a variety of mechanical, electrical, thermal, and hydraulic systems. Differential equations allow us to translate the dynamic behavior of a physical system into an equivalent mathematical model. It should be emphasized that the process of obtaining a differential equation requires a basic knowledge of the related physics rather than mathematics ability per se. Thus in deriving a variety of differential equations, we will rely upon basic and familiar principles of physics. A facility in mathematics is required when one wishes to obtain a solution to a differential equation. As will be discussed in Sec. 2-9, the solution of a differential equation is an equation that relates the variables but does not contain derivatives.

2-2 LINEARITY

The concept of linearity is extremely important in dealing with physical systems. *Linearity* in a device implies behavior such that one variable is closely proportional to another within some range of the variables. For example, doubling the current in a resistor doubles the voltage drop; when graphed, this relationship is a straight line. A spring may also be considered linear; if one end is fixed, the displacement of the other end is proportional to the force applied. However, all systems ultimately become nonlinear as the variables are increased without limit. The resistor will eventually burn out, and the spring will finally break.

The principle of linearity is especially useful in the study of control systems. When devices having a linear relationship between the input and output variables are connected in cascade, the overall input-output relationship is linear. The elements are said to form a *linear system*.

The principle of *superposition* is applicable to linear systems. This principle states that the total response of a linear system having two or more inputs may be found by separately considering the response of each input with all other inputs removed and then summing the individual responses. The implication of this statement is that an input $x_1(t) + x_2(t)$ gives a response $y_1(t) + y_2(t)$, where $y_1(t)$ is due to $x_1(t)$ and $y_2(t)$ results from $x_2(t)$.

From a mathematical standpoint, a linear system is described by equations in which none of the terms are higher than first degree in the dependent variables. This is true whether the system equations are algebraic or differential in nature. For example, the algebraic equation

$$w = 3x + 2y - 4$$

is linear because all terms are first-degree. The equation

$$w = 3x^2 + 2xy - 4$$

is nonlinear due to the two second-degree terms. Examples of linear and nonlinear differential equations will be deferred until the next section.

The distinction between linear and nonlinear equations is important since the former are much more readily solved than the latter. This is especially true insofar as differential equations are concerned. Any rigorous analysis of systems described by nonlinear differential equations is beyond the scope of this text. However, a nonlinear problem may be approached by replacing nonlinear functions by linear expressions that describe the behavior sufficiently well for purposes of analysis and design. One common approach to such linearization is to assume linear behavior over a limited operating range. For example, the small-signal operation of electron tubes and transistors is generally analyzed using linear parameters whose values are specified at a particular dc operating point. Nonlinear behavior and methods of linearization are the subject of Chap. 9.

2-3 DIFFERENTIAL EQUATIONS

Equations containing terms in an unknown dependent variable (or variables) that are derivatives of an independent variable (or variables) are called *differential equations*. It is the purpose of this section to introduce some terminology associated with differential equations so that we may describe the behavior of simple physical systems in terms of such equations.

If a differential equation involves only one independent variable, the derivatives are ordinary derivatives and the equation is called an *ordinary differential equation*. In the study of control systems, one is primarily concerned with system behavior as a function of time. Hence, the system differential equations contain time as the independent variable. For example, the equation

$$\frac{d^2y}{dt^2} + 6\frac{dy}{dt} + 5y = 0 \tag{2-1}$$

is an ordinary differential equation in which t is the independent variable and y is the dependent variable. The equation states that the behavior of y as a function of t is such that a weighted summation of the instantaneous values of y and its first and second derivatives must always equal 0. Unfortunately Eq. (2-1) does not allow us to readily visualize the manner in which y will vary with t. If an equation expressing y as a function of t without the derivatives was available, a visualization would be possible. Such an equation can be derived from the differential equation and, as was previously mentioned, is a solution of the differential equation.

When more than one independent variable is involved (such as both time and position), the derivatives are partial derivatives and the equation is known as a *partial differential equation*. Since ordinary differential equations are applicable to most control systems (as well as being much easier to handle), we will limit our consideration to this type.

The *order* of a differential equation is the order of the highest derivative that occurs in the equation. The *degree* of the equation is the exponent to which the highest derivative is raised. Thus Eq. (2-1) is second-order and first-degree. Equation (2-2) below is third-order and second-degree.

$$\left(\frac{d^3y}{dx^3}\right)^2 - \left(\frac{dy}{dx}\right)^3 + y = 0 \tag{2-2}$$

If all terms in a differential equation are first-degree, the equation is linear and the principles of linear systems are applicable. Thus Eq. (2-1) is a linear equation. If any term appears in a power other than the first, the equation is nonlinear, as is illustrated by Eq. (2-2). It should be noted that a first-degree differential equation is not necessarily linear. As an example, consider Eq. (2-3), which although first degree is nonlinear by virtue of the squared term.

$$\frac{d^2x}{dt^2} + \left(\frac{dx}{dt}\right)^2 + 3x = 0 \tag{2-3}$$

We will next distinguish between ordinary differential equations with *constant coefficients* and those with *variable coefficients*. The coefficients are the coefficients of the terms in the equation. Thus, the coefficients in the preceding equations are all constants. Variable coefficients are coefficients that are themselves functions of the independent variable.

Finally, we will define the term *homogeneous* as applied to differential equations. From a mathematical standpoint we say that a differential equation is homogeneous if the summation of terms containing the dependent variable and its derivatives is equated to 0. Thus Eqs. (2-1) to (2-3) are homogeneous.

When the summation equals a constant or a function of the independent variable, the equation is classed as *nonhomogeneous*. Equation (2-4) is an example of a nonhomogeneous equation.

$$3\frac{dy}{dt} + 5y = 10 \sin t \tag{2-4}$$

An important physical significance can be associated with the concept of homogeneity in differential equations. A homogeneous equation defines the behavior of a system independent of any external energy source or drive. The equation is based only on the parameters or constants of the elements making up the system. On the other hand a nonhomogeneous equation includes a driving force or excitation acting upon the physical system described by the equation. This input is frequently identified as either a *forcing function* or a *driving function* since it forces or drives the system to some response. Thus Eq. (2-4) contains a sinusoidal forcing function. Other forcing functions that are frequently used include step, ramp, and exponential signals.

Figure 2-1 summarizes the classification of differential equations

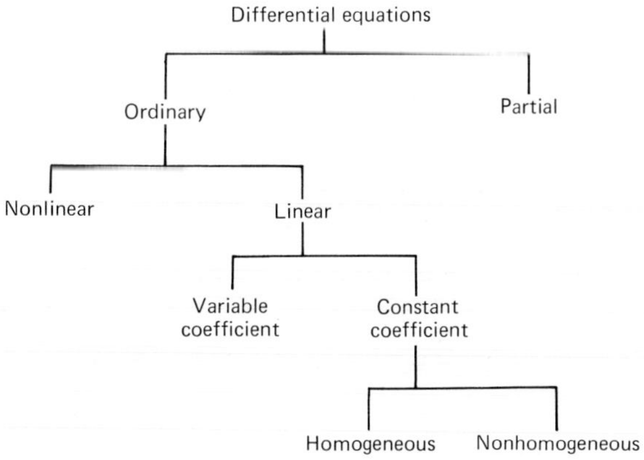

Figure 2-1 Classification of differential equations

according to the types that have been discussed. In this text we will consider only constant-coefficient linear ordinary differential equations. *These equations are fundamental to the study of automatic control.*

2-4 MECHANICAL SYSTEMS

The first kinds of physical systems that we shall consider are mechanical in nature. Mechanical systems include both *translational*[1] and *rotational* types.

Translational mechanical systems obey Newton's second law of motion, which states that the sum of the forces acting on a rigid body is equal to the product of the mass and the linear acceleration of the body. This law may be expressed by the equation

$$\sum F = Ma \tag{2-5}$$

where F = force, lb
M = mass, lb-s^2/ft(slugs)
a = acceleration, ft/s^2

Although familiar English units have been used, other systems of units are equally valid.

A simple mechanical system is the spring-mass system shown in Fig. 2-2. The mass M rests on a surface and is connected to a wall by a spring. At some equilibrium position, the spring is neither stretched nor compressed and exerts no force on the mass. If this position is taken as zero displacement, displacements to the right may be considered as positive and those to the left as negative. Once a sign convention has been adopted, it must be applied consistently to all the variables

1 *Translational means motion in a straight line.*

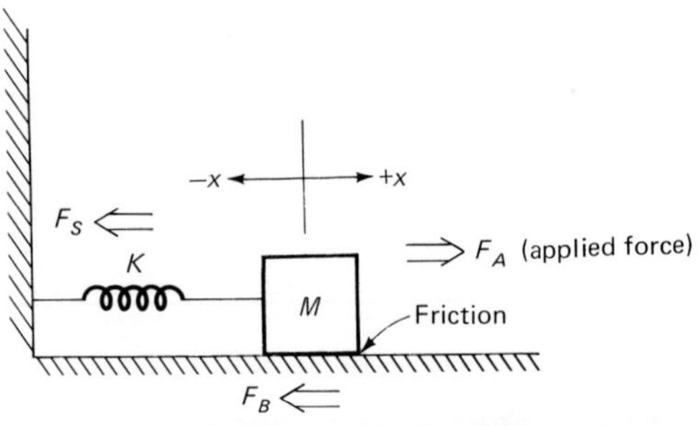

Figure 2-2 Spring-mass system

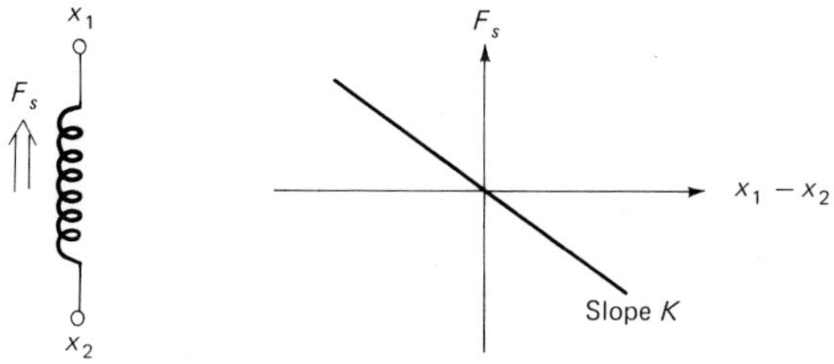

Figure 2-3 Spring and characteristic

in the system. Thus, velocity, acceleration, and force directed to the right must be taken as positive quantities. Obviously the reverse sign convention could have been used equally well. In addition to the mass and spring, friction that acts in opposition to motion is present in the system.

If an external force F_A is applied to the mass, the spring produces a reacting force proportional to the relative displacement of the ends from their equilibrium displacement. Figure 2-3 shows this characteristic. If stretched, the spring tries to contract; if compressed, it seeks to expand. The proportional relationship between restoring force and displacement may be expressed as

$$F_s = -K(x_1 - x_2) \tag{2-6}$$

where F_s = restoring force, lb
$x_1 - x_2$ = relative displacement from equilibrium, ft
K = spring constant, lb/ft

The negative sign indicates that the force opposes the displacement. If one end of the spring is fixed, the equation reduces to $F_s = -Kx$. The spring constant K is also referred to as the *stiffness* in the system.

Friction in the system is an energy-absorbing effect and tends to damp the motion of the mass. Hence, this effect is frequently referred to as system *damping*. Friction that introduces a restraining force proportional to velocity is known as *viscous friction*. The characteristic of viscous friction is shown in Fig. 2-4*a*.

The dashpot shown in Fig. 2-4*b* is a mechanical component that is frequently used to represent or actually provide damping. A dashpot consists of a piston inside a fluid-filled cylinder. A familiar example is the automobile shock absorber. Relative motion between the piston and the cylinder is resisted by the fluid which must flow from one side of the piston to the other through small openings in or around the

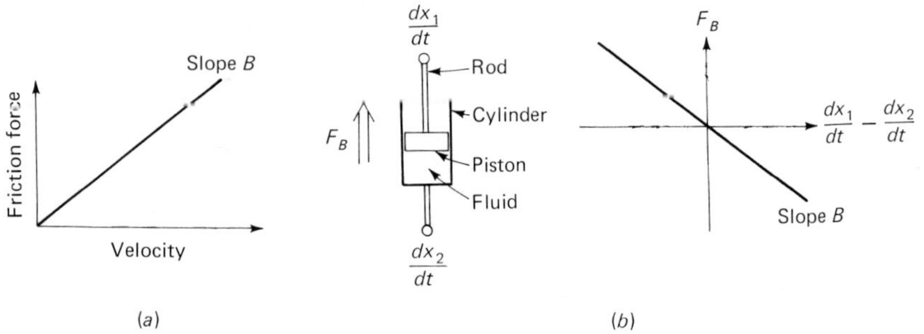

Figure 2-4 (a) Viscous friction characteristic, (b) dashpot and characteristic

piston. The greater the relative velocity between rod and housing, the greater must be the fluid flow rate and the greater the resultant force. Thus, the equation defining the damping characteristic is

$$F_B = -B\left(\frac{dx_1}{dt} - \frac{dx_2}{dt}\right) \tag{2-7}$$

where F_B = damping force, lb

$\frac{dx_1}{dt} - \frac{dx_2}{dt}$ = relative velocity, ft/s

B = damping coefficient, lb/(ft/s)

If one end of the dashpot is fixed, Eq. (2-7) becomes $F_B = -B(dx/dt)$.

Another form of friction that occurs in mechanical systems is called *coulomb friction*. Coulomb friction produces a force that is independent of velocity as shown in Fig. 2-5. Since coulomb friction is a nonlinear effect (force is not proportional to velocity), it will not be included at this time but will be considered in Chap. 9.

We may now write a differential equation for the motion of the

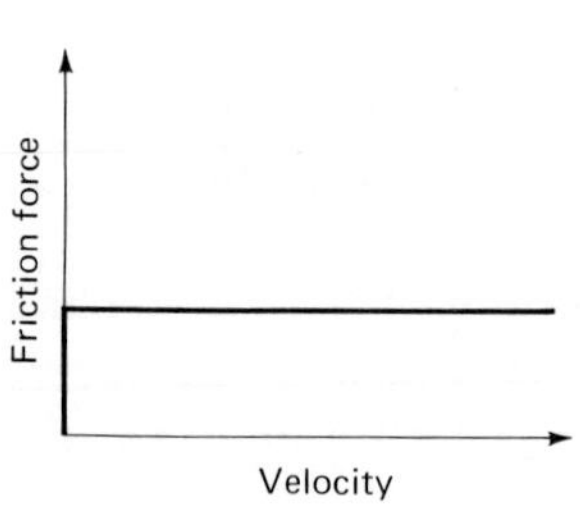

Figure 2-5 Coulomb friction characteristic

spring-mass system that incorporates the elements of mass, stiffness, and damping. The solution of this differential equation will yield an expression for the displacement of the mass as a function of time. Recalling that acceleration is the second derivative of displacement with respect to time, Eq. (2-5) may be written as

$$\sum F = M\frac{d^2x}{dt^2} \tag{2-8}$$

There are three forces to be included in the summation: the restoring force of the spring, the force due to viscous friction, and the external applied force F_A. Thus, Eq. (2-8) becomes

$$\sum F = F_A - Kx - B\frac{dx}{dt} = M\frac{d^2x}{dt^2} \tag{2-9}$$

or

$$M\frac{d^2x}{dt^2} + B\frac{dx}{dt} + Kx = F_A \tag{2-10}$$

Equation (2-10) is the differential equation for the spring-mass system of Fig. 2-2. More generally, the equation describes translational motion in a simple mechanical system.

In developing Eq. (2-10), we have assumed that the stiffness and damping are linear effects so that the system itself is linear. Thus, the equation of motion is a constant-coefficient, linear, ordinary differential equation of second order. Since the applied force F_A represents a forcing function, the equation is nonhomogeneous.

EXAMPLE 2-1

A 12-lb weight, when attached to the spring shown in Fig. 2-6, causes the spring to stretch 3 in. A dashpot having a damping coefficient $B = 0.6$ lb/(ft/s) is also coupled to the weight. Write the differential

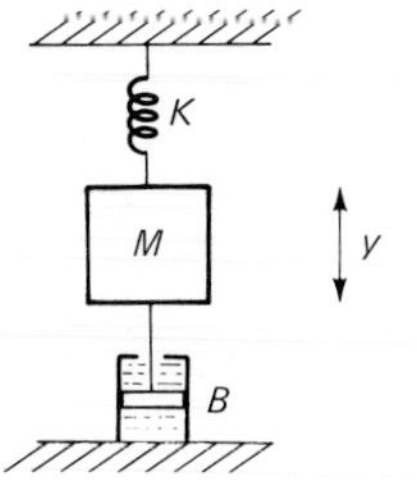

Figure 2-6 Spring-mass system for Example 2-1

equation for the motion of the weight if it is pulled below the equilibrium point and then released.

Solution

The spring constant K may be calculated using the initial displacement.

$$K = \tfrac{12}{3} = 4\ \text{lb/in.} = 48\ \text{lb/ft}$$

Using Eq. (2-10), the differential equation is

$$\frac{12}{32}\frac{d^2y}{dt^2} + 0.6\frac{dy}{dt} + 48y = 0$$

which simplifies to

$$\frac{d^2y}{dt^2} + 1.6\frac{dy}{dt} + 128y = 0$$

Note that the weight in pounds has been converted to mass in slugs by dividing by the acceleration due to gravity $g = 32\ \text{ft/s}^2$. The right-hand side of the equation is 0 since there is no driving force acting on the moving mass. In this case, motion is due to the initial displacement of the mass from its point of equilibrium.

It should also be noted that the weight of the mass in Example 2-1 has no effect on the differential equation, even though the weight does establish a force within the system. When the equilibrium position is chosen as the reference, weight is not a factor in the equation.

Rotational systems are similar to translational systems, and the principles of motion can be extended to set up their differential equations. The basic differences are that torque equations are written in place of force equations and displacement, velocity, and acceleration are now angular quantities.

Newton's law of rotation states that the sum of the torques acting on a rotating body is equal to the product of the moment of inertia of the body and its angular acceleration. Stated mathematically, this is

$$\sum T = J\alpha \tag{2-11}$$

where T = torque, lb-ft
J = inertia, lb-ft-s^2(slug-ft^2)
α = angular acceleration, rad/s^2

Figure 2-7 shows a simple torsional spring system. A body having

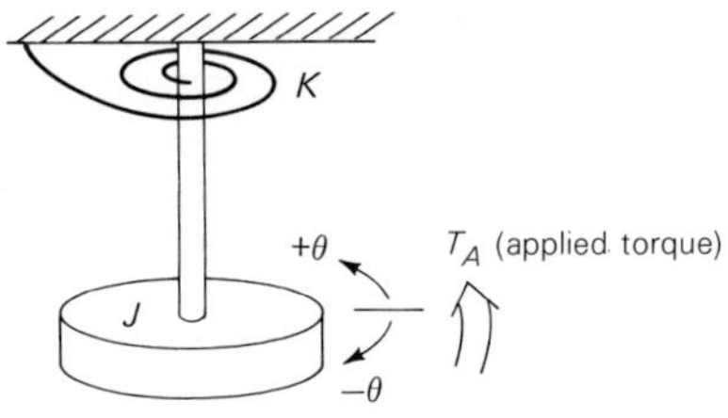

Figure 2-7 Rotational system

moment of inertia J is free to rotate under the constraint of a torsion spring K. As the angle of rotation θ increases in the positive direction, the spring winds up and generates a reaction moment that opposes the input moment. Similarly, a negative rotation causes an unwinding of the spring which likewise generates an opposing moment. The proportional relationship between restoring torque and angular displacement may be stated as

$$T_s = -K\theta \tag{2-12}$$

where T_s = restoring torque, lb-ft
θ = angular rotation from equilibrium, rad
K = spring constant, lb-ft/rad

As with a translational system, viscous friction is present and introduces a restraining torque proportional to angular velocity. Rotational damping may be expressed as

$$T_B = -B\frac{d\theta}{dt} \tag{2-13}$$

where T_B = damping torque, lb-ft
$\frac{d\theta}{dt}$ = angular velocity, rad/s
B = damping coefficient, lb-ft/(rad/s)

Using Eqs. (2-11) to (2-13), the differential equation for the rotational system is

$$\sum T = T_A + T_s + T_B = T_A - K\theta - B\frac{d\theta}{dt} = J\frac{d^2\theta}{dt^2} \tag{2-14}$$

or

$$J\frac{d^2\theta}{dt^2} + B\frac{d\theta}{dt} + K\theta = T_A \tag{2-15}$$

where T_A = applied torque, lb-ft

As we might expect, the equations describing translational and rotational mechanical systems are of the same form but differ in their variables and parameters.

EXAMPLE 2-2

For the torsional system shown in Fig. 2-8, write the differential equation for the rotation of the mass in response to an input rotation θ_i. The mass is a solid cylinder weighing 8 lb and having a radius of 6 in.

Solution

The inertia of a solid cylinder about its own axis is equal to $\frac{1}{2}Mr^2$. Substituting values,

$$J = \tfrac{1}{2} \times \tfrac{8}{32} \times (\tfrac{6}{12})^2 = \tfrac{1}{32} \text{ slug-ft}^2$$

The torque applied through the torsional spring is

$$T_A = K(\theta_i - \theta_o)$$

Applying Eq. (2-11), the differential equation is then

$$K(\theta_i - \theta_o) - B\frac{d\theta_o}{dt} = J\frac{d^2\theta_o}{dt^2}$$

Inserting the system parameters, this becomes

$$\frac{1}{32}\frac{d^2\theta_o}{dt^2} + \frac{1}{10}\frac{d\theta_o}{dt} + \theta_o = \theta_i$$

or

$$\frac{d^2\theta_o}{dt^2} + 3.2\frac{d\theta_o}{dt} + 32\theta_o = 32\theta_i$$

which is the differential equation of motion.

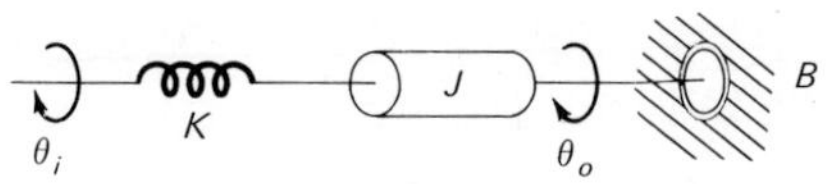

$K = 1$ lb-ft/rad
$B = 0.1$ lb-ft/(rad/s)

Figure 2-8 Torsional system for Example 2-2

2-5 ELECTRICAL NETWORKS

A second kind of physical system is electrical in nature and is commonly identified as an *electrical circuit* or *network*. An electrical network is composed of elements such as resistors, inductors, and capacitors connected together in some manner. If the network contains no energy sources, such as batteries or electrical generators, it is known as a *passive network*. On the other hand, if one or more energy sources are present, the resultant combination is an *active network*. In studying the behavior of an electrical network, we are interested in determining the voltages and currents that exist within the circuit. Since a network is composed of passive circuit elements, we must first define the electrical characteristics of these elements.

In the case of a resistor, the voltage-current relationship is given by Ohm's law, which states that the voltage across the resistor is equal to the current through the resistor multiplied by the value of the resistance. Mathematically, this is expressed as

$$v = iR \tag{2-16}$$

where v = voltage, V
i = current, A
R = resistance, Ω

The voltage across a pure inductor is defined by Faraday's law, which states that the voltage across the inductor is proportional to the rate of change with time of the current through the inductor. Thus we have

$$v = L\frac{di}{dt} \tag{2-17}$$

where v = voltage, V
$\frac{di}{dt}$ = rate of change of current, A/s
L = inductance, H

The voltage developed across a capacitor is proportional to the electric charge q accumulating on the plates of the capacitor. Since the accumulation of charge may be expressed as the summation, or integral, of the charge increments dq, we have the equation

$$v = \frac{1}{C}\int dq \tag{2-18}$$

where the capacitance C is the proportionality constant relating voltage and charge. By definition, current equals the rate of change of charge with time and is expressed as $i = dq/dt$. Thus an increment of charge dq is equal to the current multiplied by the corresponding time increment, or $dq = i\,dt$. Equation (2-18) may then be written as

$$v = \frac{1}{C}\int i\,dt \tag{2-19}$$

where C = capacitance, farads
v = voltage, V
i = current, A

A summary of Eqs. (2-16), (2-17), and (2-19) for the three forms of passive circuit elements is given in Fig. 2-9. Note that conventional current flow is used; hence the current in each element is shown in the direction of decreasing voltage.

Active electrical devices involve the conversion of energy to electrical form. For example, the electrical energy in a battery is derived from its stored chemical energy. The electrical energy of a generator is a result of the mechanical energy of the rotating armature.

Active electrical elements occur in two basic forms: *voltage sources* and *current sources*. In their ideal form, voltage sources generate a constant voltage independent of the current drawn from the source. The aforementioned battery and generator are regarded as voltage sources since their voltage is essentially constant with load. On the other hand, current sources produce a current whose magnitude is independent of the load connected to the source. Although current sources are not as familiar in practice, the concept does find wide use

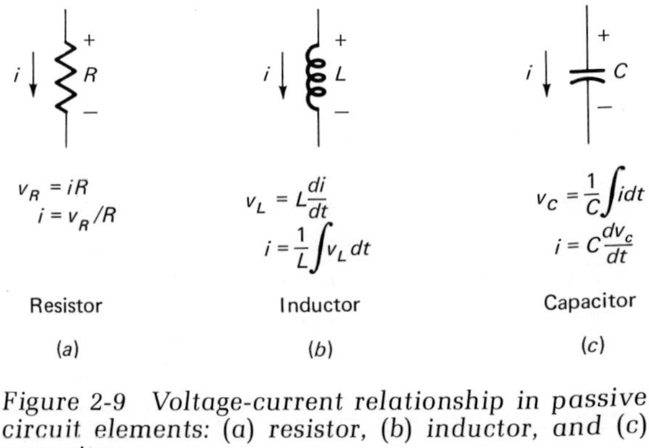

Figure 2-9 Voltage-current relationship in passive circuit elements: (a) resistor, (b) inductor, and (c) capacitor

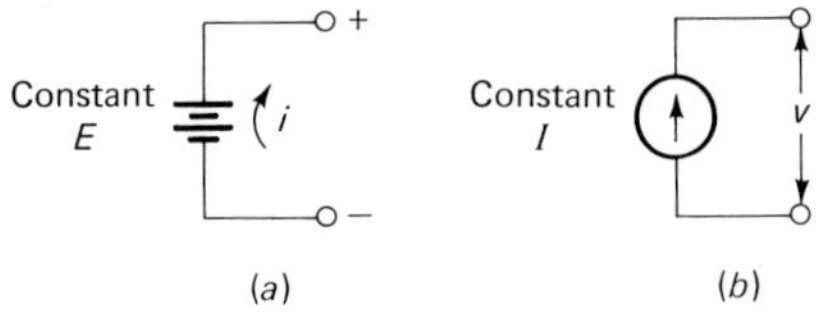

Figure 2-10 (a) Voltage source and (b) current source

in representing an amplifying device, such as the transistor, by means of an equivalent electrical circuit. Symbolic representations of voltage and current sources are shown in Fig. 2-10.

A common method of analyzing an electrical network is *mesh* or *loop analysis*. The fundamental law that is applied in this method is Kirchhoff's first law, which states that the algebraic sum of the voltages around a closed loop is 0, or, in any closed loop, the sum of the voltage rises must equal the sum of the voltage drops. Mesh analysis consists of assuming that currents—termed *loop currents*—flow in each loop of a network, algebraically summing the voltage drops around each loop, and setting each sum equal to 0.

Consider the circuit shown in Fig. 2-11*a*, which consists of an inductor and resistor connected in series to a voltage source e. Assuming a loop current *i*, the voltage drops summed around the loop are

$$-e + v_R + v_L = 0 \tag{2-20}$$

The input voltage is summed negatively since, in the direction of assumed current, it represents an increase in voltage. The drop across each passive element is positive since the current is in the direction of the developed voltage.

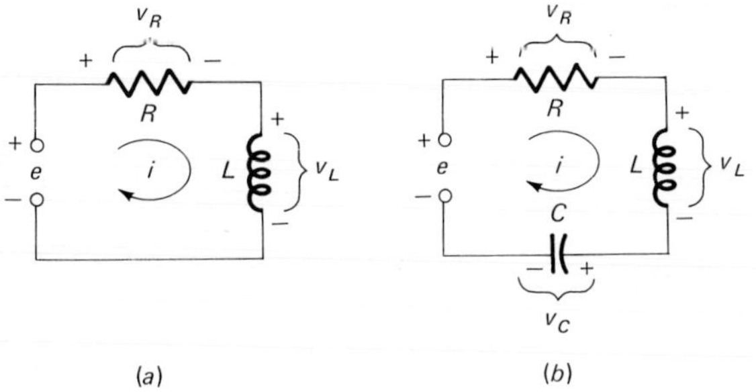

Figure 2-11 Loop analysis method

Using the equations for the voltage drops in a resistor and inductor, we have

$$L\frac{di}{dt} + Ri = e \tag{2-21}$$

Equation (2-21) is the differential equation for the current in the circuit.

It may be that the inductor voltage rather than the current is the variable of interest in the circuit. As noted in Fig. 2-9, $i = 1/L \int v_L\, dt$. Substituting this integral for i in Eq. (2-21) gives

$$v_L + \frac{R}{L}\int v_L\, dt = e \tag{2-22}$$

After differentiation with respect to time, Eq. (2-22) becomes

$$\frac{dv_L}{dt} + \frac{R}{L}v_L = \frac{de}{dt} \tag{2-23}$$

which is the differential equation for the inductor voltage.

Figure 2-11*b* shows a series circuit containing a resistor, inductor, and capacitor. Following the mesh-analysis method outlined above, the circuit equation is

$$L\frac{di}{dt} + Ri + \frac{1}{C}\int i\, dt = e \tag{2-24}$$

Recalling that current $i = dq/dt$, a substitution of this variable may be made to eliminate the integral from the equation. The result is the second-order differential equation

$$L\frac{d^2q}{dt^2} + R\frac{dq}{dt} + \frac{q}{C} = e \tag{2-25}$$

EXAMPLE 2-3

For the circuit shown in Fig. 2-12, write the differential equation for

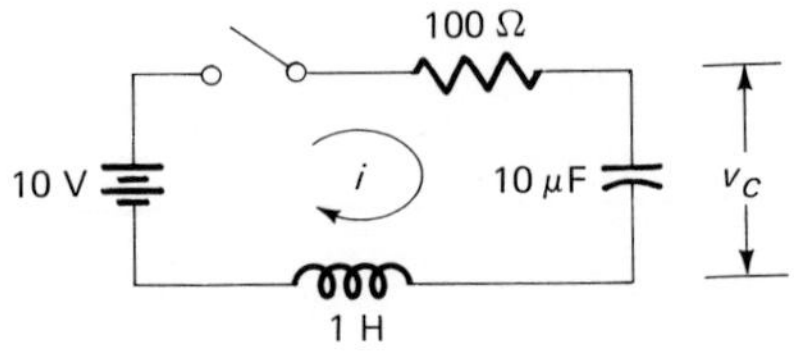

Figure 2-12 Circuit for Example 2-3

the voltage across the capacitor after closure of the switch. Assume that the capacitor is initially uncharged.

Solution

Equation (2-24) contains the current i as the dependent variable. Since we are interested in the capacitor voltage, the current in the equation can be replaced by v_c using the relation $i = C(dv_c/dt)$. Noting that $di/dt = C(d^2v_c/dt^2)$ and $\int i\,dt = Cv_c$, Eq. (2-24) can be rewritten as

$$LC\frac{d^2v_c}{dt^2} + RC\frac{dv_c}{dt} + v_c = e$$

Substituting the circuit parameters gives

$$10^{-5}\frac{d^2v_c}{dt^2} + 10^{-3}\frac{dv_c}{dt} + v_c = 10$$

We see that this is a nonhomogeneous linear differential equation in which the battery voltage is the forcing function.

2-6 THERMAL SYSTEMS

The basic thermal processes are *mixing* of hot and cold fluids; *exchange* of heat through adjoining bodies; *generation* of heat by combustion, chemical reaction, or atomic disintegration; and direct *induction* of heat by molecular or atomic agitation. These thermal processes are used in a variety of ways, and the particular designs of apparatus are literally without number.

Two laws of thermodynamics are applicable to the study of thermal systems. The first governs the manner in which heat energy is generated and determines the amount generated. The second law of thermodynamics governs the flow of heat.

Temperature change in a body is a consequence of the first law of thermodynamics. For a given body, heat input raises the internal energy and the rate of change of body temperature will be proportional to the heat flow to the body. The proportionality constant relating temperature change and heat flow is the *thermal capacity* of the body. Thus

$$C\frac{dT}{dt} = q \qquad \text{(2-26)}$$

where C = thermal capacity, Btu/°F

$\frac{dT^*}{dt}$ = rate of change of temperature, °F/s

q = heat flow, Btu/s

The thermal capacity of a body may be found by multiplying the specific heat of the material by the weight of the body. Thus

$$C = WS \tag{2-27}$$

where W = weight of body, lb
S = specific heat of material, (Btu/lb)/°F

Thermal capacity is analogous to the electric capacitance of a capacitor in an electric circuit. For example, in Fig. 2-13*a*, heat flow into a block of thermal capacitance C causes the temperature to rise above the ambient value T_o. A symbolic representation of this system is shown in Fig. 2-13*b*. Heat flow and charge flow (electric current) as well as temperature and electric voltage are analogous quantities, as can be seen by comparing Figs. 2-9*c* and 2-13*b*.

Heat transmission takes place by *conduction, convection*, and *radiation*. Conduction involves transmission through adjoining bodies, convection involves transmission and mixing, and radiation of heat is the same phenomenon as light radiation.

The rate of heat flow through a body is determined by its *thermal resistance*. This is defined as the change in temperature resulting from

* *This is an exception to the use of lowercase letters to designate time-varying quantities as stated in Sec. 2-1. Capital letter T will be used to represent temperature as a time-varying quantity since lowercase t is reserved for designating time.*

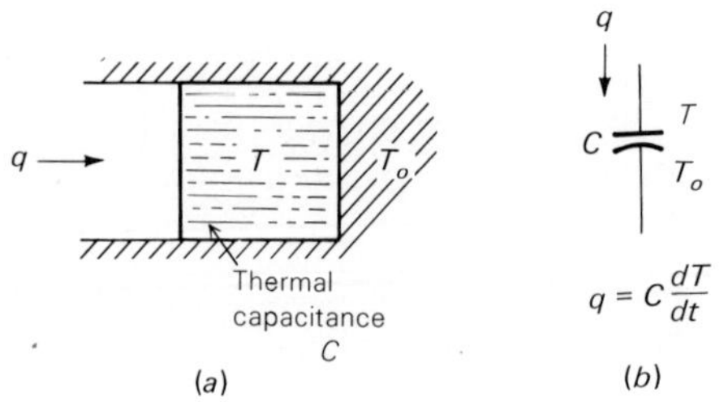

Figure 2-13 (a) Thermal capacitance and (b) symbolic representation

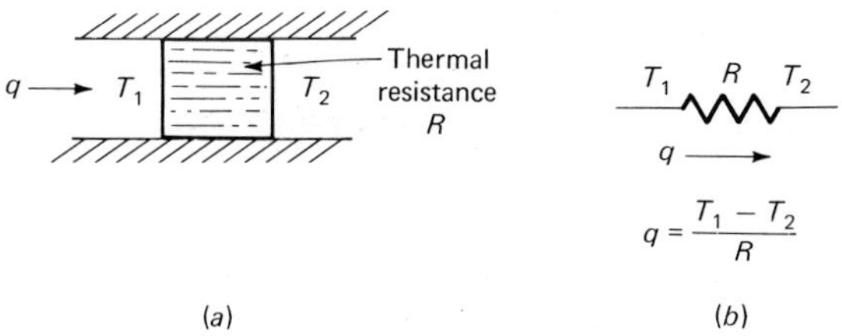

Figure 2-14 (a) Thermal resistance and (b) symbolic representation

a unit change in heat flow rate. Thermal resistance is usually considered to be linear, in which case

$$R = \frac{T_1 - T_2}{q} \tag{2-28}$$

where R = thermal resistance, °F/(Btu/s)
$T_1 - T_2$ = temperature change, °F
q = heat flow rate, Btu/s

Thermal resistance may be considered analogous to the electric resistance of a resistor in an electric circuit. This is illustrated in Fig. 2-14*a*, in which heat flow through thermal resistance R causes the temperature to decrease from T_1 to T_2. A symbolic representation of this effect is shown in Fig. 2-14*b*.

Depending upon the mode of heat transmission, the thermal resistance of a body may be calculated using appropriate transmission coefficients and the dimensions of the body.

If the temperature of a body is considered to be uniform throughout, its thermal behavior can be described by a linear differential equation. This assumption is generally valid for small bodies, or for gases or liquids where perfect mixing takes place. For such a system thermal equilibrium requires that, at any instant, the heat added to the system equals the heat stored plus the heat removed. Thus

$$q_i = q_s + q_o \tag{2-29}$$

where q = heat flow, Btu/s
i = input
s = storage
o = output

EXAMPLE 2-4

Figure 2-15 shows an insulated tank of water heated by an electric element. Develop a differential equation for the water temperature T_w if the rate of heat flow from the heating element is q_h. Assume that all the water is at a uniform temperature and there is no heat storage in the insulation.

Solution

Using Eq. (2-26), the heat stored by the water is

$$q_s = C\frac{dT_w}{dt}$$

where C is the thermal capacity of the water. From Eq. (2-28) the heat loss through the insulation is

$$q_o = \frac{T_w - T_a}{R}$$

where T_a is the temperature of the air around the tank. Applying Eq. (2-29), we have

$$q_h = C\frac{dT_w}{dt} + \frac{T_w - T_a}{R}$$

which may be rewritten as a first-order differential equation

$$\frac{dT_w}{dt} + \frac{T_w}{RC} = \frac{q_h}{C} + \frac{T_a}{RC}$$

where T_w is the variable of interest.

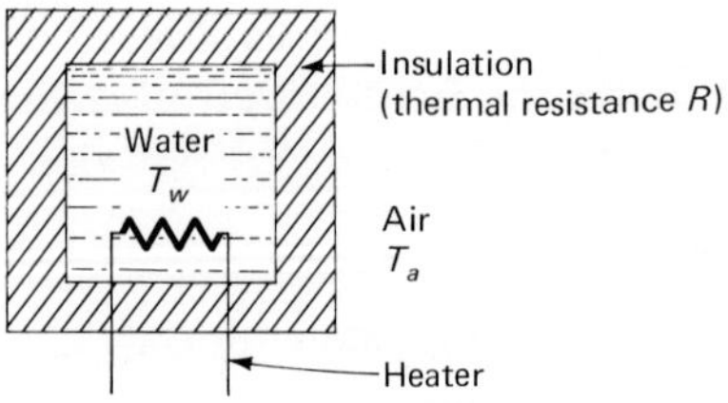

Figure 2-15 Thermal system for Example 2-4

2-7 LIQUID SYSTEMS

Liquid systems are made up of liquid-filled tanks or vessels connected through pipes, valves, and other flow-restricting devices. Such systems are analyzed using the fundamental laws of fluid mechanics. Since we will consider only the most basic aspects of this subject, certain simplifying assumptions are made. Tanks are considered to have a free surface of liquid, connecting pipes are assumed to be full of liquid, and fluid accelerations are considered negligible.

Figure 2-16 shows a simple fluid system consisting of a tank with inflow q_i and outflow q_o, where q is the volumetric flow rate expressed in cubic feet per second or other appropriate units. The liquid level h in the tank is referred to as the *head* and would be measured in feet or another unit of length.

The head provides a pressure which causes the fluid to flow from the tank. This flow is opposed by a property of pipes, or other restrictions, that is called *hydraulic resistance*. For a tank discharging under the pressure of its head, the hydraulic resistance is defined as the change in head required to cause a unit change in flow. Mathematically this is expressed as

$$R = \frac{dh}{dq} \tag{2-30}$$

where R = hydraulic resistance, s/ft^2
h = head, ft
q = flow rate, ft^3/s

Two different types of flow can take place in liquid conducting elements: *laminar flow* and *turbulent flow*. Laminar flow occurs where the fluid velocity is relatively low and the liquid flows in layers. The flow is directly proportional to the head as shown by the straight line

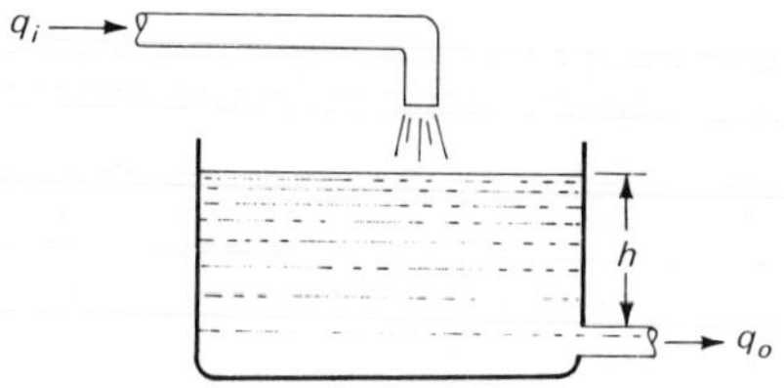

Figure 2-16 Simple fluid system

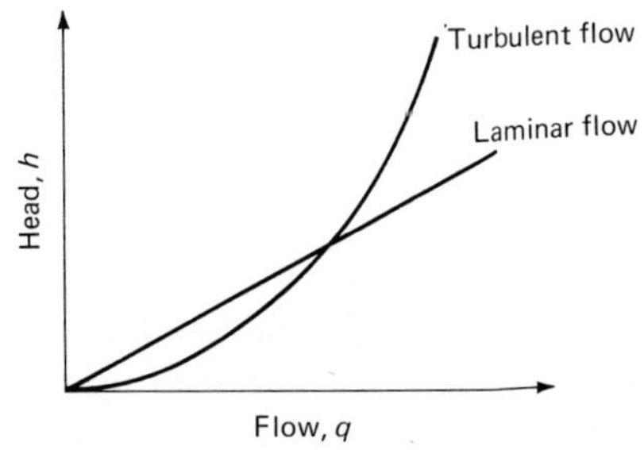

Figure 2-17 Characteristics of turbulent and laminar flow

in Fig. 2-17. The slope of the line is the laminar flow resistance, and Eq. (2-30) becomes simply

$$R = \frac{h}{q} \tag{2-31}$$

Unfortunately, laminar flow is not always encountered in actual practice and usually occurs only with very viscous fluids such as oils.

Turbulent flow occurs when the fluid velocity is relatively high and has the characteristic that the velocity of the liquid at any point varies irregularly. When turbulent flow occurs from a tank discharging under its own head, the flow is found using Bernoulli's law. This may be stated as

$$q = C_d \sqrt{2gh} \tag{2-32}$$

where C_d = coefficient of discharge, ft^2
g = acceleration due to gravity, 32 ft/s^2

Thus we see that turbulent flow varies as the square root of the head. This characteristic is shown in Fig. 2-17. Since this graph is not a straight line, the turbulent resistance is not constant but depends upon the flow rate and head. The slope of the flow curve represents the resistance at a particular value of flow and head. This resistance may be found by solving Eq. (2-32) for h and then differentiating with respect to q. Thus

$$R = \frac{dh}{dq} = \frac{2q}{2C_d{}^2 g} = \frac{2h}{q} \tag{2-33}$$

Over a narrow operating range, a linear flow characteristic may be assumed and the resistance approximated using Eq. (2-33).

Let us now consider the relationship between inflow and liquid level in the system shown in Fig. 2-16. The law for the conservation of mass in a tank containing a liquid may be stated as

$$\text{Inflow} - \text{outflow} = \text{rate of accumulation}$$

or

$$q_i - q_o = A\frac{dh}{dt} \tag{2-34}$$

where A = cross section of tank, ft²

Assuming laminar flow from the tank, $q_o = h/R$ and Eq. (2-34) becomes

$$q_i = \frac{h}{R} + A\frac{dh}{dt} \tag{2-35}$$

Equation (2-35) is a first-order linear differential equation which expresses liquid level as a function of time with the inflow q_i as a forcing function.

We can again draw an analogy between two different physical systems. Comparing the variables in the liquid system with electrical quantities, we see that head, flow, and hydraulic resistance are analogous to voltage, current, and electrical resistance. Further, the storage of liquid in the tank is akin to the accumulation of charge on the plates of a capacitor. Since electrical capacitance is defined as the change in charge per unit change in voltage, we may define *hydraulic capacitance* C as the change in volume per unit change in head. Thus, by definition

$$C = \frac{\Delta V}{\Delta h} = \frac{A\Delta h}{\Delta h} = A \tag{2-36}$$

This equation shows that the hydraulic capacitance is equal to the cross-sectional area of the tank taken at the liquid surface.

EXAMPLE 2-5

A single tank such as shown in Fig. 2-16 has a nominal operating head of 4 ft and a normal outflow of 0.125 ft³/s. The cross-sectional area of the tank is 10 ft².

Determine the differential equation for the outflow as a function of time assuming turbulent flow.

Solution

Using Eq. (2-33), the resistance to flow over the nominal operating range is

$$R = \frac{2h}{q_o} = 64 \text{ sec/ft}^2$$

Since $dh = R\,dq$, $dh/dt = R(dq/dt)$. From Eq. (2-34),

$$q_i - q_o = A\frac{dh}{dt} = AR\frac{dq_o}{dt}$$

and so

$$q_i = q_o + AR\frac{dq_o}{dt}$$

$$= q_o + 640\frac{dq_o}{dt}$$

2-8 GENERAL SYSTEM CHARACTERISTICS

Certain likenesses or analogies in the characteristics of physical systems have been pointed out in this chapter. This section will present a tabulation of these characteristics so that the similarities are more apparent. By applying these analogous properties, the mathematical model of a given system may be analyzed in terms of methods familiar to another type of system.

The behavior of a physical system may be characterized by a cause-and-effect relationship. The system contains material or energy as a quantity or potential. A driving function applied to the system causes a change or an effect in this quantity. The effect is generally in the form of a flow of material or a change in potential, which is opposed by certain properties of the system.

Three identifiable properties of a physical system are resistance, capacitance, and inductance.

Resistance in a physical system is an opposition to the flow of material or energy. It can be defined as the change in potential required to cause a unit change in the flow of the reference quantity.

Capacitance describes the storage properties of a system. It may be defined as the change in the quantity being stored per unit change in the potential.

Inductance is a property that opposes change in the flow of material or energy. It can be defined as the change in potential required to cause a unit change in the rate at which the flow is taking place.

Table 2-1 summarizes the characteristics of the four most common types of physical systems. The English system of units is shown since it is followed in this text. Although the fundamental unit of time (the second) is used, other units of time may be employed as appropriate in a given system.

To further illustrate the analogy in physical systems, Table 2-2 com-

TABLE 2-1 CHARACTERISTICS OF PHYSICAL SYSTEMS

	Type of System			
Property	*Electrical*	*Mechanical*	*Thermal*	*Liquid*
Quantity	Charge, C	Length, ft	Heat, Btu	Volume, ft^3
Time	Second, s	Second, s	Second, s	Second, s
Potential	Voltage, V	Force, lb	Temperature, °F	Head, ft
Flow	Current, A	Velocity, $\frac{ft}{s}$	Rate, $\frac{Btu}{s}$	Rate, $\frac{ft^3}{s}$
Resistance	$\frac{V}{A} = \Omega$	Damping, $\frac{lb}{ft/s}$	$\frac{°F}{Btu/s}$	$\frac{ft}{ft^3/s} = \frac{1}{ft^2/s}$
Capacitance	$\frac{C}{V}$ = farad	Elasticity, $\frac{ft}{lb}$	$\frac{Btu}{°F}$	Area, $\frac{ft^3}{ft} = ft^2$
Inductance	$\frac{V}{A/s}$ = henry	Mass, $\frac{lb}{ft/s^2}$ = slug	Not applicable	Inertance, $\frac{ft}{ft^3/s^2} = \frac{1}{ft^2/s^2}$

pares first-order systems of the type described in this chapter. In the first three systems listed in the table the input is a potential and the response is a flow. By contrast the input to the other systems is a flow and the response is a change in potential. Note that the system differential equation is of the same form in all cases although the variables and constants differ.

2-9 SOLUTIONS OF DIFFERENTIAL EQUATIONS

This chapter has been concerned with writing differential equations to describe physical systems. However, a differential equation does not provide a ready visualization of system behavior since the equation contains one or more derivatives of the dependent variable. On the other hand the *solution* of a differential equation is a relation between the variables which does not contain derivatives and which satisfies the differential equation. Thus the solution of a differential equation is an equation that shows the behavior of the dependent variable as a function of the independent variable.

If nonlinear effects are disregarded, control systems can generally

TABLE 2-2 COMPARISON OF FIRST-ORDER PHYSICAL SYSTEMS

System		Input	Response	Physical Principle	Differential Equation
Mechanical, translational	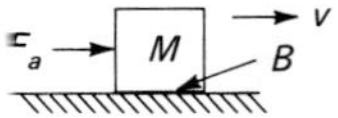	Force, F_a	Velocity, v	$\sum F = 0$	$M\frac{dv}{dt} + Bv = F_a$
Mechanical, rotational	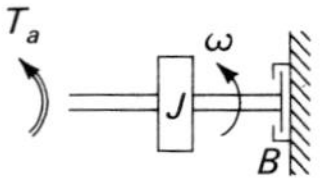	Torque, T_a	Angular velocity, ω	$\sum T = 0$	$J\frac{d\omega}{dt} + B\omega + T_a$
Electrical, series	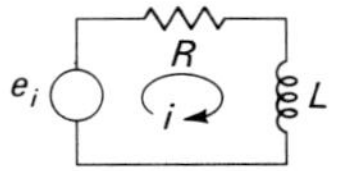	Voltage, e_i	Current, i	$\sum v = 0$	$L\frac{di}{dt} + Ri = e_i$
Electrical, parallel	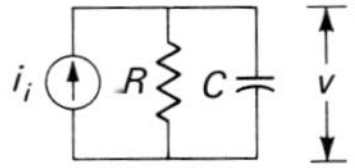	Current, i_i	Voltage, v	$\sum i = 0$	$C\frac{dv}{dt} + \frac{v}{R} = i_i$
Liquid	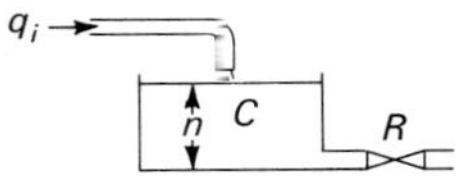	Inflow, q_i	Head, h	$\sum q = 0$	$C\frac{dh}{dt} + \frac{h}{R} = q_i$
Thermal	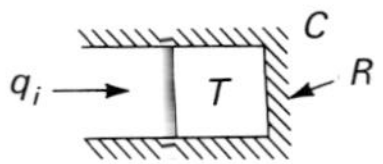	Heat, q_i	Temperature, T	$\sum q = 0$	$C\frac{dT}{dt} + \frac{T}{R} = q_i$

be described by ordinary differential equations with constant coefficients. The solution of these equations yields information concerning the performance of the system and provides a basis for system design. In Chap. 4 we will examine the response of physical systems by considering the solutions of certain differential equations without regard to a method of solution. This will permit a study of system response with a minimum use of mathematics.

Solving a differential equation involves the process of integration in order to eliminate the derivatives contained in the equation. Since a constant of integration is introduced each time we integrate, the solution of a differential equation will contain a number of arbitrary constants corresponding to the highest derivative contained in the equation. For example, a second-order differential equation will have a solution containing two arbitrary constants. The solution containing these arbitrary constants is called the *general solution* of the differential equation.

When specific values are assigned to the arbitrary constants in the general solution, the resulting expression is called a *particular solution* of the differential equation. To arrive at the particular solution, additional information is required concerning the system described by the differential equation. The additional information which allows values to be assigned to the arbitrary constants is generally known as the *boundary conditions*. In physical problems where time is the independent variable, the boundary conditions are frequently the initial values of the quantities involved in the equation. In this case, since the boundary conditions are known at time $t = 0$, they are referred to as *initial conditions*.

At this point we should distinguish between the initial conditions which lead to the particular solution and the problem constants which define the general solution. The problem constants which describe a system are known quantities in the original differential equation and are often referred to as *parameters*. As an example, consider a differential equation which shows how the current in an electrical circuit varies with time, starting at a time identified as $t = 0$. The initial conditions might involve such factors as (1) the quantities of charge on the various capacitors when $t = 0$, (2) the rates of change of current when $t = 0$, and (3) the amounts of current flowing when $t = 0$. On the other hand, the parameters would be the values of the resistors, capacitors, and inductors which make up the circuit.

There are several basic approaches to solving differential equations, and the various methods are extensively treated in higher-level mathematics texts. The so-called classic methods of solution involve integral calculus, and their application is usually an arduous task for all but the simplest equations. Depending upon the form and complexity of the differential equation, any one of a number of classic techniques may be applied. In addition the problem of solution is compounded when the physical system is described by a set of differential equations rather than by a single equation. Classic methods of solution will not be used in this text since they find little application in practice.

In addition to the classic methods of solution, techniques known as *transform methods* are applicable in solving certain linear differential equations. The most common of these is based upon a mathematical relationship known as the Laplace transformation. In particular this method has found extensive application in the study of control systems. With the Laplace transform method a differential equation may be transformed into an algebraic equation. The algebraic equation is solved for the desired variables. This solution is then inversely transformed to obtain the solution of the differential equation. A feature of the Laplace transform method is that initial conditions can be introduced into the differential equation while it is in algebraic form so that the inverse transformation directly yields a particular solution. Moreover simultaneous equations can be transformed and combined using standard substitution or matrix methods. The Laplace transformation is an extremely powerful method for solving differential equations and can be used without a prior understanding of classic methods. It will be presented in Chap. 5 and will be applied in subsequent chapters in analyzing control systems.

A somewhat different approach to solving differential equations has found extensive use during the past two decades (particularly in the study of control systems) and is worthy of mention. We have seen that various dissimilar physical systems are described by differential equations having the same form. In terms of mathematical models a likeness or analogy exists between such systems even though the components and variable quantities may be quite different. If a model of a system can be constructed from elements with the same mathematical characteristics as the elements of some actual system, the model will exhibit behavior of the same form as the system itself. It is important to note that the model need bear no physical resemblance to the system it represents; rather it will behave in a manner similar to the system. For this reason it is known as an *analog simulation* or, using the more common term, an *analog computer*. By observing the time response of the simulation, we are in effect obtaining the solution of the differential equation of the system upon which the simulation is based.

The most common form of analog computer is based on electrical circuits incorporating electronic amplifiers. The circuit voltages represent the variables in the system under simulation. The electronic analog computer has wide versatility and has proven capable of solving very complex differential equations, including those that are nonlinear.

2-10 SIMULTANEOUS EQUATIONS

In closing this chapter it should be emphasized that setting up mathematical models for all physical systems is not a simple matter. The examples in this chapter have been chosen to present the basic concepts and are rather obvious compared with the problems usually encountered in practice. Nevertheless they do illustrate the fact that differential equations are written from a knowledge of the physical laws underlying the system being considered.

The mathematical models described in this chapter involve a single first- or second-order differential equation. Since physical systems are invariably more complex than these examples, we should be aware of the approach commonly used in modeling such systems.

In practice, systems are usually described by a set of simultaneous differential equations of either first or second order. There are a number of familiar examples that may be cited. For instance, in analyzing an electrical network by means of Kirchhoff's laws, a series of loop equations may be written. The equation describing each loop can be no higher than second-order no matter how complex the network may be. In a multiloop network, coupling exists between loops because of common branches or mutual inductance. Since the current in one loop affects the current in other loops, the network represents what is known as a *coupled system*. In the mechanical realm, an example of a coupled system would be two spring-mass systems in which there exists some form of mechanical coupling between the masses.

The reader acquainted with basic circuit theory will recall that network analysis involves the simultaneous solution of loop equations to determine the loop currents. For direct or alternating currents, these equations are algebraic in form (although complex numbers are involved in the ac case). The same approach of simultaneously solving loop equations can be followed for the dynamic case even though the equations are differential in nature. In the preceding section, it was noted that a differential equation may be transformed into an algebraic equation by means of the Laplace transformation. Using this transformation, a set of differential equations can be replaced by a set of algebraic equations and the algebraic equations can then be solved for the variables of interest. This method of simultaneous solution of differential equations will be applied in Chap. 5.

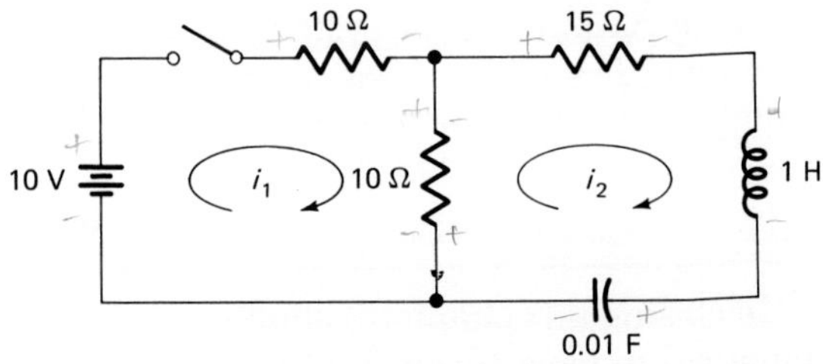

Figure 2-18 Circuit for Example 2-6

EXAMPLE 2-6

For the circuit shown in Fig. 2-18, write the loop equations for the currents after the switch is closed.

Solution

The two loop equations for the circuit are

$$10 = 20i_1 - 10i_2$$

$$0 = -10i_1 + \frac{di_2}{dt} + 25i_2 + 100 \int i_2 \, dt$$

PROBLEMS

2-1. Write the loop equations for the currents in the circuit shown below after the switch is closed.

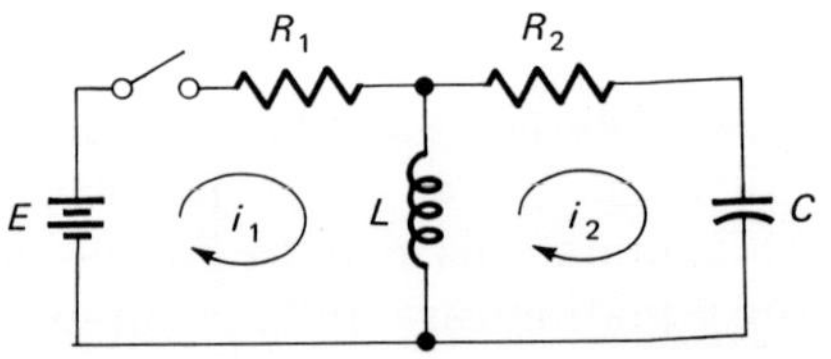

Prob. 2-1

2-2. Write the differential equation for each of the systems shown below.

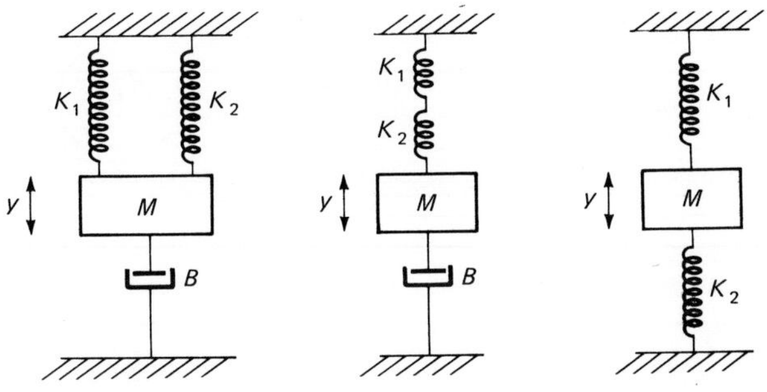

Prob. 2-2

2-3. Obtain the differential equation for the system shown below, where x is the displacement input and y is the displacement response.

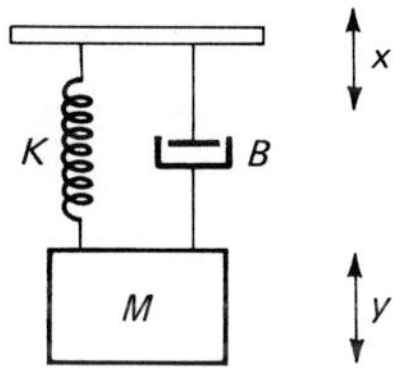

Prob. 2-3

2-4. Write the differential equations that define the system shown below.

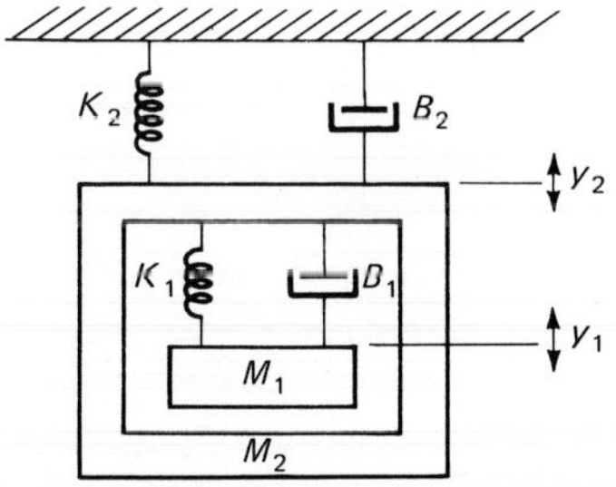

Prob. 2-4

2-5. Write the differential equation for the system shown below, where f is the force input and θ is the angular response.

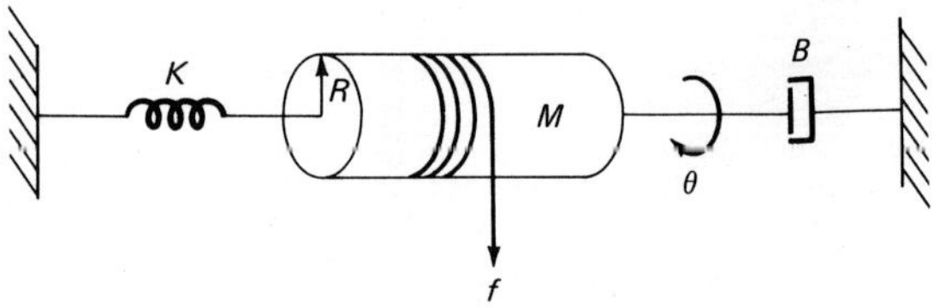

Prob. 2-5

2-6. Write the loop equations for the circuit shown below.

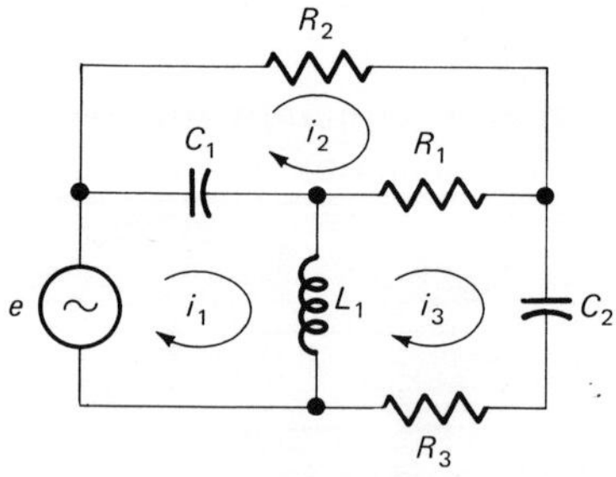

Prob. 2-6

2-7. The number n of bacteria in a certain culture grows at a rate proportional to n. Write a differential equation for this biological system.

2-8. Write a differential equation for the angular position of the pendulum shown below. Neglect friction. Hint: The equation will be nonlinear. Under what conditions might a linear equation be possible?

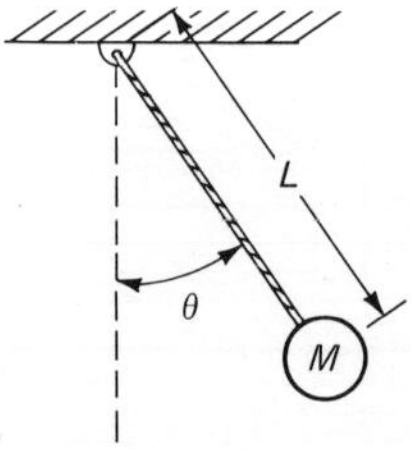

Prob. 2-8

2-9. For the liquid system shown below, write differential equations where q_i is the input and h_2 is the response.

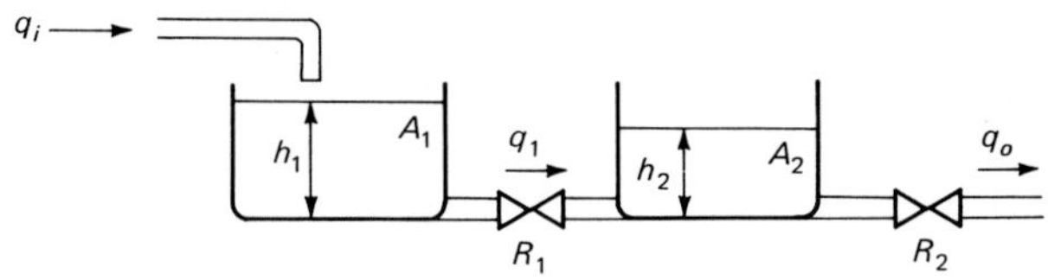

Prob. 2-9

2-10. An insulated water tank contains 1,000 lb of water at an initial temperature of 40°F. Water at 100°F flows into the tank at the rate of 200 lb/min. Assume that perfect mixing occurs, the heat loss through the insulation is negligible, and the outflow is also 200 lb/min. Write a differential equation for the temperature of the water in the tank. Hint: Express heat flow in terms of the flow of the water.

2-11. A heated metal block having thermal capacity C is cooled in a water bath. Assume that the water temperature T_w remains constant, the temperature of the block is uniform throughout, and thermal resistance R occurs only at the surface of the block. Write a differential equation for the temperature of the block.

2-12. In the circuit below, switch 1 is closed long enough to completely charge the capacitor. Write a differential equation for the capacitor current after switch 2 is closed.

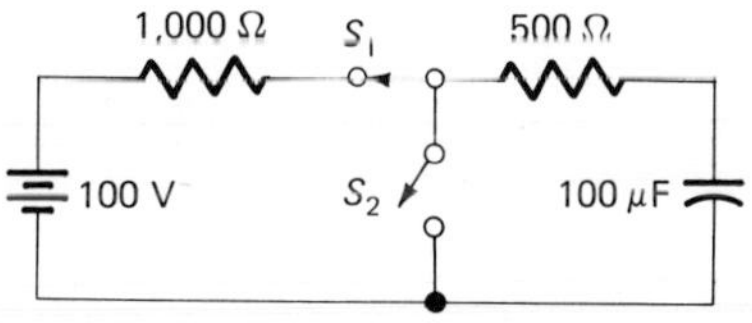

Prob. 2-12

2-13. The tank shown below is a vertical cylinder with a cross section of 5 ft². It initially contains water to a depth of 10 ft. Water is flowing in at a rate of 2 ft³/min and out at a rate of 4 ft³/min. Write a differential equation which shows the water level as a function of time.

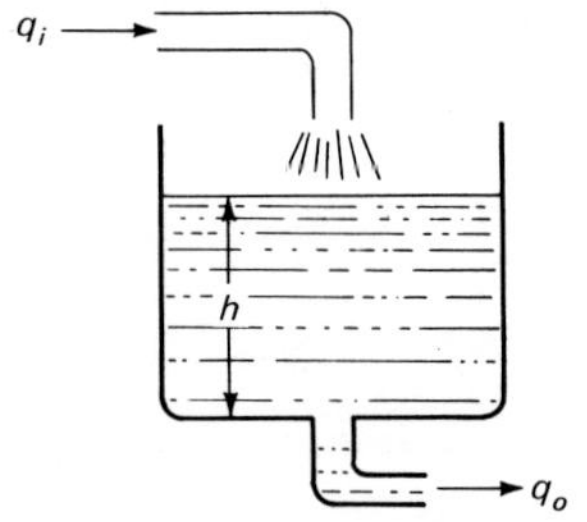

Prob. 2-13

2-14. A flow-restricting device operates according to the equation $q = h^n$. Calculate the hydraulic resistance of the device.

2-15. For the circuit below, derive a differential equation relating the output voltage e_o to the input voltage e_i.

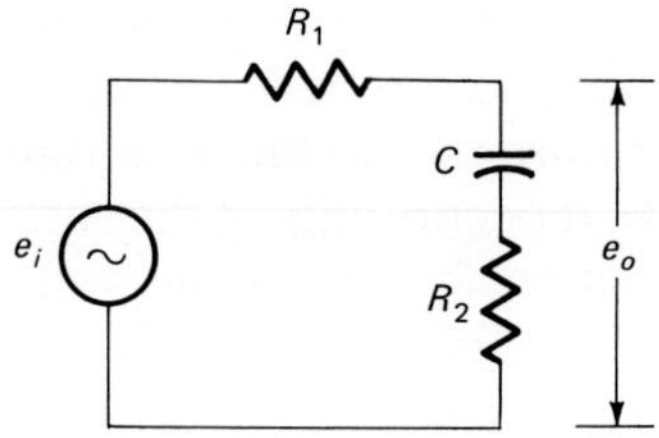

Prob. 2-15

Chapter Three

control system components

3-1 INTRODUCTION

Before proceeding further with the analysis of automatic control systems, we will examine some of the components or devices that are found in actual systems. The various drawings and block diagrams in previous chapters were intended to illustrate principles, and they bear little resemblance to actual components. Although we will continue to use block-diagram representation in future chapters, such diagrams should be more meaningful with some knowledge of the devices being represented.

This chapter is not intended to provide a detailed or exhaustive coverage of control system components. Rather the purpose is to identify some of the more common devices and to set forth their characteristics.

It has been pointed out that the function of an automatic control system is to maintain control over an output variable. It is therefore necessary that the system contain some means of measuring the controlled variable. This is done by a device known as a *transducer*. A transducer may be defined as an element which converts a signal from one form to another—that is, it receives information in the form of a physical quantity such as position, pressure, temperature, etc., and transmits this information in the form of another physical quantity which is usually electrical. Since the behavior of a transducer in a control system is generally unaffected by the closed loop, its operation, accuracy, and stability are major requirements. In many cases the output quantity of a transducer is directly proportional to the input quantity, although there are exceptions. Part A (of this chapter) will describe various transducers used in control applications.

A second necessary part of a control system is a device which measures the error between the actual output of the system and the desired output. Such a device is known as an *error detector*. A number of transducers also incorporate the error-detection function within the device.

The output or final control element is the part of the system that actually controls the flow of energy to the controlled variable. It is sometimes referred to as the *power actuator*. Typical actuators include electric motors, control valves, pumps, and hydraulic cylinders. Obviously the output element must have sufficient power capability to drive the load in the desired manner. A number of common power actuators are discussed in Part B (of this chapter).

In almost all cases the power from the error detector is inadequate to drive the final control element. Since the function of the control system is to drive the output to reduce the error to near 0, it is necessary to provide a *power amplifier* between the error detector and the output. Conventional electronic power amplifiers using transistors are widely used for this purpose. However, other forms of power amplification are also employed.

Control systems using electrical devices frequently incorporate a *voltage amplifier* preceding the power amplifier. In particular, the operational amplifier finds application in automatic control systems. Part C will deal with some common power amplifiers and the use of operational amplifiers.

The choice of control system components and their general arrangement is a design problem in which experience plays a significant part. Often the final choice of components is not made on the basis of best performance alone but includes consideration of other factors such as cost, environment, and standardization of parts.

A. Transducers

3-2 POTENTIOMETERS

A transducer that is frequently used to convert mechanical position to an electrical representation of that position is the *potentiometer*. A potentiometer is a resistor which has a sliding contact or wiper that moves along the resistance element. The wiper motion is generally produced by the rotation of a shaft; however, some potentiometers are designed to operate with a translational motion. A physical representation of a potentiometer is shown in Fig. 3-1*a*. The schematic representation in Fig. 3-1*b* is commonly used because of its simplicity.

Potentiometers are widely used in electrical equipment where an

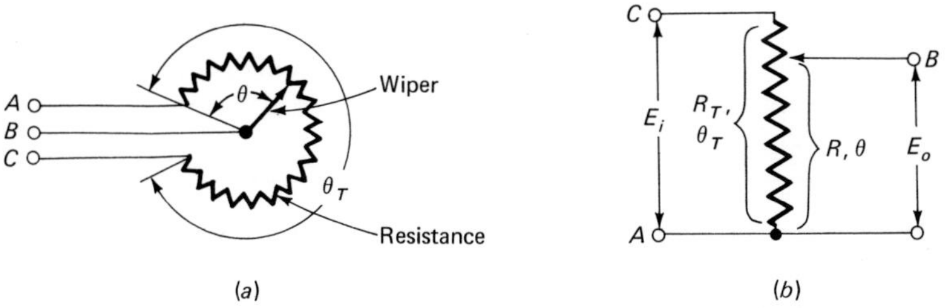

Figure 3-1 Potentiometer: (a) physical representation and (b) schematic representation

adjustable voltage level is desired. A familiar example of potentiometer use is the volume control of a radio receiver. In a servomechanism, the potentiometer serves as a reference input element and as a feedback element.

A potentiometer is referred to as *linear* if the resistance per unit length is a constant. In this case, the resistance is proportional to distance, and referring to Fig. 3-1*b*, we have

$$\frac{R}{R_T} = \frac{\theta}{\theta_t} \tag{3-1}$$

If no current flows through the wiper, the potentiometer is said to be *unloaded*. Since the current through the resistance element is uniform, voltage drops are proportional to resistance, and so

$$\frac{E_o}{E_i} = \frac{R}{R_T} = \frac{\theta}{\theta_T}$$

or

$$E_o = \left(\frac{\theta}{\theta_T}\right) E_i \tag{3-2}$$

Thus the voltage from the potentiometer wiper provides a direct measure of the shaft position.

Servo potentiometers differ from other potentiometers in that they are high precision devices with low shaft torque and inertia and long rotational life. In a potentiometer, torque is required to overcome friction between the wiper shaft and the bearings, between the wiper contact and the resistance element, and between the slip rings and their contacts. The starting torque of servo potentiometers typically ranges from 0.5 to 2 oz-in. Once the wiper shaft is moving, less torque is required to maintain motion, and the running torque may be about one-half the starting value. The moment of inertia depends largely on the diameter of the potentiometer and varies from 0.005 to 0.5 oz-in.^{2} for different sizes. As the potentiometer wiper moves, various types of extraneous signals or *noise* are generated. This noise tends to increase with wear and often limits the useful operating life of the potentiometer. Life specifications of 10^6 to 10^7 revolutions are typical of those quoted by manufacturers.

The most common form of potentiometer has a resistance element made of a length of wire that is wound around a cylindrical core known as a *mandrel*. Such *wire-wound* potentiometers can be fabricated in a wide range of resistance values and power ratings. A second basic type is the *film* potentiometer, which has a resistance element made of a thin film of metal or carbon deposited on a nonconducting base. Since the wiper in a film potentiometer moves along a relatively smooth surface, the torque and wear are generally less than in a wire-wound unit. Thus longer life and improved reliability are obtained. However, film potentiometers do have the disadvantages of higher cost, lower wattage ratings, and greater temperature sensitivity in terms of resistance changes.

If the resistance element is bent into a circular shape as suggested by the diagram in Fig. 3-1*a*, a *single-turn* potentiometer is obtained. Single-turn potentiometers have a usable rotation of slightly less than 360° (although the wiper may have continuous mechanical rotation) since 10° or so are required to provide electrical insulation between the ends of the resistance. Single-turn potentiometers are commonly available with diameters between $\frac{1}{2}$ and 5 in. or more.

As the diameter of a single-turn potentiometer is increased, improved accuracy is obtained as a result of the greater length of the resistance element. By shaping the element in the form of a helix and permitting the slider to move along the helix as it rotates, the length can be increased further with a corresponding increase in accuracy. The resultant unit is a *multiturn* potentiometer, and it is frequently made in 3-, 10-, and 15-turn sizes. Obviously, the shaft in a multiturn unit cannot be rotated continuously, and mechanical stops are necessary at the limits of travel.

For a linear potentiometer, accuracy is determined by how well its characteristic fits an ideal straight-line relationship. For a precision potentiometer, the largest deviation tends to occur at the ends of the windings, where the resistance between wiper and terminal introduces what is known as *end resistance*. This effect creates a step in the actual characteristic as is indicated in Fig. 3-2. In specifying potentiometer accuracy, manufacturers generally quote the *independent linearity*. This is defined as the maximum deviation of the actual characteristic from the best straight line that can be drawn through the actual curve, the deviation being expressed as a percentage of the total resistance

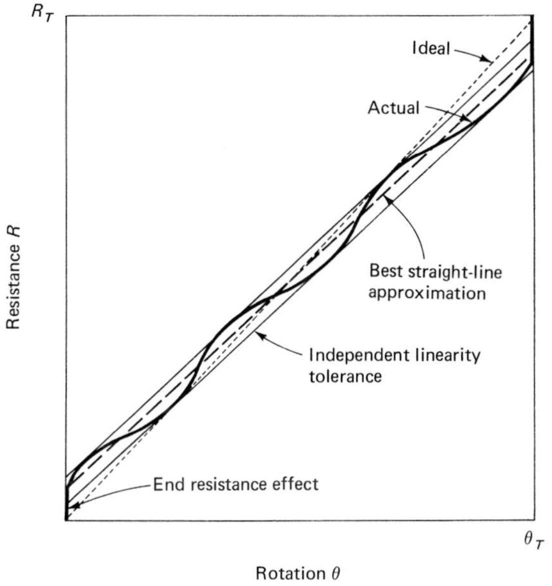

Figure 3-2 Linearity characteristic of potentiometer

of the device. This form of linearity is shown in Fig. 3-2. Since the straight line is not necessarily through the origin, a better figure for linearity is obtained. The linearity tolerance generally improves as the resistance of the element, the diameter, and the number of turns are increased. Typical values are 0.1 to 1.0 percent for single-turn potentiometers and 0.01 to 0.1 percent for multiturn units.

As indicated in Fig. 3-2, the output voltage of a potentiometer varies continuously and smoothly as the shaft angle is changed. This is actually not the case for wire-wound potentiometers since the voltage will vary in steps as the wiper moves from turn to turn of the resistance element. Figure 3-3 shows this characteristic with the steps exaggerated for illustration. For example, if the resistance element had 1,000 turns on the mandrel, the voltage would vary in 1,000 steps as the shaft is fully rotated. Each step would be $\frac{1}{1,000}$ or 0.1 percent of the total range. These increments are referred to as the *resolution* of the potentiometer. Resolution is generally expressed as a percentage of total resistance and can be calculated as

$$\text{Percent resolution} = \frac{100}{\text{number of turns of wire}} \tag{3-3}$$

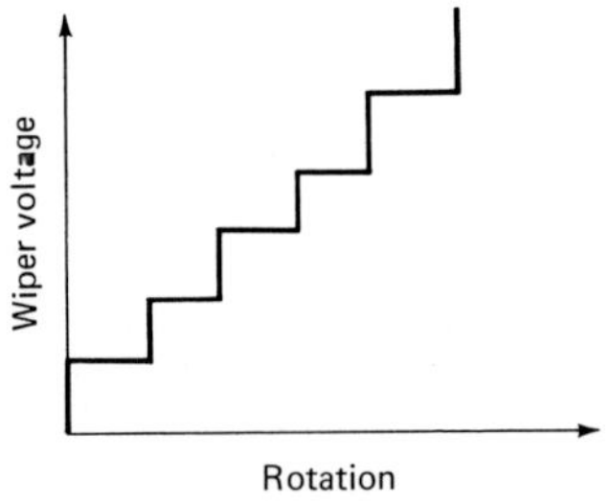

Figure 3-3 Potentiometer resolution

The resolution is actually somewhat better than that indicated by Eq. (3-3) since the wiper usually contacts several turns of wire at the same time. The resultant shorting of turns tends to improve the resolution by a factor of 2 or so. Better resolution is obtained with high-resistance fine wire and a longer resistance element such as are found in a multiturn potentiometer. Wire-wound potentiometers usually have a resolution between 0.001 and 0.1 percent. It should be noted that film potentiometers do not exhibit finite resistance steps and therefore have a resolution that is practically 0.

The resolution of potentiometers is important in servomechanisms because it determines a minimum value of the error signal. If the error signal is too large, the system may tend to oscillate between adjacent wires of the potentiometer and thereby cause greatly increased wear at this point.

In deriving Eq. (3-1) we assumed that no current was drawn from the potentiometer wiper. In practice, however, the potentiometer output voltage must be applied to a load resistance such as the input of an amplifier. When a potentiometer is operated with a load, there is a *loading effect* due to the fact that an output current flows. This is shown in Fig. 3-4. The load current reduces the output voltage so that a nonlinear rather than linear characteristic is obtained. As we shall see, the deviation from linearity depends on both the wiper position and the magnitude of the load.

With the wiper at some angular position, the resistance R that the loaded potentiometer places on the excitation is

$$R = (1 - \alpha)R_P + \frac{\alpha R_P R_L}{R_L + \alpha R_P} \tag{3-4}$$

where R_P = potentiometer resistance
R_L = load resistance
α = normalized shaft setting $= \dfrac{\theta}{\theta_T}$

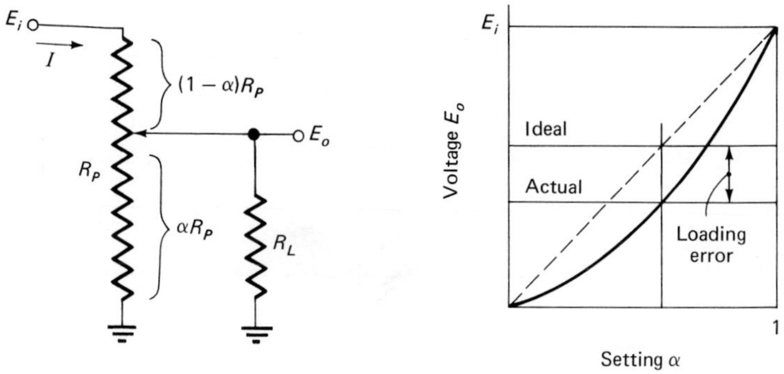

Figure 3-4 Potentiometer with load

Since the current I to the potentiometer is E_i/R, we may express the output voltage as

$$E_o = I\left(\frac{\alpha R_P R_L}{R_L + \alpha R_P}\right) = \frac{E_i}{R}\left(\frac{\alpha R_P R_L}{R_L + \alpha R_P}\right) \tag{3-5}$$

Substituting Eq. (3-4) in Eq. (3-5), we obtain for the output voltage

$$E_o = \frac{\alpha E_i}{1 + \alpha(1-\alpha)R_P/R_L} \tag{3-6}$$

EXAMPLE 3-1

A 10-kΩ servo potentiometer is driving an amplifier whose input resistance is 50 kΩ. For an excitation of 10 V and a shaft position of $\alpha = 0.6$, determine the output voltage.

Solution

From Eq. (3-6), the output is

$$E_o = \frac{(0.6)(10)}{1 + (0.6)(0.4)(10{,}000)/(50{,}000)} = 5.73 \text{ V}$$

For a given shaft position, the error due to loading will be the difference between the ideal and actual outputs as given by Eqs. (3-2) and (3-6), respectively. Thus

$$\text{Loading error} = \alpha E_i - \frac{\alpha E_i}{1 + \alpha(1-\alpha)R_P/R_L} \tag{3-7}$$

$$= \left[\frac{\alpha^2(1-\alpha)}{\alpha(1-\alpha) + R_L/R_P}\right]E_i$$

From Eq. (3-7) we see that the error depends on both the ratio α and the resistance ratio R_L/R_P. For R_L large compared to R_P, maximum loading error occurs at $\alpha \cong 0.67$.

In servomechanisms, the error in shaft position for a given wiper voltage is frequently of interest. For small voltage errors, the position error to obtain the correct voltage is proportional to the magnitude of the voltage error. Thus, if the error factor in Eq. (3-7) is multiplied by the total rotation, the positional error may be calculated.

EXAMPLE 3-2

When the potentiometer in Example 3-1 is driven to null, 6 V appear on the wiper. Determine the positional error caused by loading if the total rotation is 350°.

Solution

The shaft setting α is approximately $E_o/E_i = 0.6$. Using Eq. (3-7), the error factor is

$$\frac{0.6^2\,(1 - 0.6)}{0.6 \times 0.4 + 50{,}000/10{,}000} = 0.0275$$

Thus

$$\theta_E = 0.0275 \times 350 = 9.63°$$

So far we have considered a single-ended potentiometer circuit; that is, one end of the potentiometer is common to both the input and output circuits. Frequently a servo potentiometer is operated in a double-ended arrangement with voltages of opposite polarity applied to its ends. In this case, the resistance element may have a connection or tap at its physical midpoint. If this center tap is grounded as shown in Fig. 3-5,

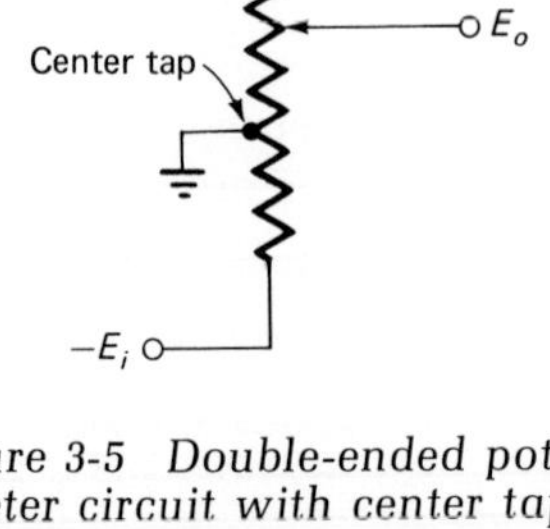

Figure 3-5 Double-ended potentiometer circuit with center tap

a zero-voltage position of the wiper is precisely established. For purposes of analysis, this circuit can be viewed as composed of two single-ended circuits with their ground ends in common. For a given load resistance, we should note that the loading error is reduced with a grounded center tap since the effective potentiometer resistance is halved. Additional taps on the resistance element are sometimes provided in order to establish particular voltage drops across segments of the potentiometer.

Since a potentiometer acts as an attenuating element ($E_o \leq E_i$), it is convenient to define a gain constant K_P for the potentiometer. This is taken as the ratio of the change in output voltage to the change in shaft angle and may be calculated as

$$K_P = \frac{E_i}{\theta_T} \tag{3-8}$$

where K_P = potentiometer constant
E_i = excitation voltage
θ_T = total angle of rotation

3-3 SYNCHROS

Another type of transducer employed in automatic control systems to measure position is the *synchro*. This term designates a range of ac electromechanical devices which are widely used in data transmission and computing systems. Essentially a synchro is a variable transformer that converts angular position to voltage and vice versa. Synchros are frequently identified by trade names such as Selsyn, Autosyn, Telesyn, and others.

There are four basic types of synchros: transmitters, receivers, transformers, and differentials. Although synchros differ in size, power level, and specific details of construction, their general arrangement and the basic principles governing their performance are similar.

Control Transmitter (CX)

The synchro control transmitter (designated CX) functions as a mechanical-to-electrical transducer and is shown symbolically in Fig. 3-6. It has a stationary member or stator made of transformer-core iron. A second iron core member called the rotor is free to rotate within the stator. The stator includes three equally spaced poles 120° apart on which are placed stator windings. One end of each stator winding is connected to a common point, giving a Y configuration. The common

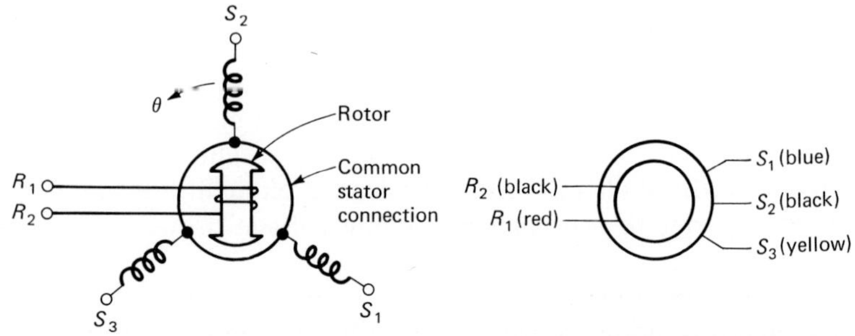

Figure 3-6 Synchro control transmitter (CX)

connection is not accessible, but the outer ends of the windings are brought to external terminals identified as S_1, S_2, and S_3.

The rotor carries a single winding with external connections R_1 and R_2. Connections to the rotor are made through slip rings to permit unlimited rotation, although these are omitted from the drawing for simplicity.

On the right Fig. 3-6 shows a symbol for a control transmitter that is frequently used in system diagrams. The color coding on the connections is standard for synchros which provide pigtail leads rather than labeled terminals.

An ac reference voltage, usually either 60 or 400 Hz, is connected to the rotor winding which functions as the primary winding. The stator windings serve as the secondaries. Recall that the voltage induced in a secondary depends on the angle at which the magnetic lines of flux cut across the turns and will vary from a maximum when the windings are parallel to 0 when the windings are orthogonal. Thus the voltage induced in any one of the secondary windings is dependent upon the physical position of the rotor with respect to that winding.

The zero position of the rotor is generally defined as that position in which the pole axis of the S_2 winding is lined up with the pole axis of the rotor winding. At this angle the voltage induced in S_2 will be a maximum while the voltages induced in S_1 and S_3 will be less. It should be noted that the voltages in S_1 and S_3 will now be equal in magnitude to each other, although differing in electrical phase from S_2 by 180°. For convenience a voltage in phase with the excitation or reference is designated as positive and an out-of-phase voltage as negative.

If the rotor is moved away from the zero position, the induced voltage in the S_2 winding will vary according to the equation

$$E_{S_2} = KE_R \cos \theta \tag{3-9}$$

where E_{S_2} = rms voltage induced in S_2 winding
E_R = rms rotor excitation voltage
K = proportionality constant based on the turns ratio and magnetic coupling between the rotor and stator
θ = rotor angle measured in counterclockwise direction from zero position

This induced voltage is shown in Fig. 3-7. Note that at an angle of 180° the induced voltage is the same magnitude as at 0°. However, the phase relationship between the excitation and the induced voltage is changed by 180 electrical degrees.

It should be pointed out that the sinusoidal waveform in Fig. 3-7 is not an ac voltage in the ordinary sense. Although the voltages induced in the stator windings are ac voltages, the magnitude of the curve represents the rms amplitude of the ac voltage as a function of rotor angle. The polarity of the curve indicates the phase relationship of the induced voltage to the excitation voltage.

The voltages induced in the other two stator windings will vary in a sinusoidal fashion according to the equations

$$E_{S_3} = KE_R \cos (\theta - 120°) \tag{3-10}$$

$$E_{S_1} = KE_R \cos (\theta + 120°) \tag{3-11}$$

These voltages are also shown in Fig. 3-7.

The voltages in Fig. 3-7 are plotted with respect to the common stator connection. Since this point is not generally available, the voltages between the external stator terminals taken by pairs must be considered. Note that for any particular rotor angle, the voltage between two stator terminals may be found by measuring the vertical distance

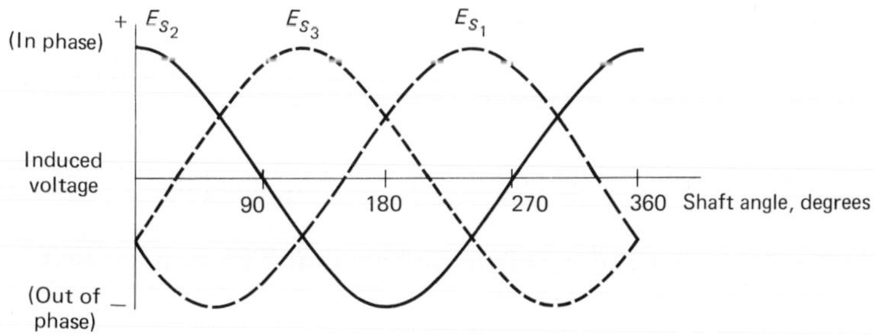

Figure 3-7 Voltage induced in stator windings of control transmitter as a function of shaft angle

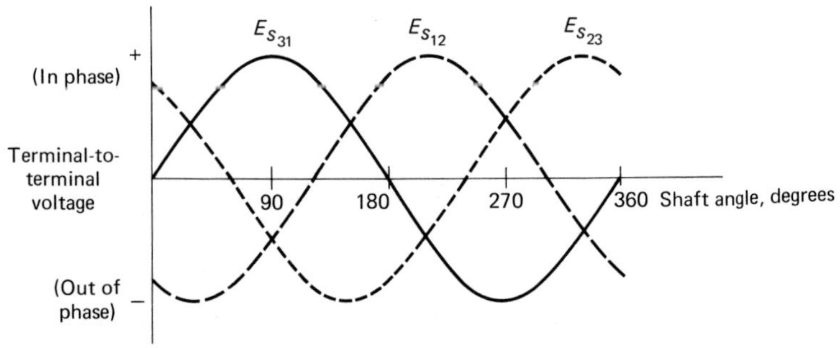

Figure 3-8 Control-transmitter terminal-to-terminal voltage as a function of shaft angle

between the appropriate curves in Fig. 3-7. By means of trigonometric identities the voltages given by Eqs. (3-9) to (3-11) may be combined and the following equations obtained for the voltages between terminals.

$$E_{S_{31}} = \sqrt{3}KE_R \sin \theta \tag{3-12}$$

$$E_{S_{12}} = \sqrt{3}KE_R \sin (\theta - 120°) \tag{3-13}$$

$$E_{S_{23}} = \sqrt{3}KE_R \sin (\theta + 120°) \tag{3-14}$$

These voltages are shown in Fig. 3-8. It should be apparent that the three terminal-to-terminal voltages uniquely identify the shaft angle, and for no two positions of the shaft will these voltages be identical.

Synchros are high-precision transducers which have good reliability. Synchro accuracy is usually specified in terms of minutes of error. Thus a synchro rated as having 10 minutes accuracy (a typical value for synchros) will produce the correct stator voltages for a given angle with no more than 10 minutes error in the angle.

Synchro Receiver (TR)

We have seen that the synchro control transmitter is a position-to-voltage transducer in which the mechanical position of the rotor is represented by a set of terminal-to-terminal stator voltages. By contrast, the synchro torque receiver (designated TR) is a voltage-to-position transducer. For given voltages applied to its stator terminals, the receiver will position its rotor to the corresponding angle.

The physical structure of the synchro receiver is similar to that of the transmitter. Figure 3-9 shows the method of interconnecting a transmitter and receiver. The corresponding stator terminals of the two devices are electrically connected together, and the two rotors are connected to the same excitation. This excitation causes voltages to be induced in the various stator windings as described above. When the rotor angles θ_i and θ_o are the same, identical voltages are induced between corresponding pairs of stator terminals and no stator currents flow in either device. Since no stator magnetic fields are generated, there is no interaction between stator and rotor fields to exert a torque on the rotor.

If the rotor of the transmitter is turned with respect to the rotor of the receiver, the voltages induced in corresponding stator windings are not equal. The voltage differences will cause currents to flow between the stator windings. The resulting stator fields interact with the rotor field to exert a torque on the rotor. The shaft of the transmitter is constrained from turning by the device setting the input angle. However, the shaft of the receiver is free to turn, and the torque will rotate the rotor in the direction such that the displacement between the rotors approaches 0. As the rotors near correspondence, the stator currents decrease. The torque thus approaches 0, and the receiver rotor will stop in a position corresponding to the transmitter-rotor position. Since the receiver duplicates or repeats the transmitter angle, the arrangement is known as a *synchro repeater*.

A synchro receiver generally includes some form of mechanical damping. Without damping the receiver rotor would move too rapidly and overshoot the correct position, then reverse and overshoot in the opposite direction. The result would be a continuous oscillation around the desired position. Another possibility is that the momentum of the rotor may carry it through a complete revolution. Under this condition, the rotor will spin continuously in true motor fashion.

The use of synchro transmitter-receiver pairs is limited to relatively low-torque loads, and the overall accuracy is determined by the friction of the receiver bearings and the load. In ordinary applications, such as driving an indicator pointer at a remote location, the receiver shaft

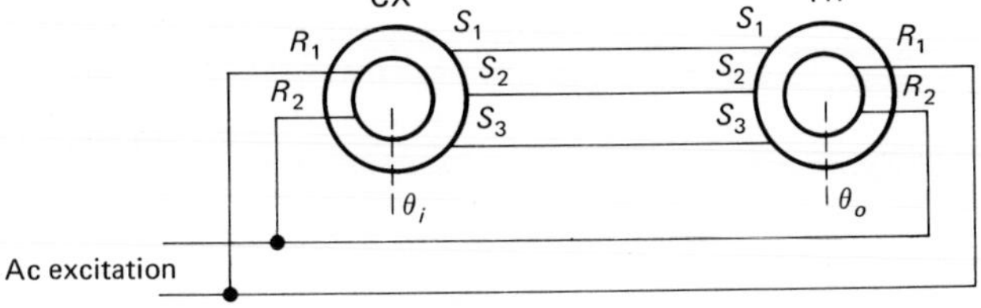

Figure 3-9 Synchro transmitter-receiver pair

will follow the transmitter shaft with less than a 1° displacement. Synchro receiver ratings are based on a degree displacement known as the *torque gradient*. For example, if a receiver is specified as having a torque gradient of 0.5 oz-in./°, a torque of 0.5 oz-in. would pull the rotor 1° away from the desired position.

Control Transformer (CT)

A synchro control transformer is specifically designed for use as a position transducer and error detector in a servomechanism. As indicated in Fig. 3-10, a control transformer is constructed similarly to a transmitter or receiver in that the stator consists of three Y-connected windings spaced at 120° intervals. However, the rotor is cylindrical, which results in a uniform air gap with minimum magnetizing current. In addition the control transformer has relatively higher winding impedances so that the power flowing into it is as low as practical.

Three-wire synchro data from a control transmitter are applied as an input to the stator windings of the control transformer, and the resulting current flow generates a magnetic field within the device. The direction of this field depends on the magnitude and phase of the terminal-to-terminal stator voltages, which in turn depend upon the position of the transmitter rotor. A second input in the form of a mechanical angle is applied to the rotor shaft. The voltage induced in the rotor winding depends on the angle at which the turns are cut by the magnetic field of the stator windings. For example, if the winding axis of the rotor is perpendicular to the magnetic field vector dictated by the transmitter-shaft position, zero voltage is induced in the rotor winding. The zero or null position of the control transformer is then defined as that position for which the rotor output voltage is 0 when

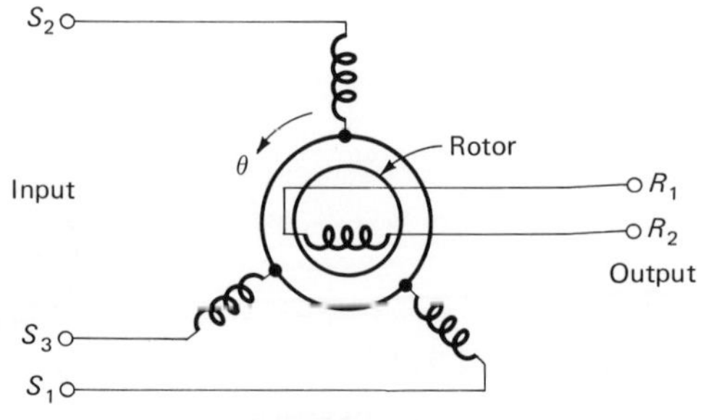

Figure 3-10 Synchro control transformer (CT)

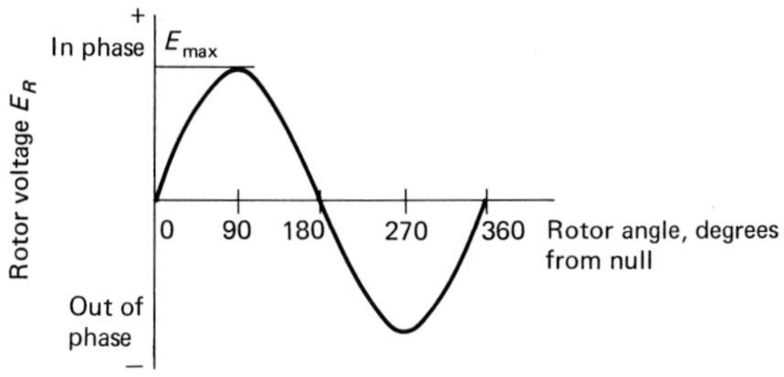

Figure 3-11 Control-transformer rotor voltage as function of rotor angle

synchro data representing 0° are applied to the stator windings. This zero position is indicated in Fig. 3-10.

If the angular position of the control-transformer rotor is turned away from the null condition, the voltage induced in the rotor winding E_R will vary as shown in Fig. 3-11. Note that voltages above the horizontal axis are of one phase, while voltages below the axis are of the opposite phase. The magnitude and phase of the rotor voltage vary as the sine of the angular displacement from null.

If the transmitter shaft is turned, the transformer null will appear at an angle corresponding to the transmitter angle. Thus we see that the induced voltage depends upon the angular difference between the shafts of the transmitter and transformer and is given by the equation

$$E_R = E_{max} \sin(\theta_o - \theta_i) \tag{3-15}$$

where E_{max} = maximum rotor voltage
$\theta_o - \theta_i$ = angular difference between transformer and transmitter shafts

For angles of 20° or less around the null, the voltage-angle curve is nearly linear. The sensitivity of the device may then be expressed as volts per degree around the null point. For typical control transformers, sensitivities from 100 to 500 mV/° are common.

Control Differential Transmitter (CDX)

In the basic synchro transmitter-receiver pair shown in Fig. 3-9 the receiver shaft reproduces the transmitter-shaft rotation. In some applications, however, it is desired to indicate the difference between two shaft positions. The synchro differential transmitter (CDX) is designed to do this.

The differential transmitter has three windings spaced 120° on both the stator and rotor, as indicated in Fig. 3-12. These windings are similar

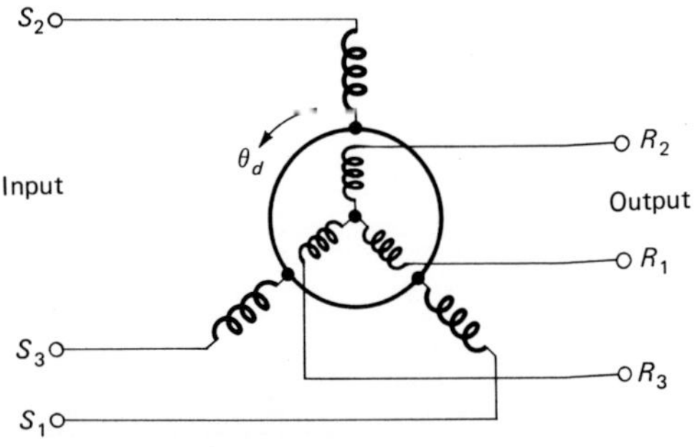

Figure 3-12 Synchro control differential transmitter (CDX)

to the stator windings of the synchro transmitter. This arrangement provides two inputs to the device: the synchro transmitter data applied to the stator windings and the mechanical input applied to the rotor shaft. As with the control transformer the voltages induced in the rotor windings depend upon the orientation of the rotor with respect to the stator. For example, if the rotor windings are aligned with the corresponding stator windings, the induced rotor voltages would duplicate the stator input voltages (assuming a unity turns ratio as is standard practice). This condition is designated as an angle of 0° for the differential shaft. Now assume that the rotor is advanced 120°. The induced rotor voltages will be the same as if the stator input voltages had been altered by moving the transmitter rotor back 120°. We see that the effect of turning the differential shaft is the same as that of turning the transmitter shaft by the same amount but in the opposite direction. Thus the stator voltages of the differential synchro represent the difference between the transmitter-shaft angle and differential-shaft angle.

Figure 3-13 shows the interconnection of a synchro transmitter, differential, and receiver. With the differential inserted between the

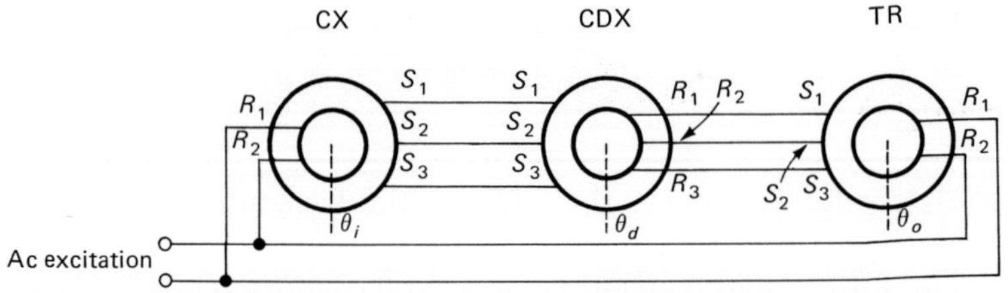

Figure 3-13 Synchro transmitter-differential-receiver chain

transmitter and receiver, the receiver angle θ_o is given by the equation

$$\theta_o = \theta_i - \theta_d \tag{3-16}$$

where θ_i and θ_d are the transmitter and differential angles, respectively.

If the S_1 and S_3 connections of the transmitter and differential are interchanged, the output will be the sum of the inputs. Thus

$$\theta_o = \theta_i + \theta_d \tag{3-17}$$

It is possible to use more than one differential transmitter where more than one input angle must be added into the system.

The synchro arrangements shown in Figs. 3-9 and 3-13 are used for remote indicating systems. Synchro devices are also used with an amplifier and motor as a servomechanism to provide more accuracy than the simple repeater. A servomechanism of this type will be described in Sec. 3-5.

3-4 RESOLVERS

In control applications that involve angular motions, trigonometric functions are invariably used. This is especially true in navigation and trajectory problems. The importance of trigonometric functions has led to the development of devices known as *resolvers*. Basically a resolver is a device that produces electrical output signals that are trigonometric functions of an angular input. Its name is derived from the trigonometric operation of resolving a vector into rectangular coordinates.

One type of resolver that finds wide use is the induction resolver shown in Fig. 3-14. In principle of operation and physical construction, this device is much like the synchro devices discussed in the previous section. The stator and rotor each mount two windings that are spaced at 90 mechanical degrees with respect to each other. If an input voltage

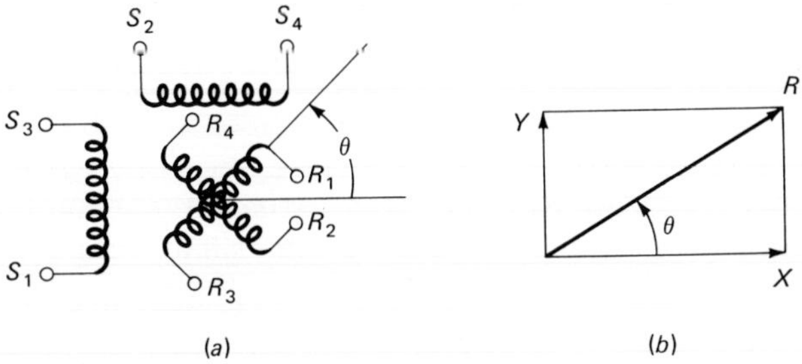

Figure 3-14 (a) Induction resolver and (b) geometry of polar-to-rectangular transformation

is applied to one of the stator windings (with the other stator winding unenergized), the voltages induced in the rotor windings are a function of the applied voltage and the angle of the rotor shaft. Referring back to Fig. 3-7, we note that the voltage induced in a rotor winding will vary as a sinusoidal function of the shaft angle. Since the two rotor windings are displaced by 90°, their output voltages will be proportional to the sine and cosine of the angle. With both stators excited the equations for the output voltages are

$$E_{R_{24}} = K(E_{S_{13}} \cos \theta + E_{S_{24}} \sin \theta) \tag{3-18}$$

$$E_{R_{13}} = K(E_{S_{24}} \cos \theta - E_{S_{13}} \sin \theta) \tag{3-19}$$

where K is a proportionality constant between the stator and rotor voltages.

In polar coordinates, a vector quantity is defined by a magnitude R and an angle θ measured with respect to a reference axis. The geometry of transforming the vector from polar to rectangular coordinates is shown in Fig. 3-14*b*. The rectangular coordinates are then expressed as

$$Y = R \sin \theta \qquad \text{and} \qquad X = R \cos \theta \tag{3-20}$$

Equation (3-20) may be implemented using the induction resolver with one stator input energized with a voltage representing the vector magnitude R. An application of this transformation is in radar operation where the slant range and elevation angle of a target (polar coordinates) are converted into altitude and ground distance (rectangular coordinates).

To perform the rectangular-to-polar transformation a resolver can be used in a servomechanism as will be described in Sec. 3-5.

The functional accuracy of a resolver defines the maximum deviation from a true sinusoidal function of the output. This is usually stated as a percentage of the maximum output voltage. For example, a functional accuracy of 0.05 percent implies that a sine output is

$$E_o = E_i K(\sin \theta \pm 0.0005) \tag{3-21}$$

It should be noted that the angular error represented by a specified amplitude error will vary because the slope of the sine function varies as θ varies. For example, at 0° a 0.05 percent amplitude error represents only 1.7 minutes, while at 90° the same error represents 101 minutes!

3-5 SYNCHRO USE IN SERVOMECHANISMS

The accuracy and load capabilities of synchros are limited when functioning by themselves (e.g., transmitter-receiver pair). For this reason synchros are most commonly used as transducers in servomechanisms.

A servomechanism incorporating a control transformer is shown in Fig. 3-15. The input to the servo is synchro data from a control transmitter whose rotor is positioned to an angle θ_i. The rotor of the control transformer is driven by the servo output, and the device functions as both a feedback element and error detector. As noted earlier the output from the rotor winding of a control transformer is a voltage whose magnitude is proportional to the difference between the desired angle θ_i and the rotor angle θ_o and whose phase is determined by the sign of this difference. This difference or error signal is amplified and drives a servomotor to position a mechanical load connected to the output shaft. With the proper phase connection the motor turns the shaft in the direction which will reduce the error to 0; that is, $\theta_i = \theta_o$. Since ac voltages are used in this servomechanism, the amplifier and motor must respond to signals of the frequency of the excitation voltage. A servo of this type is known as an *ac* or *carrier-type* system.

It can be noted from Fig. 3-11 that there are two null positions of the control-transformer rotor. One of these positions represents a stable null in that the error signal will drive the output toward that null. The other position, 180° away from the first, is an unstable null since the error voltage will drive the servo away from that null. Ambiguity in the output is theoretically possible because of the two nulls. However, in practice the servo will not settle on the unstable null because any slight disturbance, including system noise, will move the output from the unstable position toward the stable null.

In Sec. 3-4 we saw how a resolver provides transformation of a vector from polar coordinates to rectangular form. A second form of coordinate transformation is required when the rectangular coordinates are known and the polar coordinates (magnitude and phase angle) are desired. To perform this transformation, it is necessary to position the resolver shaft to the phase angle. But, since the angle is one of the quantities to be determined, it is not directly available as an input as in the case of the polar-to-rectangular transformation. However, the

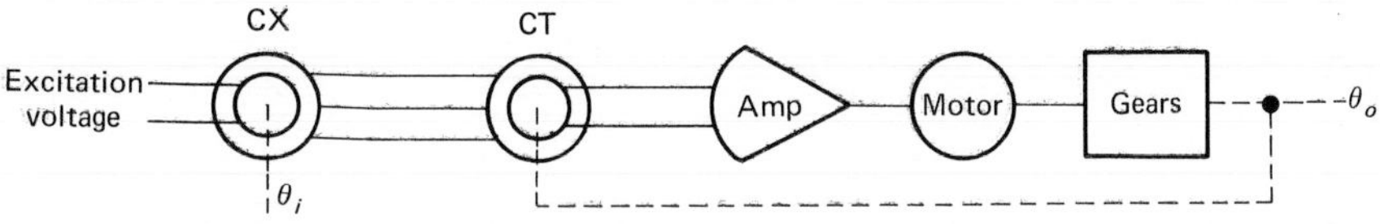

Figure 3-15 Positional servomechanism using control transformer

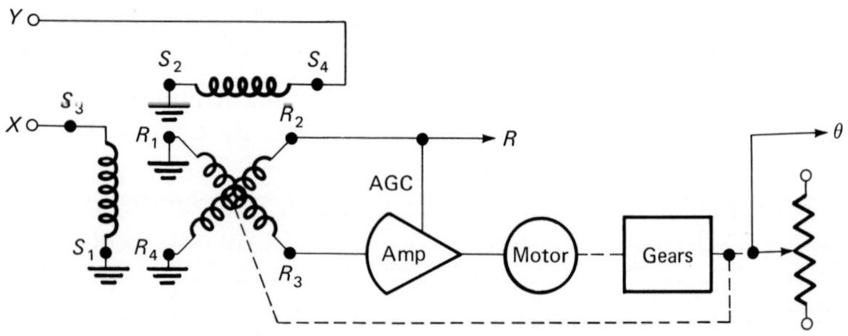

Figure 3-16 Servomechanism for performing rectangular-to-polar transformation

transformation may be accomplished using the servomechanism shown in Fig. 3-16.

Let us first examine the geometry of the transformation with the aid of Fig. 3-17. If X and Y are the rectangular coordinates, the corresponding polar coordinates may be obtained by applying the equations

$$\theta = \tan^{-1}\left(\frac{Y}{X}\right) \qquad \text{and} \qquad R = \sqrt{X^2 + Y^2} \tag{3-22}$$

Two additional equations can be developed as follows:

$$X\cos\theta + Y\sin\theta = R \tag{3-23}$$

and

$$Y\cos\theta - X\sin\theta = 0 \tag{3-24}$$

Rearranging Eq. (3-24), we obtain

$$Y\cos\theta = X\sin\theta$$

$$\frac{Y}{X} = \frac{\sin\theta}{\cos\theta} = \tan\theta \tag{3-25}$$

Thus we see that Eq. (3-24) will be valid for $\theta = \tan^{-1} Y/X$. This equation suggests a means for nulling the servo since the two terms sum

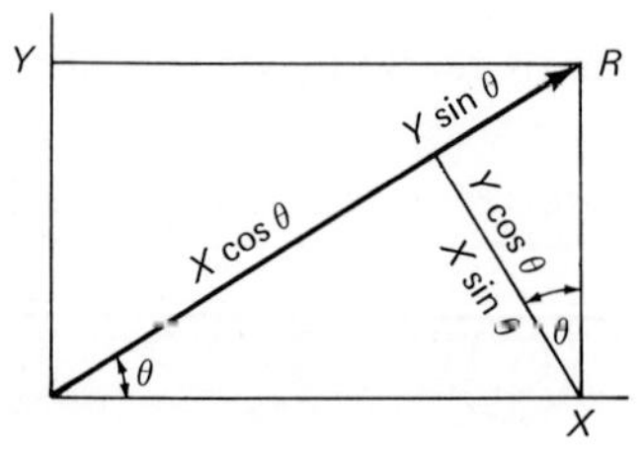

Figure 3-17 Geometry of rectangular-to-polar transformation

to 0 only at the desired angle. Comparing Eqs. (3-19) and (3-24), we see that the R_{13} winding voltage may be used as the error signal to the servo amplifier. The motor will then position the resolver to a null at the angle θ. Comparing Eqs. (3-18) and (3-23) shows that a voltage proportional to the magnitude of R can be obtained at the R_{24} winding. The angle θ can be obtained as an electrical output by connecting a potentiometer to the output shaft.

Unfortunately an additional factor complicates the operation of the servo in performing the rectangular-to-polar transformation. The error voltage applied to the servo amplifier is proportional to both the magnitude of the error in the angle θ and the magnitude of the vector R. The latter may vary over a wide range of values resulting in the gain of the servo loop varying in a like manner. To overcome this undesirable effect, it is common practice to provide automatic gain control (AGC) of the servo amplifier. The voltage $E_{R_{24}}$ which is proportional to the magnitude of R may be used as a gain control signal in a manner similar to the automatic volume control circuits in radio receivers. As the voltage representing R increases, the amplifier gain is decreased, and vice versa.

3-6 LINEAR POSITION TRANSDUCERS

We have seen that synchros and resolvers are rotary position transducers in which the coefficient of coupling between primary and secondary windings is varied by turning one set of windings with respect to the other set. It is also possible to vary the coupling using motion along a straight line (linear motion). Of the various devices that have been designed for this purpose, two, the linear variable differential transformer (LVDT) and the Inductosyn, will be described in this section.

Figure 3-18a illustrates the construction of a linear differential transformer. A single primary coil is wound between two secondary coils on a hollow fiber, aluminum, or other nonmagnetic cylindrical form. The cylinder for various transformers ranges from 1 to 6 in. in length and between one-half and several inches in diameter. The two secondary windings are connected so that the voltages induced in them by the primary winding oppose each other.

A low-reluctance magnetic core is positioned inside the nonmagnetic coil form. When the core is in an exact center position, the magnetic coupling to the two secondary windings is equal. Since the induced secondary voltages are equal in magnitude but opposite in phase, the net output voltage is 0. When the core is moved downward, there is

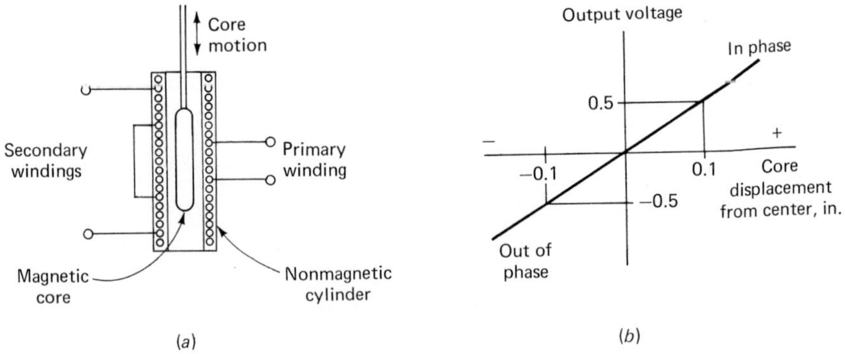

Figure 3-18 Linear variable differential transformer: (a) construction and (b) characteristic

more coupling to the lower secondary and less to the upper. The result is a net output voltage of one phase. When the core is moved upward, a net voltage of the opposite phase appears. For small deviations from the zero position the magnitude of the output is closely proportional to the core displacement. A typical characteristic is shown in Fig. 3-18*b*.

This linear differential transformer has the advantage that no brushes or slip rings are required. However, it can be used only over a limited range of core displacement. A typical device will be linear to within 1 percent over a range of $\pm\frac{1}{4}$ in. from the center position. The sensitivity of the differential transformer is extremely high, and it can detect position changes of less than 0.001 in.

A linear version of the resolver, known as the Inductosyn, is used in precision measuring systems. As shown in Fig. 3-19 this device includes a scale in the form of a glass base on which a conducting material is deposited in a pattern that repeats every 0.1 in. A slider carrying two windings with patterns as shown in the figure is mounted in close proximity to the scale. When the scale is excited by an ac

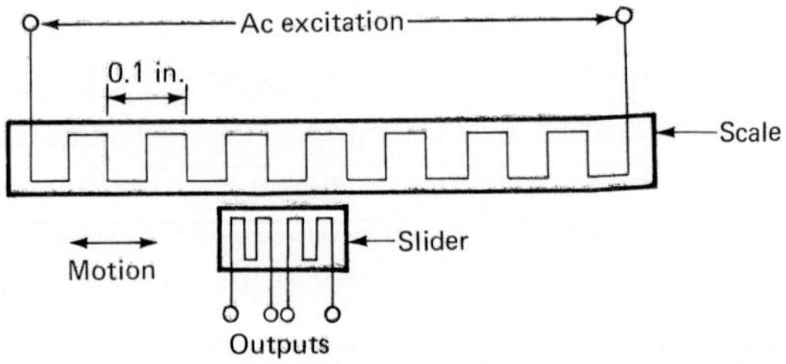

Figure 3-19 Inductosyn construction

voltage of a suitable frequency, the voltage induced in the slider windings will vary in a sinusoidal fashion with each 0.1-in. movement. The physical spacing of the slider windings is such that when the coupling to one winding is at a maximum, the other is at a minimum. By comparing the relative amplitudes of the induced voltages, the slider position can readily be determined with an accuracy of ±0.001 in.

It should be noted that the induced voltages locate the slider only within a given 0.1-in. increment. To measure position over an extended distance, means must be provided to accumulate the number of cycles of induced voltage (0.1-in. increments) through which the slider is moved.

Because of the low inductance and magnetic coupling associated with the windings, the excitation frequency is higher than with the more conventional synchro devices. Frequencies in the range of 10 to 20 kHz are typical.

3-7 TACHOMETERS

An electric generator is a device that converts mechanical energy to electrical energy by means of electromagnetic induction. Electromagnetic induction is a physical phenomenon in which an electric voltage is induced in a conductor when either a moving conductor cuts across the flux in a magnetic field or a moving field cuts across a conductor. The induced or generated voltage is proportional to the number of turns in the conductor as well as the rate of change with time of the magnetic flux with respect to the conductor.

Figure 3-20*a* shows the construction of a generator. It consists of a stationary iron structure called the *field magnet* that creates a magnetic field and a rotating member called the *armature* that carries conductors in the form of a loop or coil. The armature is rotated by an external mechanical drive. As the armature rotates through 360°, an ac voltage is induced in the coil as shown in Fig. 3-20*b*. When the conductor is moving normal to the magnetic field (coil in Fig. 3-20*a* is horizontal), the rate of change of flux is maximum and the induced voltage is maximum. When the conductor is moving parallel to the field (coil is vertical), no flux is being cut and there is no induced voltage. The induced voltage over 360° of rotation is sinusoidal in nature and is repetitive with continuous rotation. It should be noted that both the amplitude and frequency of this ac voltage will depend upon the speed of rotation of the armature.

The polarity of the induced voltage depends on the relative direction of motion of the conductors with respect to the magnetic field. To

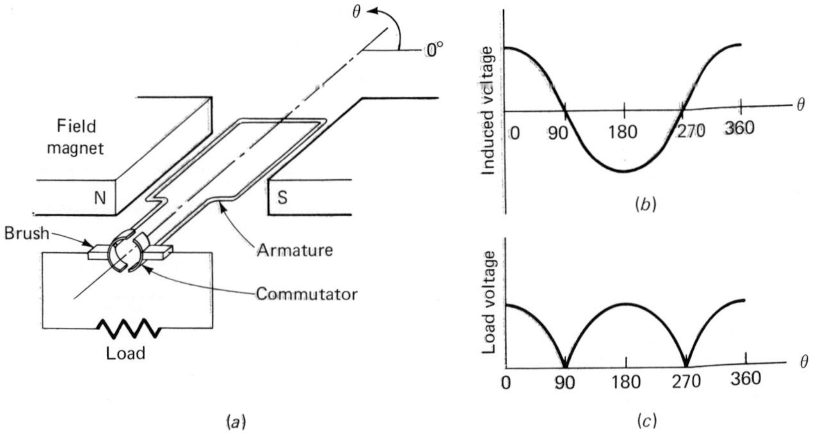

Figure 3-20 Generator (a) construction, (b) voltage induced in armature, (c) load voltage with commutator

provide a unidirectional or dc voltage from the rotating coil as shown in Fig. 3-20*c*, the polarity must be reversed at appropriate points. This reversal may be accomplished by a segmented connection ring called a *commutator*. Electrical contact with the commutator is provided by brushes as indicated in Fig. 3-20*a*. There are two segments on the commutator for each coil on the armature. By increasing the number of coils in the armature (and correspondingly the number of commutator segments) the pulsating nature of the voltage is reduced and the variation or ripple about the dc level is less.

The most extensive use of generators is as a source of electrical energy. In such applications the armature is rotated at a constant speed by an external source called the *prime mover*. The magnetic field is created by electromagnets which are excited either by an external source of voltage or by the generated voltage itself. The basic equation for the generated voltage is

$$V_g = K\phi\omega \tag{3-26}$$

where V_g = generated voltage
ϕ = magnetic flux
ω = armature speed
K = generator constant

A second use of generators, especially in servomechanisms and other control systems, is as a transducer to provide a signal proportional to velocity. In this application the generator is called a *tachometer* or *rate generator*. A primary requirement is that a highly linear relationship exist between the generator voltage and the speed of shaft rotation.

When a dc signal proportional to speed is required, a tachometer employing the principle of a basic dc generator may be used. Because of the small power required, a permanent magnetic field is used for better stability. Dc tachometers are available with linearities in the range of 0.1 to 1.0 percent of full output. A major problem with dc tachometers is the ripple voltage and noise associated with the commutation. This is typically about 1 percent of the maximum output voltage. For this reason it is desirable to operate the tachometer at its highest possible speed, which gives a larger output signal and improves the signal-to-noise ratio. This also helps to keep the noise frequency as high as possible so that the noise is more easily filtered from the signal. A change in the direction of rotation changes the polarity of the dc output voltage.

The sensitivity of a tachometer is frequently specified by the manufacturer in terms of the output voltage at a speed of 1,000 rpm. Values from 1 to 10 V/1,000 rpm are typical for dc tachometers.

When an ac velocity signal is required, the basic ac generator cannot be used because the frequency of the output, as well as the amplitude, is proportional to shaft speed. Since most ac control systems operate at a fixed frequency, this type of generator is not practical.

A type of ac tachometer that is widely used is known as a *drag-cup tachometer*. This device has two stator windings that are mechanically positioned at an angle of 90° with respect to each other. As shown in Fig. 3-21, one winding is excited by an ac reference voltage of fixed amplitude and frequency. The other winding functions as the output

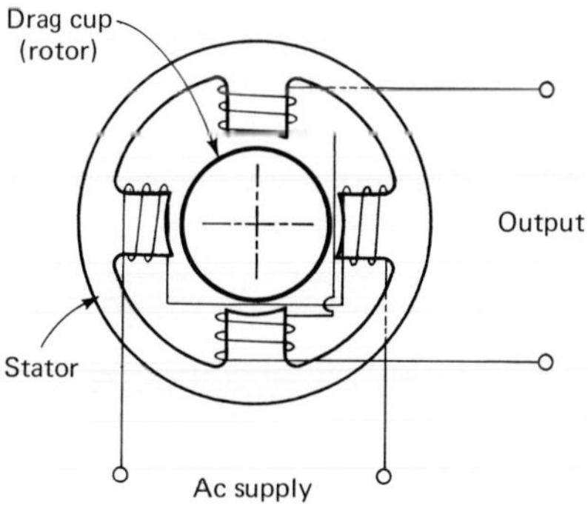

Figure 3-21 Drag-cup tachometer

of the tachometer. The rotor is made in the form of a thin cup of aluminum or some other nonmagnetic metal.

When the cup is not rotating there is no magnetic coupling between the stator windings since they are at right angles to one another. Hence no voltage is induced in the output due to the reference excitation. However, as the cup rotates, eddy currents are induced that generate a rotor flux. This rotor flux adds to the reference flux so that the direction of the total flux is shifted. This results in an induced output voltage that is proportional to cup velocity. The phase of the output voltage is 0° or 180° with respect to the reference voltage, depending on the direction of shaft rotation.

Drag-cup tachometers exhibit good linearity as well as a low residual output voltage when the shaft is not moving. Accuracies of 0.1 percent may be obtained over a wide range of speeds.

Since tachometers of both dc and ac types are operated with constant field, Eq. (3-26) may be reduced to

$$V_g = K_g \omega \tag{3-27}$$

where V_g = output voltage
K_g = tachometer constant
ω = shaft speed

3-8 ACCELEROMETERS

An accelerometer is a transducer that produces a signal proportional to the linear acceleration along its sensitive axis. The operation of most accelerometers is based on Newton's second law of motion. The acceleration is measured indirectly by measuring the force required to accelerate a known mass.

Figure 3-22 depicts a common type of accelerometer that is basically a spring-mass system similar to that described in Sec. 2-4. The accelerometer is attached to the object whose acceleration is to be measured and undergoes the same acceleration as that object. A small mass is supported by cantilever springs attached to the accelerometer case. When the entire unit is accelerated along the direction of the sensing axis, the inertia of the sensing mass causes it to be displaced from its equilibrium until the springs exert a force large enough to accelerate the mass at the same rate as the frame. When this occurs, the spring force equals the accelerating force. Thus

$$F = Kx = Ma$$

or

$$x = \frac{M}{K} a \tag{3-28}$$

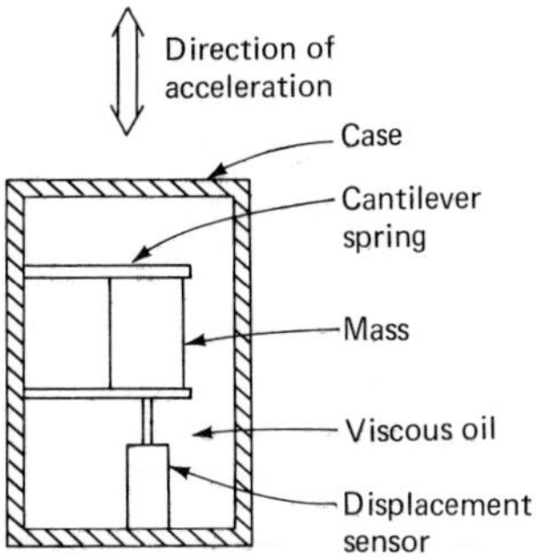

Figure 3-22 Basic accelerometer

where x = displacement of the mass
M = mass
K = spring constant
a = acceleration

Equation (3-28) indicates that the displacement of the mass is proportional to the magnitude of the acceleration. This displacement may be measured using a position transducer such as a linear potentiometer or LVDT. The transducer output is then an electrical signal proportional to acceleration along the sensing axis.

Damping is required to dissipate vibrational energy. This is frequently provided by a viscous oil surrounding the mass.

Accelerometers of the type described above are accurate to within 1 percent over a range from 1 to 100 g (g = gravitational unit = 32 ft/s^2). A more accurate type of accelerometer incorporates negative feedback in a force-balance arrangement that corrects inherent inaccuracies. Such accelerometers can measure accelerations as low as 0.001 g at an accuracy of 0.1 percent.

It should be noted that an accelerometer cannot distinguish between the acceleration of gravity and other accelerations. Thus, if an accelerometer is to be made insensitive to gravity, it must be mounted on an accurately horizontal platform.

3-9 STRAIN GAGES

A transducer that is used to measure force or weight conditions in machine tools, rolling mills, pipes, aircraft structures, etc., is the *strain gage*. The most common of the various available types is the resistance strain gage.

The resistance strain gage depends on the fact that a metal wire when stretched undergoes a change in resistance that bears a definite relation to the change in length. The dimensions of such a gage may be made small so as to allow the stress at a specific location to be determined with high accuracy.

A typical commercial strain gage consists of a grid of fine wire (several mils in diameter) bonded to a thin backing piece as shown in Fig. 3-23. This assembly, which may be about the size of a postage stamp, is cemented onto the surface of the member under test at the point where the stress is to be measured. When the member is stretched by a force, the wire is then stretched by an equal amount.

The resistance of the wire is given by the equation

$$R = \frac{\rho L}{A} \tag{3-29}$$

where R = resistance
L = wire length
A = wire cross-sectional area
ρ = resistivity of the material

When the wire is stretched, the increase in length is accompanied by a decrease in cross section. Both of these changes increase the resistance of the wire. The ratio of resistance change to change in length is known as the *gage factor*. Thus

$$G = \frac{\Delta R/R}{\Delta L/L} \tag{3-30}$$

where G = gage factor
ΔR = change in resistance R
ΔL = change in length L

For the various metals used in strain-gage wire, the gage factor is generally between 2 and 5.

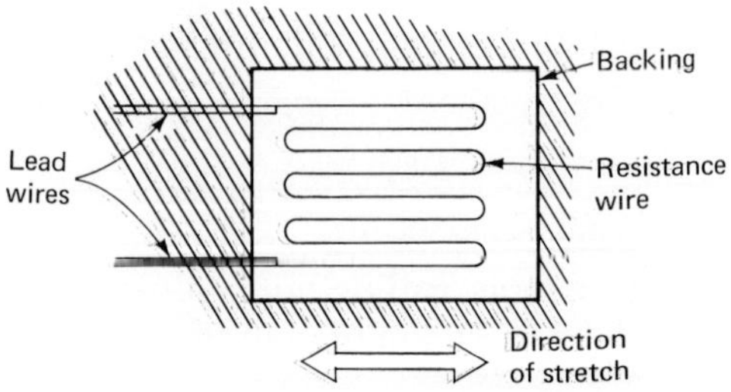

Figure 3-23 Resistance strain gage

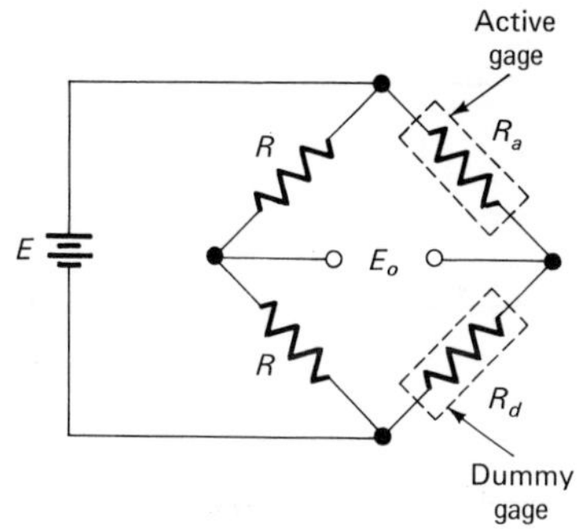

Figure 3-24 Resistance-strain-gage bridge circuit

Resistance also varies with temperature. If the temperature coefficients of the gage wire and the material under test differ, a change in temperature will give an apparent stress. The common method of alleviating this difficulty is to mount a dummy gage close to the actual one but in a position free of strain. The two gages are then connected as adjacent arms of a Wheatstone bridge as shown in Fig. 3-24. Assuming a high-impedance load on the bridge, the output voltage E_o is

$$E_o = \left(\frac{1}{2} - \frac{1}{1 + R_d/R_a}\right) E \qquad (3\text{-}31)$$

where R_a = resistance of active strain gage
R_d = resistance of dummy gage
E = bridge excitation voltage

Note that like variations in R_a and R_d due to temperature will not affect the bridge output. However, a change in the active gage resistance will cause an output.

Since the change of length to be measured may be less than 0.001 in., the corresponding change in wire resistance and the bridge output are also quite small. Considerable amplification may therefore be required to bring the bridge output to a usable level. Since ac amplifiers are generally simpler, the excitation to the bridge may frequently be an ac voltage rather than the dc voltage shown in Fig. 3-24.

3-10 TEMPERATURE TRANSDUCERS

The temperatures found in industrial processes range from about −200 to 2000°C, and the precise measurement of temperature is often necessary for controlling the chemical and physical properties of ma-

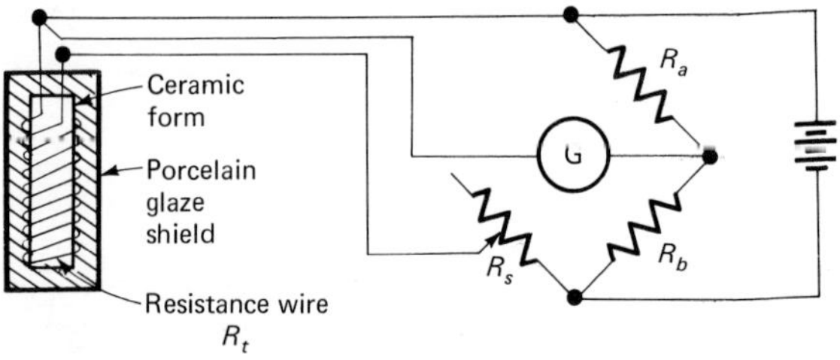

Figure 3-25 Resistance thermometer and bridge circuit

terials. This range of temperatures can be measured by various transducers that convert temperature to an electrical signal. The method selected for a given application will depend on not only the temperature but also the accuracy of measurement, the speed of response, and the cost. Two of the most common methods are the resistance thermometer and the thermocouple, and these will be briefly described in this section.

A *resistance thermometer* indicates temperature by measuring the change in resistance of a metal as the temperature changes. The materials most often used are nickel, which is inexpensive and has a relatively large change in resistance with temperature, and platinum, which is a very stable material and is highly resistant to corrosion and oxidation.

A length of fine resistance wire is wound on a nonconducting form as shown in Fig. 3-25. The form is then generally protected by a shield or sheath. Where the temperature of a corrosive fluid or conducting medium is being measured, it is common practice to enclose the resistance element in a sealed bulb.

The equation relating resistance and temperature is

$$R_1 = R_0\,[1 + \alpha(T_1 - 20)] \tag{3-32}$$

TABLE 3-1 CHARACTERISTICS OF RESISTANCE-THERMOMETER BULBS

Wire Material	*Bulb Resistance, Ω*	*Useful Range, °C*	*Temperature Coefficient of Resistance at 20°C*
Platinum	10 to 35	−150 to 700	0.00393
Nickel	100 to 300	−150 to 300	0.0038
Copper	10	−40 to 150	0.006

where R_1 = resistance at T_1°C
R_0 = resistance at 20°C
α = temperature coefficient of resistance at 20°C

Table 3-1 shows the characteristics of platinum, nickel, and copper resistance-thermometer bulbs.

For temperature measurement, the resistance thermometer may be used as an arm of a conventional Wheatstone bridge as indicated in Fig. 3-25. When the bridge is balanced, the resistance R_t of the thermometer is

$$R_t = \frac{R_s R_a}{R_b} \tag{3-33}$$

where R_s is the slide-wire resistance and R_a and R_b are fixed resistors. An error may be introduced into the measurement due to a resistance change in the conductors connecting the thermometer to the bridge. A three-wire system as shown in Fig. 3-25 is frequently used for correction of this error. Note that any change in the resistance of the leads is balanced out by including one conductor in each side of the bridge.

One of the transducers most frequently used to convert temperature to an electrical signal is the *thermocouple*. Basically a thermocouple consists of two dissimilar wires welded together at one end. This junction forms the *measuring* or *hot junction*. The other two ends are known as the *reference* or *cold junction* and connect to a device that measures electrical voltage. A thermocouple works on the principle that a small voltage is generated at the junction of two dissimilar materials, and this voltage is closely proportional to the temperature of the junction. Table 3-2 gives characteristics of four of the most commonly used thermocouples.

The thermocouple voltage can be measured by a galvanometer, another sensitive meter, or a bridge circuit. Although the hot junction is frequently located some distance from the measuring instrument, in practice the thermocouple leads themselves are kept fairly short and low-resistance copper lead wires are used to extend the distance. In

TABLE 3-2 CHARACTERISTICS OF THERMOCOUPLES

Materials	*Useful Range, °C*	*Sensitivity, mV/°C*
Copper–constantan	−200 to 350	0.059
Iron–constantan	150 to 750	0.063
Chromel–alumel	250 to 1250	0.039
Platinum, 10% rhodium–platinum	500 to 1400	0.011

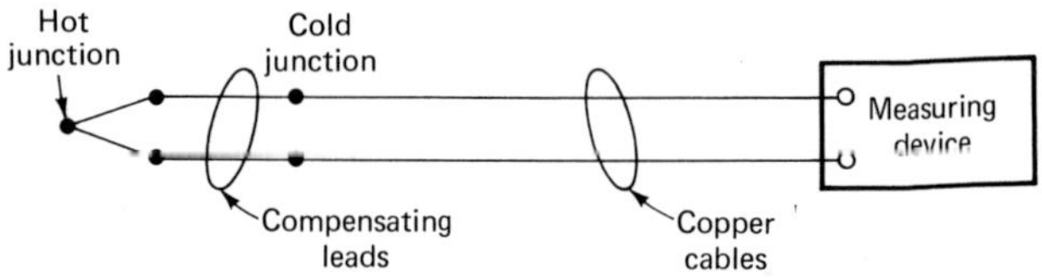

Figure 3-26 Thermocouple circuit

some cases cables having the same thermoelectric properties as the thermocouple, but of less expensive materials, are used to extend the thermocouple to a point where a steady temperature is available for the cold junction. These cables are known as *compensating leads*. This arrangement is shown in Fig. 3-26. The cold junction must either be maintained at a constant known temperature or have corrections made if its temperature changes since thermal voltages are also generated at this junction.

In the temperature range for which they are suitable, resistance thermometers are more accurate, give a larger signal, and have no cold junction problem as compared to thermocouples. On the other hand thermocouples are quite simple, cover a wide range of temperatures, and exhibit fast response.

3-11 SUMMARY OF TRANSDUCERS

The transducers described in this chapter are frequently found in automatic control systems, especially servomechanisms. However, they represent only a cross section of the many devices available for measuring physical quantities. A more extensive discussion of transducers is beyond the scope of this chapter; in fact, a text in itself would be required for a complete treatment of the subject.

Table 3-3 provides a partial listing of physical variables that are of interest in control and the basic transducers commonly applied to each. In addition to these variables, quantities such as pH, humidity, smoke density, gas composition, and specific gravity are measured in industrial processes by transducers in the form of analytical instruments.

B. *Power Actuators*

3-12 DC MOTORS

An electric motor is a device that converts electrical energy to mechanical energy. In many industrial processes rotary or linear motion

TABLE 3-3 PHYSICAL VARIABLES AND TRANSDUCERS

Variable	*Transducer*	*Operating Principle*
Acceleration	Seismic mass	Acceleration causes displacement of mass relative to case
Attitude (relative orientation)	Gyroscope	Change in orientation causes angular displacement with respect to fixed axis of rotating wheel on axle
Displacement	Potentiometer	Sliding contact on resistance element produces change in voltage ratio
	Synchro	Rotating armature produces change in ac voltages
	Linear variable differential transformer	Change in magnetic coupling with core position causes change in mutually induced voltage
Fluid flow	Orifice plate	Pressure differential across small opening in plate is proportional to flow rate
	Venturi tube	Pressure differential across constriction in tube is proportional to flow rate
	Turbine	Bladed rotor turns at speed proportional to flow rate
	Electromagnetic flowmeter	Electrically conductive fluid in magnetic field has electromotive force induced that is proportional to flow rate
Humidity	Resistance hygrometer	Change in ambient relative humidity produces variation in resistance of sensing material
Light	Photovoltaic cell	Illumination of junction between two dissimilar metals produces output voltage

TABLE 3-3 *(Continued)*

Variable	*Transducer*	*Operating Principle*
	Photoconductive cell	Resistance of semiconductor material changes with incident light level
	Photodiode or phototransistor	Resistance across junction in semiconductor device changes with light level
Liquid level	Float	Position of float on liquid surface is mechanically coupled to displacement transducer
	Resistance probe (electrically conductive liquid)	Resistance between electrodes changes with level
	Capacitive probe	Variable dielectric with level change causes capacitance change between electrodes
	Optical	Interaction of liquid surface with light beam is detected by light sensor
Nuclear radiation	Ionization chamber (Geiger counter)	Incident radiation generates ion pair in gas or solid material which produces ionization current
	Scintillation detector	Radiation generates light emission
Pressure	Bourdon tube	Internal pressure causes straightening deflection in curved or twisted tube sealed at one end
	Bellows	Axial deflection produced in thin-walled tube formed in convolutions
	Diaphragm	Deflection produced in circular plate fastened continuously around its edge

TABLE 3-3 (*Continued*)

Variable	*Transducer*	*Operating Principle*
	Manometer	Displacement of liquid column in U-shaped tube
	Piezoelectric crystal	Voltage produced by mechanical distortion of crystal
Strain	Strain gage	Deformation of sensing material causes change in electrical resistance
Temperature	Resistance thermometer	Resistance of sensing material changes with temperature
	Thermocouple	Electromotive force proportional to temperature produced at junction between two dissimilar metals
Velocity	Tachometer	Armature rotating in magnetic field generates voltage proportional to speed

is imparted to a mechanical load by a motor. In particular, motors find wide use in automatic control systems since their operation can be continuously and accurately controlled over a wide range of speeds and loads.

Electric motors are available with power ratings that range from fractional amounts to hundreds of horsepower. Although a wide variety of designs are available to meet specific requirements, a basic distinction is in terms of operation from either dc or ac voltages.

The dc motor is extensively used in both high- and low-power control systems. The motor is similar to the dc generator in that it consists of a field magnet and armature. The motor operates on the principle that a force is exerted on a current-carrying conductor located in a magnetic field. If the conductor is in the form of a loop as shown in Fig. 3-27, the force acts as a torque which tends to rotate the loop. The direction of the torque is dependent on the relative direction of the magnetic flux and the loop current. To provide a unidirectional torque when the loop is rotating, the direction of the current must be reversed at appropriate points. This reversal is accomplished by a commutator and brushes. In practical motor designs the armature is wound with a

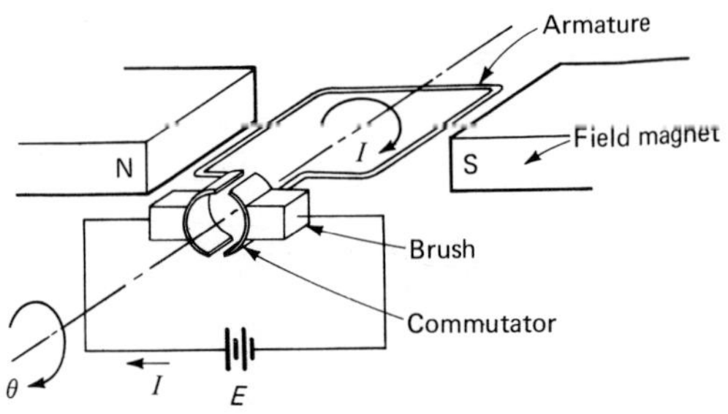

Figure 3-27 Dc motor principle

number of loops in order to provide a more uniform torque and rotation. The commutator must then have segments corresponding to the number of loops in the armature winding.

The torque developed by a motor is proportional to both the magnetic flux and the armature current. The magnetic flux may be produced by permanent magnets or electromagnetically from a winding on the field magnet. The current that establishes the magnetic field is called the *field current.*

In control-system applications, the armature and field are excited from separate sources and either may be controlled to vary the torque and speed of the motor. In the more common method, shown in Fig. 3-28*a*, a constant field current I_f is supplied and a variable control voltage V_a is applied to the armature. In this case the control source must provide the full power requirements of the motor.

Figure 3-28*b* shows an alternative arrangement in which a constant armature current is supplied and the field current is varied. This method is advantageous in that the control source must supply only the relatively small field current and the torque is closely proportional

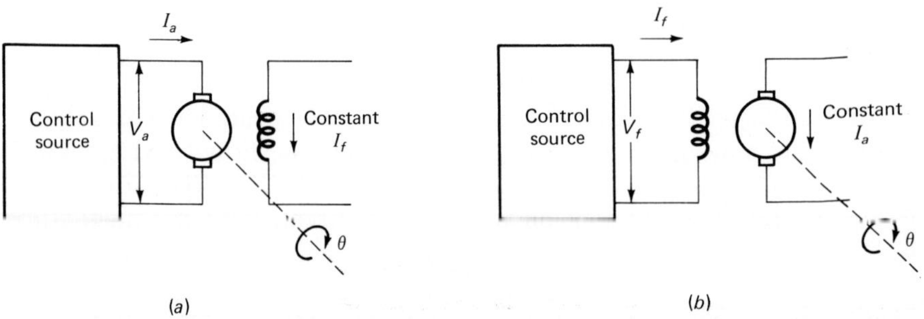

Figure 3-28 (a) Armature control and (b) field control of dc motor

to the control current. However, the speed of response of the motor to a changing input is generally slower than with armature control, and it proves somewhat difficult and inefficient to provide a constant current to the armature. For these reasons, field control is generally limited to low-power motors of up to several horsepower rating.

Dc motors intended for servo use are the same in principle as conventional motors but with several special design features. They generally incorporate the desirable characteristics of high torque and low inertia. A high torque-inertia ratio can be obtained by reducing the armature diameter and increasing its length compared to standard designs. As a further refinement, the commutation should be effective over a wide range of speeds. In particular the motor should operate smoothly at low speeds without noticeable jumping or cogging action.

A drawback in using electric motors is that their shaft speeds are invariably much higher than the desired speed of moving the load. To obtain efficient power transfer, some method of speed reduction may be employed, such as the gear train to be described in Sec. 3-14. A type of dc motor which can operate efficiently at low speeds and be directly attached to the load is the direct-drive dc torque motor. The torque motor has a wound armature and a permanent magnetic field which, acting together, convert electric current directly into torque. Most torque motors are frameless, require little space, and are relatively thin compared to their diameter. In general, torque motors find application in high-torque, "standstill" positioning systems and in slower-speed control systems.

Much of the information concerning performance can be obtained from the torque and speed characteristics of the motor. If the field is held constant, the torque developed by the motor is closely proportional to the armature current as shown by Fig. 3-29. The intercept of the line on the horizontal axis is the value of current required to overcome the static-friction torque of the motor. The slope of the line is defined as the *torque constant* of the motor. Thus

$$K_i = \frac{\Delta T}{\Delta I_a} \tag{3-34}$$

where K_i = torque constant, oz-in./A
T = torque, oz-in.
I_a = armature current, A

When the armature is rotating, a voltage proportional to speed is generated within the winding according to Eq. (3-26). This voltage is called a *back electromotive force (emf)* because it opposes the voltage applied to the armature and tends to limit the armature current. The

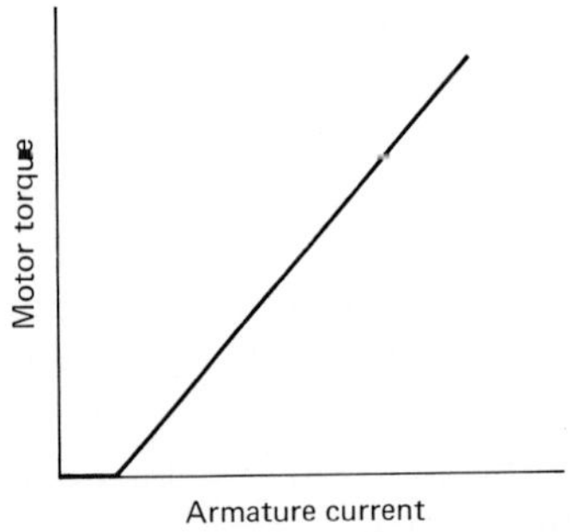

Figure 3-29 Torque characteristic of dc motor

armature speed will be such that the back emf is just enough smaller than the applied voltage to permit an armature current to drive the torque load. If the load is increased, the armature slows down. As it does, the back emf is reduced and a larger current flows. This increased current causes the motor to pull the increased load but at a lower speed. If the original speed is desired, the applied voltage must be increased.

This behavior is shown graphically in the family curves of Fig. 3-30. The curves known as *speed-torque curves* show motor torque as a function of speed for constant values of armature voltage. The zero-torque condition represents the no-load speed of the motor. At this speed no torque may be derived from the motor. Although the speed-torque curves of a motor are not ideally linear, a linear approximation is generally adequate for a dc motor.

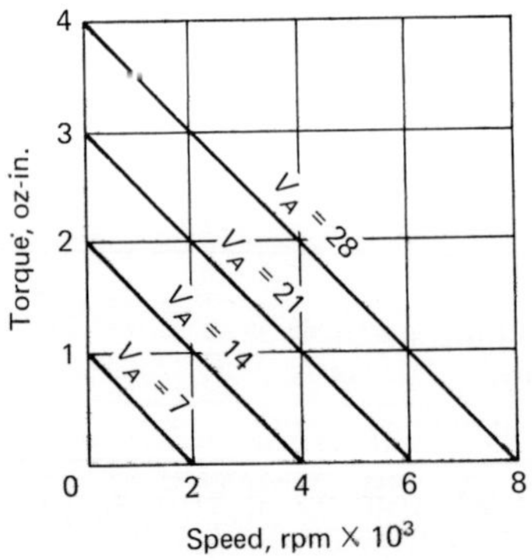

Figure 3-30 Speed-torque curves of dc motor with armature control

Although motor torque with armature control is proportional to armature current, it proves convenient to define a torque constant in terms of armature voltage. At zero speed, or stall, no back emf is generated and the armature current is limited solely by the armature resistance. Thus under stall conditions

$$K_t = \frac{\Delta T_s}{\Delta V_a} \tag{3-35}$$

where T_s = stall torque, oz-in.
V_a = armature voltage, V
K_t = torque constant, oz-in./V

The drop in motor speed with increasing torque is characterized by a damping constant that is taken as the slope of the speed-torque curves. Thus

$$B_m = \frac{\Delta T}{\Delta \dot{\theta}_s} \tag{3-36}$$

where T = torque, oz-in.
$\dot{\theta}$ = speed, rad/s*
B_m = damping constant, oz-in./(rad/s)

If the speed-torque curves are linear, as shown in Fig. 3-30, the damping constant may be determined as

$$B_m = \frac{T_s}{\dot{\theta}_{NL}} \tag{3-37}$$

where T_s = stall torque at rated voltage, oz-in.
$\dot{\theta}_{NL}$ = no-load speed at rated voltage, rad/s

EXAMPLE 3-3

Determine the torque and damping constants for the armature-controlled motor whose speed-torque curves are shown in Fig. 3-30. The armature resistance is 50 Ω.

Solution

Using Eq. (3-35),

$$K_t = \tfrac{4}{28} = 0.143 \text{ oz-in./V}$$

**A common practice that will be followed in this text is to use a notation wherein dots are placed above a dependent variable to indicate its derivatives with respect to time. Thus $\dot{\theta} = d\theta/dt$, $\ddot{\theta} = d^2\theta/dt^2$, etc.*

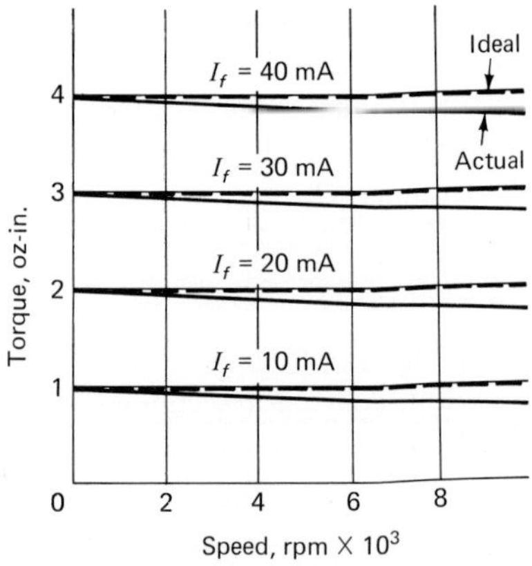

Figure 3-31 Speed-torque curves of dc motor with field control

Using Eq. (3-37),

$$B_m = \frac{4}{8{,}000 \times 2\pi/60} = 4.8 \times 10^{-3} \text{ oz-in./(rad/s)}$$

With field control of the motor, the armature current in theory is held constant and the armature current and torque are unaffected by the back emf. Hence the torque would be independent of speed as indicated by the horizontal lines in Fig. 3-31, which show the torque for various values of field current.

In practice the armature current is not obtained from a true constant-current source and consequently will be somewhat affected by the back emf. In addition, mechanical friction increases with speed, further reducing the available torque. Thus, the speed-torque curves with field control exhibit a slight droop with increasing speed. Contrary to what one might expect, the decrease in torque with speed is generally advantageous in closed-loop control systems since this damping characteristic reduces the tendency toward overshoot and instability.

EXAMPLE 3-4

Determine the torque and damping constants for the field-controlled motor whose speed-torque curves are shown in Fig. 3-31.

Solution

By definition

$$K_i = \frac{T}{I_f} = \frac{4}{0.04} = 100 \text{ oz-in./A}$$

Using Eq. (3-36)

$$B_m = \frac{4 - 3.8}{8{,}000 \times 2\pi/60} = 2.4 \times 10^{-4} \text{ oz-in./(rad/s)}$$

3-13 AC MOTORS

The qualities that make an ac motor (such as the three-phase squirrel-cage induction motor) the choice for most constant-speed industrial drives are not necessarily those needed for control-system applications. Instead a motor that can operate over a wide range of speeds with high starting torque is desired. The primary advantage of the ac motor over the dc motor is its compatibility with the ac signals of synchros, LVDTs, and other ac transducers without the necessity of demodulation.

One type of ac motor that finds extensive use in low-power control applications is the two-phase induction motor shown schematically in Fig. 3-32. This motor has two stator windings physically located at 90° with respect to each other. The rotor consists of a slotted cylinder of iron laminate mounted on the motor shaft. Solid copper or aluminum bars are laid lengthwise in the slots of the rotor and are connected together at both ends by rings. An elongated, small-diameter rotor is generally used to minimize inertia resulting in faster motor response.

The principle of operation of the motor is shown in Fig. 3-33. Stator winding B serves as the reference winding and is supplied from a sinusoidal source of fixed voltage and frequency. This is generally the power line of either 60 or 400 Hz frequency. The second winding A is the control winding and is supplied from the voltage source of the same frequency but shifted in electrical phase by 90°. The currents through the two stator windings cause lines of magnetic flux to pass

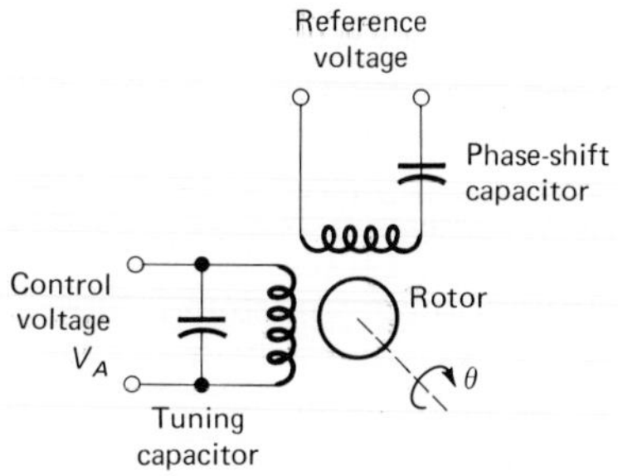

Figure 3-32 Two-phase induction motor

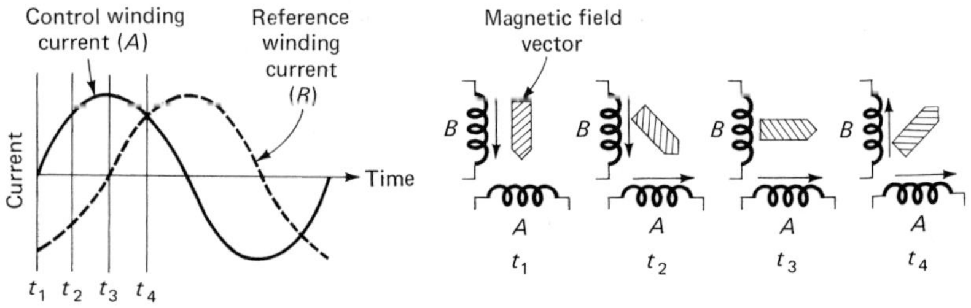

Figure 3-33 Rotation of magnetic field in two-phase ac motor

through the rotor. These flux lines combine, and at any instant the effect of all the lines can be represented by a single vector having some angular position relative to the axis of the rotor. As the magnitudes of the currents change at successive instants, the angle changes. The result is a magnetic field that rotates at the frequency of applied voltage. This is indicated in Fig. 3-33 at four successive times. As the lines of flux rotate, they cut the rotor bars which induces currents in the bars. These currents produce a magnetic flux which reacts with the stator flux and causes the rotor to turn in a smooth and continuous manner.

If the phase of the control-winding current is shifted by 180°, the phase relationship between the stator currents will change from 90° in one direction to 90° in the other. The effect is to reverse the direction of the rotating magnetic field which in turn reverses the direction of mechanical rotation.

The method of operation in control applications is to make the control-winding voltage variable. The resultant characteristics are shown in the speed-torque curves of Fig. 3-34. Note that the stall torque is closely proportional to the control voltage. However, if no torque load is present, the speed is approximately the same for various voltages. This is the synchronous speed of the motor, which is determined by the pole structure and the frequency of the ac voltage.

The effect of load torque is to reduce the speed of the rotor causing a greater speed difference or slip between the rotating magnetic field and the rotor. The induced magnetic field will increase with this difference, creating additional torque to pull the load but at a lower speed.

Although the internal consequence of rotor slowdown is somewhat more complex in the induction motor than in the dc motor, the external effects are much the same. A comparison of Figs. 3-30 and 3-34 shows that the speed-torque characteristics are similar, although those of the ac motor are quite nonlinear.

In practice the speed-torque characteristic of the ac motor is generally taken as linear over the low-voltage range as indicated by the dashed

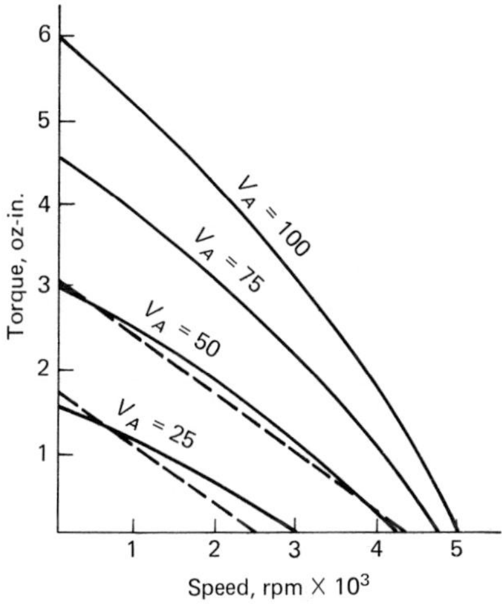

Figure 3-34 Speed-torque curves of two-phase ac motor

lines in Fig. 3-34. Using this approximation, motor constants can be derived that are similar to those of the armature-controlled dc motor.

EXAMPLE 3-5

Determine the torque and damping constants for the motor whose speed-torque curves are shown in Fig. 3-34, assuming linear operation around a nominal control voltage of 25 V.

Solution

$$K_t = \frac{1.7}{25} = 0.068 \text{ oz-in./V}$$

$$B_m = \frac{1.7}{2{,}500 \times 2\pi/60} = 6.49 \times 10^{-3} \text{ oz-in./(rad/s)}$$

To increase the impedance of the control circuit, a tuning capacitor is often added in parallel with the control winding as shown in Fig. 3-32. If the capacitor value is chosen to produce unity power factor, the parallel resonant circuit will have maximum impedance. A convenient equation for determining the parallel tuning capacitance for unity power factor is

$$C = \frac{X_L}{2\pi f Z^2} \tag{3-38}$$

where C = capacitance, farads
X_L = inductive reactance of control winding, Ω
Z = impedance of control winding, Ω
f = frequency of excitation, Hz

A common method of obtaining the required 90° phase difference between the two currents is to use an appropriate phase-shifting capacitor in series with the reference winding as indicated in Fig. 3-32. A suitable value for this capacitor is frequently given in the motor specifications.

An additional factor to be considered in using a control motor of either dc or ac type is the static friction associated with the armature. This requires that a minimum control voltage or "breakaway" voltage be applied to cause rotation of the armature. Static friction represents a basic error in servomotors which degrades system performance. It causes system insensitivity since the error-detecting device must develop enough error voltage through an amplifier to overcome the breakaway voltage. The static friction may be expressed as a torque and is generally between 1 and 10 percent of the rated torque of the motor.

3-14 GEARS

Gears are a common method of coupling rotational mechanical systems, and their use can improve the transfer of power between drive and load. For example, motors are generally high-speed, low-torque devices. A gear arrangement between the motor shaft and load shaft can reduce speed and increase torque to more nearly fit the requirements of a typical load.

The most common type of gear is the *spur gear* shown in Fig. 3-35. In a spur gear the teeth are at right angles with the gear disk.

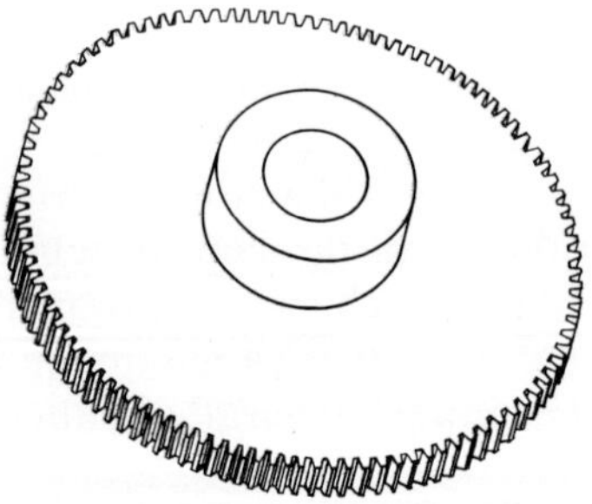

Figure 3-35 Spur gear

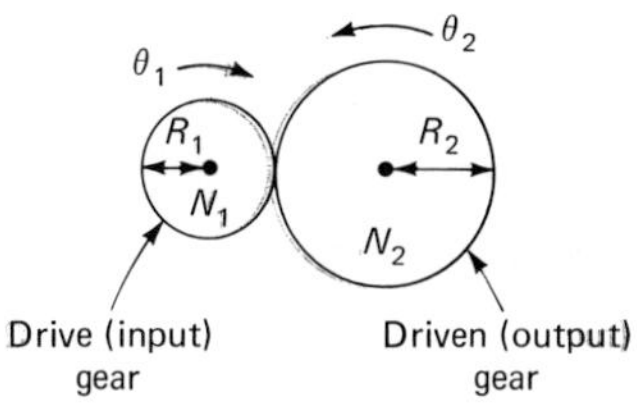

Figure 3-36 Gear mesh

Figure 3-36 indicates two spur gears transmitting motion from the lefthand or drive gear to the righthand or driven gear. Such an arrangement is called a *gear mesh.* Note that the rotational axes of the two gears are parallel and that the direction of rotation is reversed. During a given time the same linear distance must be traversed along the surface of each gear. Thus we have the relationship

$$\theta_1 R_1 = \theta_2 R_2 \tag{3-39}$$

where θ is the angle of rotation and R is the gear radius. Since the number of teeth N of a gear is proportional to its radius, Eq. (3-39) may be rewritten as

$$\theta_1 N_1 = \theta_2 N_2$$

or

$$\frac{\theta_1}{\theta_2} = \frac{N_2}{N_1} \tag{3-40}$$

The ratio N_2/N_1 is known as the *gear ratio* and is identified by the letter N. Thus

$$N = \frac{\text{number of teeth of driven (output) gear}}{\text{number of teeth of drive (input) gear}} = \frac{N_2}{N_1} \tag{3-41}$$

The velocity and acceleration ratios are also defined by the gear ratio as may be seen by taking successive time derivatives of Eq. (3-40)

$$N = \frac{\theta_1}{\theta_2} = \frac{\dot{\theta}_1}{\dot{\theta}_2} = \frac{\ddot{\theta}_1}{\ddot{\theta}_2} \tag{3-42}$$

In practice, gears operate quite efficiently and there is little power loss in a gear mesh. If we assume that the power loss is negligible, the rate of doing work is the same at both input and output shafts. Since work in a rotational system equals the product of torque and angle, the power associated with the rotating shafts in Fig. 3-36 is the product of the torque and the rate of change of the angle or angular velocity.

Thus

$$P = T_1\dot{\theta}_1 = T_2\dot{\theta}_2 \tag{3-43}$$

Using Eq. (3-42) and substituting for $\dot{\theta}_1$ gives

$$T_2\dot{\theta}_2 = T_1 N\dot{\theta}_2 \tag{3-44}$$

or

$$\frac{T_2}{T_1} = N \tag{3-45}$$

Thus, the torque ratio is proportional to the ratio of teeth of the two gears.

In general, the highest gear ratio that is practical with spur gears is about 10 for a single mesh. For larger ratios, two or more meshes may be cascaded; the resulting combination is known as a *gear train*. The overall gear ratio of a gear train is the product of the ratios of the individual meshes.

A small spur gear is commonly identified as a *pinion*. Pinions are frequently made by cutting teeth directly on a shaft. If the larger gear is laid out flat, the result is called a *rack and pinion* and is used to convert rotational motion to translational motion.

Various size gears are identified by their *pitch*. Pitch is defined as the number of teeth on a disk with a 1-in. diameter. Thus, pitch provides a measure of the fineness of gear teeth. Gear pitches commonly used with fractional-horsepower motors are 32, 48, and 64.

A difficulty associated with the use of gears, especially in precision applications, is looseness or play between the input and output of the gear train. This effect, known as *backlash*, can produce an error since the input shaft can turn without any corresponding motion of the output shaft. Backlash is commonly specified by the play between two assembled gears at their point of mesh. Typical values range from 0.001 to 0.005 in. for various classes of gears. Backlash can be minimized by setting the shafts for tight coupling or by using special spring-loaded split gears. However, in either case, increased friction results, and, with anti-backlash gears, the torque load is limited.

EXAMPLE 3-6

A motor that develops 2 oz-in. of torque at a speed of 750 rpm is coupled to a load by means of the gear train shown in Fig. 3-37. Determine the torque and speed of the output shaft.

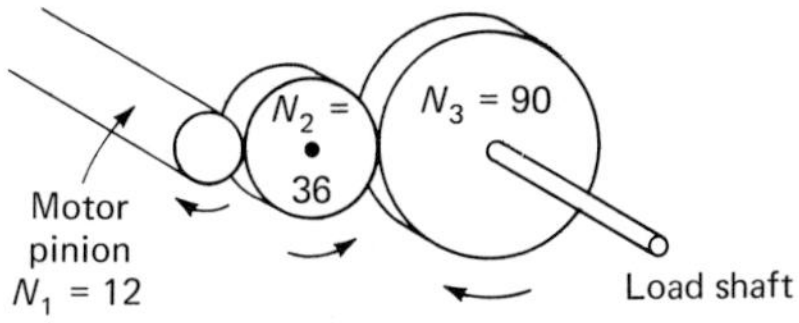

Figure 3-37 Gear train for Example 3-6

Solution

The overall gear ratio N is

$$N = \frac{N_2}{N_1} \times \frac{N_3}{N_2} = \frac{N_3}{N_1} = 7.5$$

The load torque T_L determined from Eq. (3-45) is

$$T_L = T_M N = 15 \text{ oz-in.}$$

The load velocity $\dot{\theta}_L$ determined from Eq. (3-42) is

$$\dot{\theta}_L = \frac{\dot{\theta}_M}{N} = 100 \text{ rpm}$$

In Example 3-6, we note that the load and motor shafts rotate in the same direction because of the presence of the middle gear. The same load torque and speed could have been obtained with the 90-tooth gear directly coupled to the motor pinion. However, the direction of load rotation would now be reversed.

We will next consider the relationship between the inertia associated with the output or load shaft and the effect of this inertia at the input or drive shaft. This effect is called the *reflected inertia* of the load.

Referring to the gear mesh of Fig. 3-36, the acceleration $\ddot{\theta}_1$ of the input shaft is

$$\ddot{\theta}_1 = \frac{T_1}{J_1}$$

or

$$J_1 = \frac{T_1}{\ddot{\theta}_1} \tag{3-46}$$

where J_1 and T_1 are the inertia and torque at the input. Let us initially assume that the entire inertia is that reflected from the output load. From Eqs. (3-42) and (3-45) we have

$$T_1 = \frac{T_2}{N} \qquad \text{and} \qquad \ddot{\theta}_1 = N\ddot{\theta}_2$$

Substituting in Eq. (3-46) gives

$$J_{21} = \frac{T_2/N}{N\ddot{\theta}_2} = \frac{T_2}{\ddot{\theta}_2 N^2} \tag{3-47}$$

where J_{21} is the reflected inertia. Since $T_2/\ddot{\theta}_2$ equals the inertia J_2 of the output, Eq. (3-47) may be rewritten as

$$J_{21} = \frac{J_2}{N^2} \tag{3-48}$$

Equation (3-48) states that the inertia of the load reflected back to the input is altered by a factor of $1/N^2$. The significance of this relationship lies in the fact that, with only a moderate gear ratio, the effect of load inertia is significantly reduced.

If the inertia associated with the drive is included, the total inertia J_T at the input is

$$J_T = J_1 + J_{21} = J_1 + \frac{J_2}{N^2} \tag{3-49}$$

As indicated by Eq. (3-48), the effect of load inertia becomes substantially less as N increases. Hence the inertia of the gears, particularly those near the input shaft, can be a major factor. Since gear inertia is proportional to the gear radius squared, the use of small gears is indicated. For an overall ratio up to 100, it proves desirable to use a gear train of not more than three meshes since there is little improvement in inertia effect with added meshes and a minimum number of meshes reduces the effect of backlash.

The inertia of a gear train can frequently be reduced by the choice of gears. By using punched rather than solid gear disks, the inertia can be reduced by about one-fourth. Substitution of materials can provide an even greater reduction. For example, the use of aluminum rather than steel reduces the inertia by about two-thirds. Aluminum gears are especially useful at the input of gear trains where their tensile strength and durability are compatible with the lower torques. Other materials that are used for medium- and light-duty gears are brass and nylon.

We have seen that friction or damping is also present in mechanical systems. From the relationship $B = T/\dot{\theta}$ given by Eq. (2-13), we can apply the method of analysis that was used in analyzing inertia effects. It can be shown that the damping coefficient B_2 associated with the output shaft can be reflected back to the input shaft as

$$B_{21} = \frac{B_2}{N^2} \tag{3-50}$$

where B_{21} is the reflected damping coefficient.

EXAMPLE 3-7

Referring to Example 3-6, the inertia of the motor armature is 2.5×10^{-4} oz-in.-s^2 and the inertia of the load is 7.5×10^{-3} oz-in.-s^2. Neglecting the inertia of the gears, what is the total inertia appearing at the drive shaft?

Solution

Using Eq. (3-48), the reflected inertia is

$$J_{21} = \frac{7.5 \times 10^{-3}}{7.5^2} = 1.33 \times 10^{-4} \text{ oz-in.-s}^2$$

The total inertia is then

$$J_T = 2.5 \times 10^{-4} + 1.33 \times 10^{-4} = 3.83 \times 10^{-4} \text{ oz-in.-s}^2$$

3-15 GEAR-RATIO SELECTION

In servomotor applications, the capabilities of the motor must be compatible with the requirements of the load. Because a motor is typically a high-speed, low-torque actuator, it should be geared down to provide sufficient torque to drive a load. The choice of gear ratio is an important consideration since it is desirable to use the smallest motor adequate for the job in order to minimize size, weight, and cost.

For a given application the motor and gearing must be able to drive the load at a specified speed and must supply enough torque to accelerate the load as desired. This section will outline a simple procedure for selecting an optimum gear ratio to meet those requirements.

Let us first examine the case of a constant-velocity output. In this instance only the friction torque will be considered. The friction torque reflected at the motor shaft is

$$T_{MF} = \frac{T_{LF}}{N} \qquad (3\text{-}51)$$

where T_{LF} = load friction torque
T_{MF} = reflected load friction torque
N = gear ratio

The motor speed $\dot{\theta}_M$ for a given load speed $\dot{\theta}_L$ is

$$\dot{\theta}_M = N\dot{\theta}_L \qquad (3\text{-}52)$$

Various possible gear ratios between motor and load are arbitrarily assumed. Based on each assumed value, the motor torque and motor speed are calculated using Eqs. (3-51) and (3-52). Using these data points,

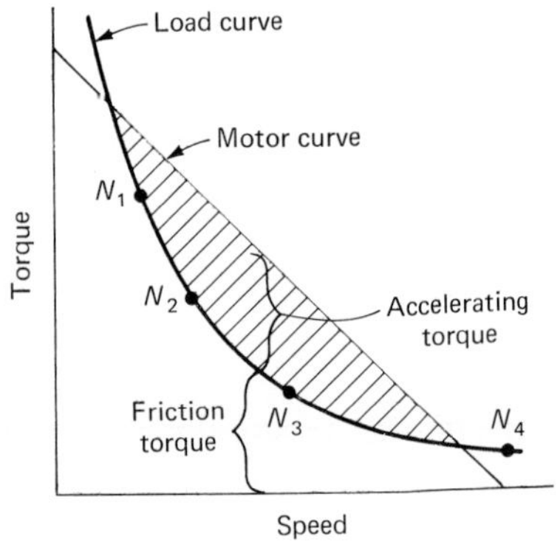

Figure 3-38 Speed-torque curves for gear-ratio selection

a load speed-torque curve is superimposed on the motor speed-torque characteristic as shown in Fig. 3-38.

All parts of the load curve that lie beneath the motor curve (shaded area) represent satisfactory gear ratios. Any gear ratios above the motor curve will not give the required load performance. If the entire load curve lies outside the motor curve, the motor is inadequate and a different unit must be selected.

The torque that is available for accelerating the load may also be determined from Fig. 3-38. The difference between the torque required to drive the friction load and the torque provided by the motor is the torque available for load acceleration. The gear ratio that provides maximum accelerating torque can be found by holding a scale parallel to the vertical axis of the speed-torque curves and moving it between the motor and load curves until the distance between them is greatest.

For the case of a motor driving a pure inertial load, there is another criterion for selecting the gear ratio for maximum acceleration. The acceleration will be a maximum when the gear ratio gives a reflected load inertia equal to the inertia of the drive. This ratio is

$$N = \sqrt{J_L/J_M} \tag{3-53}$$

where J_L and J_M are load and motor inertias, respectively.

The load acceleration can be computed as

$$\ddot{\theta} = \frac{NT_a}{J_L + N^2 J_M} \tag{3-54}$$

where T_a is the torque available for acceleration.

If friction and inertia are combined, the gear ratio for maximum acceleration will be somewhere between the two values when friction

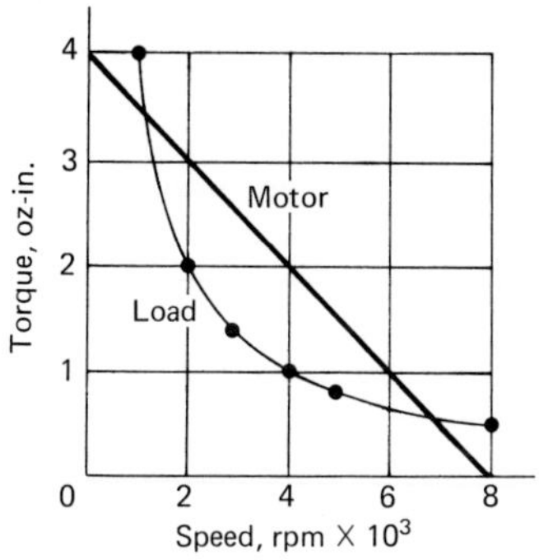

Figure 3-39 Speed-torque curves for Example 3-8

and inertia are considered independently. The optimum ratio in this case can be found as follows. The accelerating torque for a trial value of N can be read from Fig. 3-38, and the acceleration can be calculated using Eq. (3-54). This procedure is repeated on a trial-and-error basis until the optimum value is determined.

EXAMPLE 3-8

The motor whose speed-torque characteristic is shown in Fig. 3-39 is to drive a torque load of 40 oz-in. at a speed of 100 rpm. Determine (1) the range of gear ratios that will meet this specification and (2) the gear ratio that will provide maximum accelerating torque.

Solution

Values of motor speed and torque to meet the load specification are calculated for a series of assumed gear ratios using Eqs. (3-51) and (3-52). The resulting values are:

Gear Ratio	*Motor Speed, rpm*	*Motor Torque, oz-in.*
10	1,000	4.00
20	2,000	2.00
30	3,000	1.33
40	4,000	1.00
50	5,000	0.80
80	8,000	0.50

The reflected load characteristic is plotted on Fig. 3-39. From the intersections of the two curves, it can be seen that a gear ratio between approximately 12 and 70 will be satisfactory.

As a next step it can be determined that the maximum vertical spacing between the curves occurs at about $N = 30$. The available accelerating torque is then

$$T_a = 2.50 - 1.33 = 1.17 \text{ oz-in.}$$

EXAMPLE 3-9

A motor has a stall torque of 6 oz-in. and a rotor inertia of 1.0×10^{-4} oz-in.-s^2. What is the maximum acceleration that it can impart to a pure inertial load of 1.0×10^{-2} oz-in.-s^2, and what gear ratio is required to obtain this result?

Solution

Using Eq. (3-53), the gear ratio for maximum acceleration is

$$N = \sqrt{\frac{10^{-2}}{10^{-4}}} = 10$$

Using Eq. (3-54), the maximum load acceleration is

$$\ddot{\theta} = \frac{10 \times 6}{10^{-2} + 10^2 \times 10^{-4}} = 3{,}000 \text{ rad/s}^2$$

This is the value of acceleration that first occurs as the motor gains velocity starting from a stall condition.

3-16 CONTROL VALVES

In a process control system the need often exists to control the rate of flow of a fluid in such a way as to establish or maintain process operating conditions. In systems of this type the final control element may be a valve located in a flow line. Very little power may be required to position a valve that controls a large power output.

Basically a control valve varies the area through which the flowing fluid must pass. In this respect the valve may be regarded as a variable orifice. The body of the valve is a pressure-type vessel capable of withstanding the worst conditions likely to be met in a particular application. The valve body is screwed, bolted, or welded into the flow line. The most important part of the valve is the plug which is positioned within the body to vary the orifice opening. The plug is moved

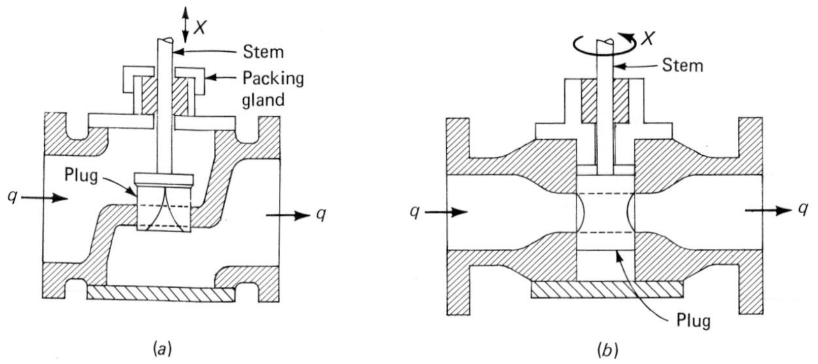

Figure 3-40 Typical control-valve construction: (a) *sliding-stem and* (b) *rotating-stem*

by a valve stem which in turn may be driven by an actuator such as a hydraulic or electric motor, an electromagnetic solenoid, or a pneumatic diaphragm.

Control valves in which the plug is operated by means of a linear motion are termed *sliding-stem valves*. Figure 3-40*a* shows a single-seated valve of this type. Other arrangements, including double-seated valves, are also available. Control valves in which the restriction is accomplished by rotation of the plug are called *rotating-stem valves*. A typical arrangement is shown in Fig. 3-40*b*. The plug is a cylindrical element with one or more transverse openings.

The relationship between valve stem position X and effective cross-sectional flow area A could have one of several forms for various types of valves. Two of the more common characteristics are shown in Fig. 3-41. With the linear characteristic, flow area is closely proportional to stem position. The second characteristic is called *equal percentage*

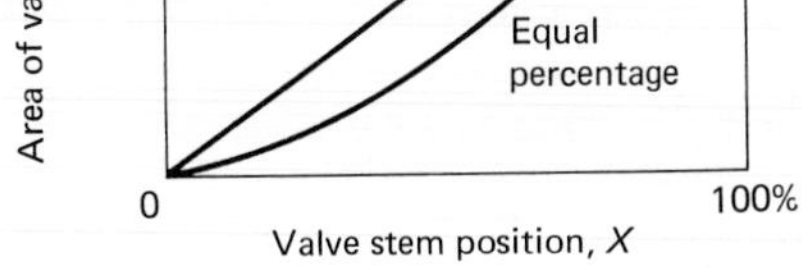

Figure 3-41 Control-valve characteristics

because an increment of stem movement gives the same percentage change in valve-opening area in the high, medium, or low flow range (although not in the very low range). Thus a given stem movement permits a fine adjustment of flow in the low range and a sizable adjustment in the high range. This valve characteristic is nonlinear but may be taken as linear over a small range of operation. In applying a control valve, it can be assumed that there is no time lag between the quantities X and A.

The proper size of control valve for a specific application is important. If the control valve is too large, the valve must operate in the lower range and the minimum controllable flow will be too large. On the other hand, if the valve is too small, the maximum desired flow may not be provided.

The flow of liquid through a control valve is defined by the basic equation

$$Q = KA\sqrt{\frac{2g\Delta P}{\gamma}} \tag{3-55}$$

where Q = flow rate, ft^3/s
K = flow coefficient
A = area of valve opening, ft^2
ΔP = pressure drop across valve, lb/ft^2
γ = fluid density, lb/ft^3
g = acceleration due to gravity = 32 ft/s^2

The flow coefficient K and the area A are different for various styles and sizes of valves. Consequently a *size coefficient* C_v has been developed by combining certain of the terms in Eq. (3-55). The size coefficient is defined as the number of gallons of water per minute that will flow through a wide-open valve with a pressure drop of one psi.
Thus

$$Q_L = C_v\sqrt{\frac{\Delta P}{G}} \tag{3-56}$$

where Q_L = liquid flow rate, gal/min
ΔP = pressure drop, psi ($lb/in.^2$)
G = specific gravity of liquid (referred to water)
C_v = size coefficient

The size coefficient for a control valve must be determined by actual test. However, the value of C_v for a control valve of the sliding-stem and plug types is approximately equal to the square of the valve size multiplied by 10.

EXAMPLE 3-10

A fully open 2-in.-size valve passes 100 gal/min of water with a pressure drop of 7.0 psi. Determine the size coefficient of the valve.

Solution

From Eq. (3-56),

$$C_v = Q_L\sqrt{\frac{G}{\Delta P}} = 100\sqrt{\frac{1}{7}} = 37.8$$

Note that this value is approximately the size squared (4) and then multiplied by 10.

A simplified analysis of flow can be made if the pressure drop across the valve is assumed to be constant and independent of valve position and if the orifice area is taken as proportional to the stem position. Under these conditions Eq. (3-56) can be simplified to

$$Q_L = C_x X \tag{3-57}$$

The flow Q_L is directly proportional to stem position X, and the quantity C_x represents a valve transfer constant. This equation will be essentially correct when the valve is installed such that most of the system's pressure drop is across the valve. When the pressure drop is not constant with flow, the resulting transfer characteristic is nonlinear. This case will not be analyzed inasmuch as we are limiting our consideration to linear systems.

C. Amplifiers

3-17 INTRODUCTION

An amplifier may be defined as a device that receives an input signal and delivers an amplified output signal related to the input in a definite way. Many kinds of amplifying devices are available, and most are adaptable for use in control systems. If the output of the amplifier delivers more power than is required at the input, the amplifier is generally classed as a *power amplifier*. Another class of amplifier is the *voltage amplifier*. In this type of amplifier, output voltage is larger than input voltage, although the output is not necessarily capable of delivering higher power.

Amplifiers in electrical systems may be as simple as an electromechanical relay, or they may consist of electronic devices such as tubes, transistors, or magnetic circuits. Pneumatic relays, valve positioners, and diaphragm valves may be used to provide amplification when compressed air serves as a medium, and fluidic amplifiers find application in liquid systems. Because our emphasis is on electromechanical systems, we will limit our consideration to amplifiers commonly found in such systems.

A relay can be regarded as an amplifier when its contacts are connected in series with a load. For example, a power change of 1 W in a relay coil could cause a change of 1 kW in the load. Thus the relay provides power amplification by a factor of 1,000. However, this control action is discontinuous since the output has only the two values associated with the contacts open or closed. For continuous control, an amplifying device must provide an output that is proportional to a change in the input.

With signals of very low frequency (including zero frequency or direct current), some form of dc amplifier must be used. Unfortunately dc amplifiers have a disadvantage in their tendency to drift. Drift is a variation in output that is independent of the signal input. It frequently originates in power-supply variations and/or variation of the amplifier-circuit parameters. Since dc amplifier stages are directly coupled, such changes are transmitted and amplified in the same manner as the signal input. Minimization of drift is a major problem in the design of dc amplifiers.

The bandwidth of an amplifier is usually far greater than that of an overall control system, which quite often includes electromechanical devices. In general the high-frequency cutoff of the amplifier need not be considered, and over the range of system frequencies the response may be taken as flat. However, we shall see in subsequent chapters that it is invariably necessary to shape or compensate the frequency response of a control system to obtain satisfactory performance. It often is convenient to incorporate this frequency compensation in the system amplifier.

Additionally, the input impedance of an amplifier should be high enough not to load the signal source, and the output impedance should be low enough to adequately drive the power element.

3-18 OPERATIONAL AMPLIFIERS

Voltage amplification in many control systems is obtained by use of a versatile electronic circuit known as an *operational amplifier*. The term operational amplifier was originated by those in the analog-computer field to designate an amplifier circuit that performs various mathematical operations such as summation, integration, and differentiation. The circuit incorporates an electronic amplifier with passive circuit elements in such a way that circuit operation is almost entirely determined by the passive elements. Operational amplifiers are still used in electronic analog computers. However, the application of these devices has been widely extended, and they are now used in many instrumentation and control applications including signal conditioning, filtering, and impedance transformation.

Solid-state operational amplifiers have almost entirely superseded amplifiers employing electron tubes. In addition to their inherent advantages in size, weight, power requirements, heat dissipation, and reliability, solid-state amplifiers have surpassed tube amplifiers in performance specifications and, in many ways, permit greater flexibility in system design.

An operational amplifier generally exhibits the following properties:

Extremely high dc voltage gain, generally in the range from 10^4 to 10^9

Wide bandwidth starting at direct current and dropping to unity gain in the frequency range from 1 to 10 MHz

Plus or minus output voltage over a wide dynamic range, typically either ± 10 or ± 100 V

High input impedance

Solid-state operational amplifiers are available in discrete-component and integrated-circuit (IC) forms. Over the past several years IC amplifiers have become extremely popular because of their excellent characteristics combined with low cost. IC amplifiers that are satisfactory for many, if not most, control applications are available for less than 1 dollar. Thus in using operational amplifiers one need not be concerned about the amplifier circuit per se but only about its characteristics and performance.

From a functional standpoint the typical operational amplifier provides two input terminals and an output terminal. The amplifier is usually represented in symbolic form by a pie-shaped wedge with the apex pointing in the direction of the output. This is shown in Fig. 3-42.

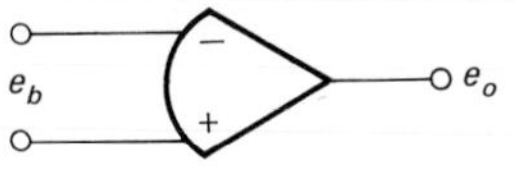

Figure 3-42 Operational amplifier symbolic representation

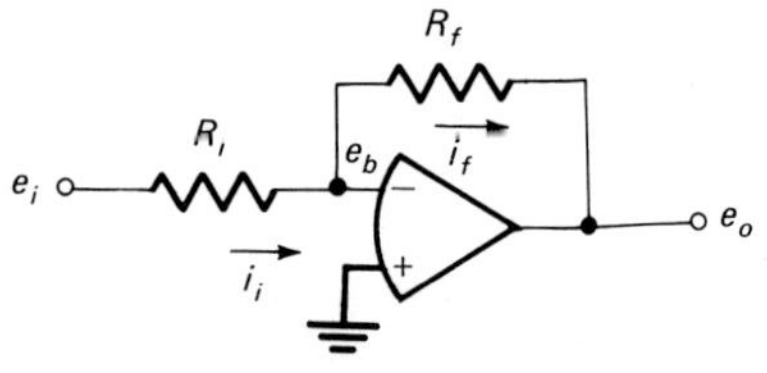

Figure 3-43 Operational-amplifier scaling circuit

The amplifier output e_o is referenced to a common or ground point. The inputs are differential in behavior; that is, the output voltage is proportional to voltage e_b, which is the difference of the voltages applied to the two terminals. The input designated by the minus sign is the inverting input since the output due to this input is reversed or inverted in polarity. The noninverting input designated by the plus sign maintains the same polarity of the output. The magnitude of the amplifier gain is designated as A.

The circuit in Fig. 3-43 is a common arrangement of an operational amplifier with resistors as the passive elements. The circuit employs the principle of negative feedback in which a portion of the amplifier output signal is coupled back to the input in an out-of-phase relationship. Note that the noninverting input is grounded and the amplifier input voltage, designated e_b, is applied to the inverting input.

The closed-loop transfer function may be derived in a manner analogous to that in Sec. 1-7. The circuit input voltage e_i is applied to the amplifier through the input resistor R_i, and the output voltage e_o is coupled back by the feedback resistor R_f. In general the input current to the amplifier is negligibly small, in which case $i_i = i_f$. The two currents may be expressed in terms of the voltage drops across their respective resistors, and the current identity may then be rewritten as

$$\frac{e_i - e_b}{R_i} = \frac{e_b - e_o}{R_f} \tag{3-58}$$

The amplifier output voltage may be written as

$$e_o = -Ae_b \tag{3-59}$$

where the negative sign is associated with the inverting input. Substituting Eq. (3-59) into Eq. (3-58), we have

$$\frac{e_i + e_o/A}{R_i} = \frac{-e_o/A - e_o}{R_f} \tag{3-60}$$

The circuit voltage gain (which we may regard as its transfer function) may be obtained by solving Eq. (3-60) for the ratio e_o/e_i. Thus

$$\frac{e_o}{e_i} = \frac{-AR_f}{R_f + R_i(1 + A)} = -\frac{R_f}{R_i}\left[\frac{1}{1 + (1 + R_f/R_i)/A}\right] \tag{3-61}$$

If the amplifier gain A is very high, then to a close approximation Eq. (3-61) becomes

$$\frac{e_o}{e_i} = -\frac{R_f}{R_i} \tag{3-62}$$

Thus we see that the high-gain operational amplifier in this circuit arrangement produces a transfer function that multiplies the input by a scaling constant. The output is essentially independent of the characteristics of the amplifier and is simply determined by the ratio of two resistors. If the two resistors are equal in value, the output is equal to the input in magnitude but opposite in sign. This case is known as an *inverting amplifier* or *inverter*.

EXAMPLE 3-11

In the circuit in Fig. 3-44 the amplifier gain is 20,000. Compare the actual output voltage with the output voltage assuming the amplifier gain to be infinitely large.

Solution

Using Eq. (3-61), the exact output voltage is

$$e_o = -\frac{100\ \text{k}\Omega}{10\ \text{k}\Omega}\left[\frac{1}{1 + (1 + 100\ \text{k}\Omega/10\ \text{k}\Omega)/20{,}000}\right]1.0 = -9.9945\ \text{V}$$

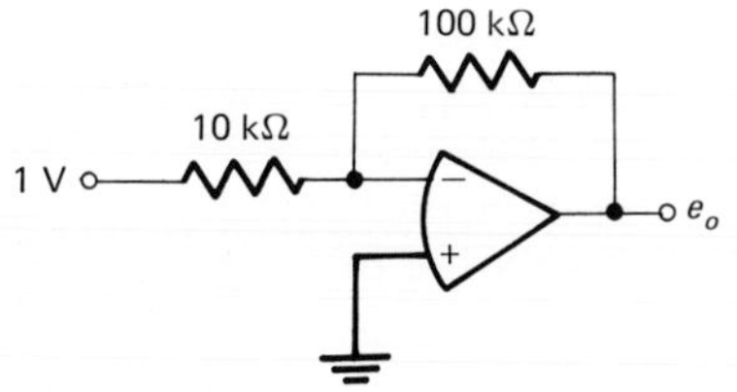

Figure 3-44 Circuit for Example 3-11

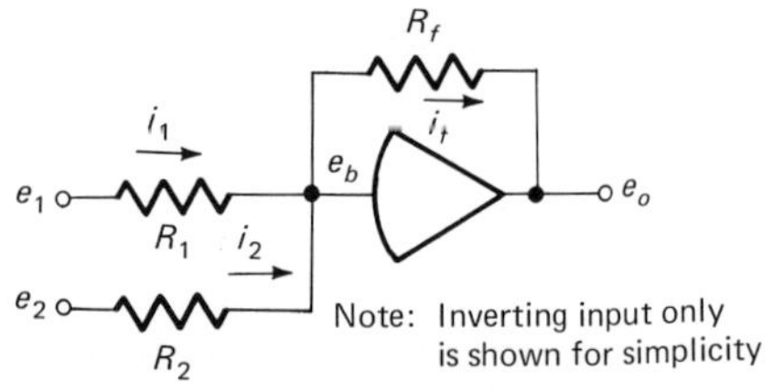

Figure 3-45 Operational-amplifier summing circuit

From Eq. (3-62) the output is

$$e_o = -\frac{100\ \text{k}\Omega}{10\ \text{k}\Omega} \times 1.0 = -10.00\ \text{V}$$

We note that the error in the output voltage calculated from the simplified equation is extremely small.

An operational amplifier may be used for the addition or summation of voltages by connecting a number of inputs in parallel to the amplifier input. The summing-amplifier circuit is illustrated in Fig. 3-45, in which two inputs are combined. The output of the circuit can be determined by extending the previous analysis of the operational amplifier with a single input. The currents in the circuit may be combined at their point of summation using Kirchhoff's current law. If the amplifier input current is assumed to be 0, we may write the equation

$$i_1 + i_2 = i_f \tag{3-63}$$

Expressing the currents in terms of voltage drops and substituting $-e_o/A$ for e_b, we can derive the relationship

$$e_o = -\left(\frac{R_f}{R_1} e_1 + \frac{R_f}{R_2} e_2\right)\left[\frac{1}{1 + (1 + R_f/R_1 + R_f/R_2)/A}\right] \tag{3-64}$$

If the amplifier gain A is again large, Eq. (3-64) may be simplified to

$$e_o = -\left(\frac{R_f}{R_1} e_1 + \frac{R_f}{R_2} e_2\right) \tag{3-65}$$

Thus we see that the amplifier output is the sum of the inputs, where each input is weighted according to the ratio of the feedback resistor and the resistor associated with that input. It is important to note that the output voltage is inverted in polarity. Inversion is inherent in operational amplifier circuits of this type because of their negative-feedback principle of operation.

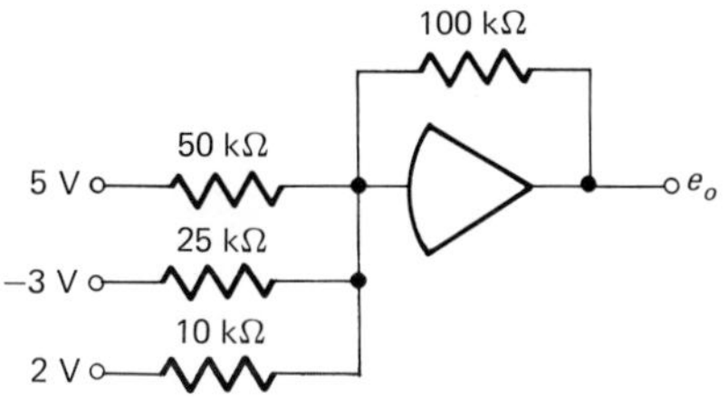

Figure 3-46 Circuit for Example 3-12

The summing circuit can be extended to include additional inputs, in which case Eq. (3-65) will incorporate additional terms. Subtraction of one or more inputs may be accomplished by introducing these inputs to the amplifier with a negative polarity.

EXAMPLE 3-12

Determine the output of the summing amplifier shown in Fig. 3-46. Assume that the amplifier gain is extremely high.

Solution

Applying Eq. (3-65) for the case of three inputs, we have

$$e_o = -\left[\left(\frac{100\ \text{k}\Omega}{50\ \text{k}\Omega}\right)(5) + \left(\frac{100\ \text{k}\Omega}{25\ \text{k}\Omega}\right)(-3) + \left(\frac{100\ \text{k}\Omega}{10\ \text{k}\Omega}\right)(2)\right]$$

$$= -(10 - 12 + 20) = -18\ \text{V}$$

As mentioned earlier, an operational amplifier can also perform the basic calculus operations of integration and differentiation. As we shall see in Chap. 7, these functions are frequently introduced into control systems in order to achieve high accuracy combined with stability.

Near-perfect integration can be obtained by combining an operational amplifier with a resistor-capacitor network. The basic integrating

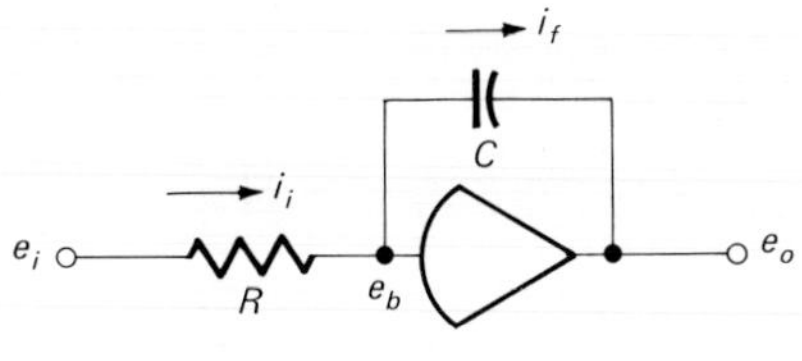

Figure 3-47 Operational-amplifier integrating circuit

circuit shown in Fig. 3-47 uses a resistor as an input element and a capacitor as the feedback component. To analyze the circuit let us again assume that currents i_i and i_f flow in the input and feedback elements, respectively. If the amplifier input current is considered negligible,

$$i_i = \frac{e_i - e_b}{R} = i_f \tag{3-66}$$

Referring to Fig. 2-9, the capacitor current may be expressed as

$$i_f = C\frac{d(e_b - e_o)}{dt} \tag{3-67}$$

If the amplifier gain is extremely high, the input voltage e_b will be negligible compared to e_i and e_o. Equation (3-66) can then be written as

$$\frac{e_i}{R} = -C\frac{de_o}{dt} \tag{3-68}$$

Integrating both sides of Eq. (3-68) with respect to time, we obtain the result

$$e_o = -\frac{1}{RC}\int e_i\,dt + E_o \tag{3-69}$$

Equation (3-69) states that the output of the integrator circuit is proportional to the input voltage integrated with respect to time. The constant $1/RC$ which multiplies the integral represents a gain factor for the circuit. The quantity RC is defined as a *time constant* and has the dimension of seconds, where R is expressed in ohms and C in farads. The factor $1/RC$ then has the dimension seconds^{-1}; and, if the variable of integration is measured in seconds, the output of the circuit is a voltage of the same units as the input voltage.

The voltage E_o in Eq. (3-69) is the value of e_o at the start of operation. This is the voltage to which the capacitor may initially be charged and represents the constant of integration associated with this operation.

More than one input may be applied to the integrating circuit by using parallel input resistors similar to the setup of a summing amplifier. For multiple inputs the output voltage is given by

$$e_o = -\int\left(\frac{e_1}{R_1C} + \frac{e_2}{R_2C} + \cdots + \frac{e_n}{R_nC}\right)dt + E_o \tag{3-70}$$

where input e_1 is applied through resistor R_1, e_2 through R_2, etc. Thus the amplifier output is the integral with respect to time of the weighted sum of the input voltages.

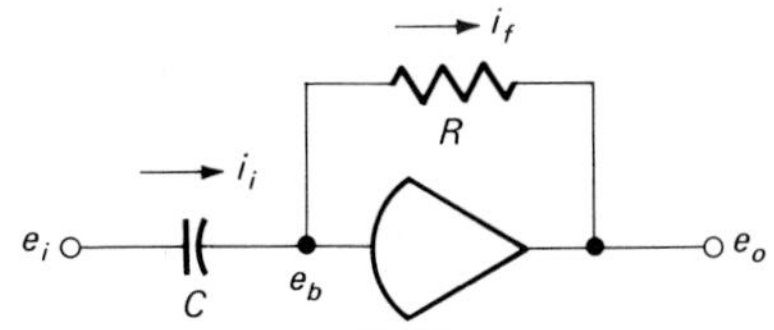

Figure 3-48 Operational-amplifier differentiating circuit

If the passive elements in Fig. 3-47 are interchanged as shown in Fig. 3-48, the circuit will perform the operation of differentiation. Expressing the input and feedback currents in terms of voltage drops and making the assumption that e_b is very near 0, we may derive the equation for the output as

$$e_o = -RC\frac{de_i}{dt} \tag{3-71}$$

In this case the output is equal to the time derivative of the input multiplied by the RC time constant of the circuit.

In practice, the differentiating circuit of Fig. 3-48 has several limitations. The circuit tends to amplify any noise that may be present in the input signal, and it is somewhat prone to oscillate. These adverse characteristics may be overcome by placing a small resistor (generally less than one-tenth the value of the feedback resistor) in series with the input capacitor. However, this additional resistor does make the differentiation function less than ideal.

EXAMPLE 3-13

Write the equation of the output of the circuit shown in Fig. 3-49 for a sinusoidal input voltage $e_i = 10 \sin 100t$ V. The capacitor is initially uncharged.

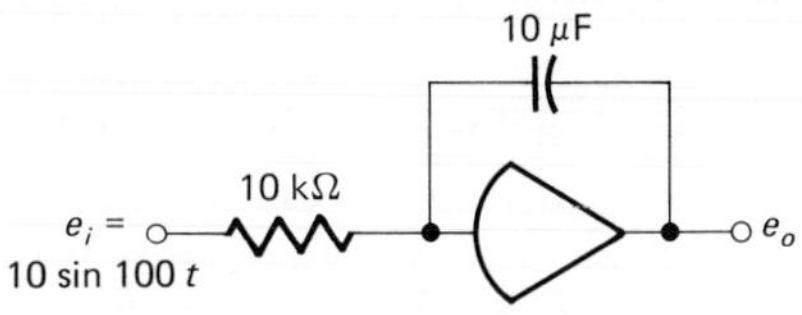

Figure 3-49 Circuit for Example 3-13

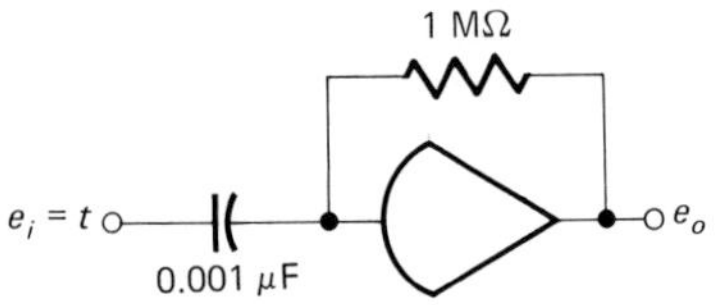

Figure 3-50 Circuit for Example 3-14

Solution

From Eq. (3-69)

$$e_o = -\frac{1}{10^4 \times 10^{-5}} \int (10 \sin 100t)\, dt$$

$$= (-10)(-\tfrac{10}{100})(\cos 100t) = \cos 100t \text{ V}$$

EXAMPLE 3-14

Write the equation of the output of the circuit shown in Fig. 3-50 for an input voltage that increases linearly with time ($e_i = t$).

Solution

Using equation (3-71),

$$e_o = -(10^{-9})(10^6)\left[\frac{d(t)}{dt}\right] = -10^{-3} \text{ V}$$

3-19 POWER AMPLIFIERS

The power actuator or output element of a control system usually requires considerable drive power. A number of electromechanical and electronic devices are available to provide needed power amplification between the error detector and the output stage.

As mentioned earlier, relays provide a simple means of amplifying electric power. Power vacuum tubes are used in some cases although heat dissipation and power-supply requirements impose an upper limit on their power-handling capability. Gas tubes such as the thyratron are used at higher power levels.

In recent years semiconductor devices have largely superseded tubes as a means of power amplification. Both power transistors and silicon-controlled rectifiers are widely used.

Magnetic amplifiers are another form of amplifying device that can be used over a wide power range. However, magnetic amplifiers have a disadvantage because of the time delay in their response as compared to electronic amplifiers.

For very large power applications, rotating machines can provide efficient power amplification. However, these devices also suffer from slow response.

In addition to these devices, pneumatic and hydraulic amplifiers find use in control applications. However, these devices generally lack the flexibility of electrical equipment and are more expensive.

3-20 TRANSISTOR AMPLIFIERS

In almost all electronic amplifiers (including amplifiers used in control systems), transistors have replaced vacuum tubes because of their advantages in efficiency, reliability and life, space and weight, and resistance to shock and vibration. However, the temperature sensitivity of transistors, especially at high power levels, does impose a limitation on their use. Careful circuit design and heat sinks for dissipating heat are required to maintain thermal stability.

This section is not intended to encompass the design of power amplifiers since this is an extensive subject in itself. Rather several basic amplifier types that are commonly found in control systems will be briefly described.

With low-power ac control systems (such as those employing an ac servomotor) conventional transformer-coupled–push-pull amplifiers are commonly used. A typical circuit is shown in Fig. 3-51. Transistor Q_1

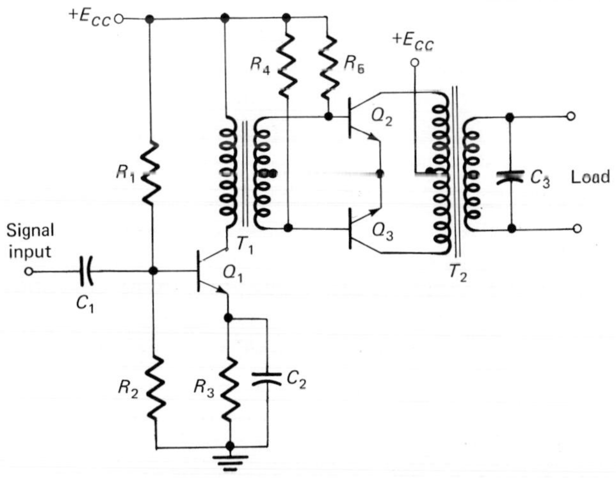

Figure 3-51 Ac power amplifier circuit

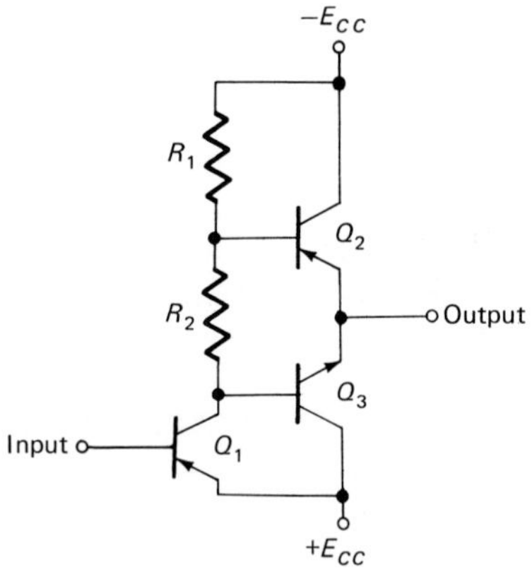

Figure 3-52 Dc power amplifier circuit

functions as a phase inverter that provides a push-pull signal through transformer T_1 to the output stage. The push-pull output comprises transistors Q_2 and Q_3. The transistors are operated class AB for improved efficiency. Forward bias to reduce crossover distortion is obtained through resistors R_4 and R_5. Capacitor C_3 and the motor control winding form a resonant circuit at the frequency of operation as described in Sec. 3-13. This arrangement improves the load power factor as well as rejecting noise at frequencies other than that of the signal.

The circuit shown in Fig. 3-52 is a conventional power amplifier used to provide a dc output in low-power applications. Transistor Q_1 is the driver for the output state. The output consists of transistors Q_2 and Q_3—each connected in a common-collector (emitter-follower) configuration. Transistor Q_2 is a PNP-type transistor, and Q_3 is an NPN type with similar characteristics. The circuit employs complementary symmetry to provide push-pull operation with a single-ended input. The NPN transistor conducts when the output of Q_1 goes positive, and the PNP transistor conducts when the output of Q_1 goes negative. The output voltage appears between the common-emitter point and ground. Resistor R_2 provides a slight forward bias for class AB operation.

Figure 3-53 shows a bridge-type power amplifier capable of driving dc loads at high power levels with good efficiency. This circuit has the advantage that only one polarity of high-power supply voltage is required. Transistors Q_1 and Q_2 are emitter-followers, and one or the other will conduct depending on the polarity of the input signal. Transistors Q_3 and Q_4 act as common-emitter switches and provide a ground return for the load current. For example, with a negative input signal, the collector voltages of the differential stage Q_7 and Q_8 will change, with the Q_7 collector going positive and the Q_8 collector negative. Transistor Q_3 will be switched to a low-resistance state to provide a ground path for the load that is being driven from Q_2. For a positive input, transistors Q_1 and Q_4 are conducting and the load current is thereby reversed. Transistors Q_5 and Q_6 are emitter-followers to minimize loading on the differential stage. A circuit of this type is capable of delivering up to 10 A of current to the load.

Although the class AB power amplifier provides nearly ideal operation, the method of amplification results in substantial power loss within the amplifier, especially with low values of load resistance.

A recent type of power amplifier that exhibits high efficiency employs switching transistors to provide a power output proportional to a dc input signal. This type of amplifier usually employs two output transistors connected in a manner similar to that shown in Fig. 3-52. However, in contrast to the class AB amplifier previously described, the output transistors in the switching amplifier are controlled by drive circuitry to be in either the fully conducting state or the fully nonconducting state. The proportional power output is provided by switching the output transistors in accordance with the input so that the *average* value of the output is proportional to the input.

In actual operation the frequency at which the transistors are switched is usually between 1,000 and 25,000 Hz. The inductive armature circuit of a motor provides a filtering effect so that the output-

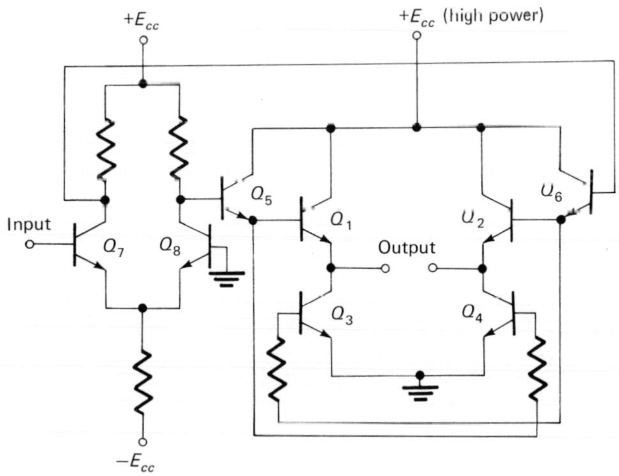

Figure 3-53 Bridge-type dc power amplifier circuit

current waveform is adequately smoothed to signal frequencies as high as one-third the switching frequency.

The switching transistor amplifier achieves excellent efficiencies because the output transistors are either fully on or off. In either of these two states, the power dissipation of the transistor is near 0.

3-21 SCR CONTROL

For very high-power control applications extending into the megawatt range, the *silicon controlled rectifier* (SCR) is widely used. The SCR, or *thyristor* as it is sometimes called, is essentially a solid-state version of the thyratron or ignitron gas tube. It can be used with a variety of electrical loads, including motor armatures, heaters, and lamp banks. Single devices are available that can control currents up to 500 A and voltages up to 500. With a forward voltage drop of tenths of a volt and an off current of several milliamperes, the SCR provides an extremely efficient (about 98 percent) power-drive system. Although SCRs have high power gains, large heat sinks or forced cooling methods are needed in practice to realize their full capacity.

The simplest circuit that uses an SCR as a linear power amplifier is the single-phase half-wave circuit shown in Fig. 3-54*a*. This circuit is similar to the familiar phase-control thyratron drive. As indicated by the waveform in Fig. 3-54*b*, a gating signal repetitively turns the SCR on at a desired time t_1 during the positive half-cycle. The reversal of the ac power line turns the SCR off at the end of the half-cycle.

Many types of firing circuits that convert an analog input signal into a time delay are available. Simplest is a resistance-capacitance integrator that is synchronized to start generating a ramp at each positive half-cycle. The analog input voltage controls the steepness of the ramp, which determines how soon the trigger level is reached and the firing pulse applied to the SCR.

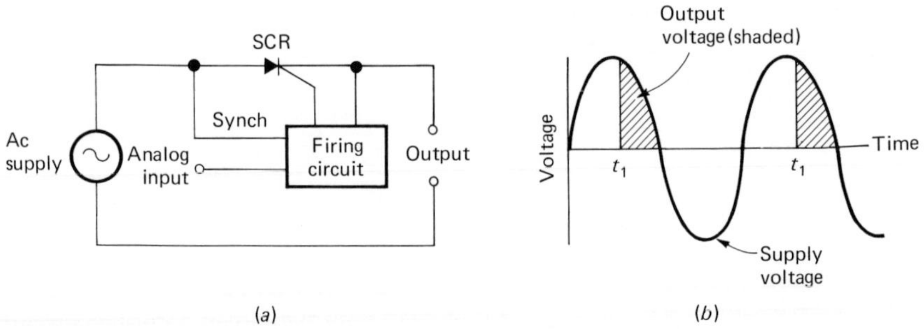

Figure 3-54 (a) Unidirectional half-wave SCR control circuit and (b) waveform

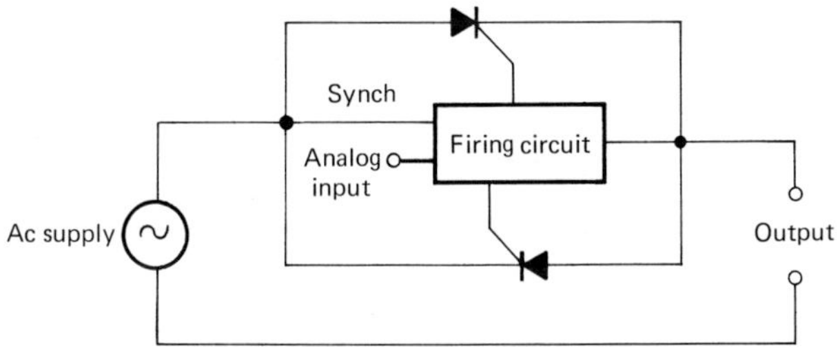

Figure 3-55 Bidirectional half-wave SCR control circuit

Electrical reversing of the load current requires two SCRs connected back to back in parallel as shown in Fig. 3-55. One SCR controls forward power flow, and the other handles reverse power. The triggering circuit for the SCRs must ensure that both do not conduct simultaneously.

Banks of SCRs may be paralleled for greater current-handling capacity, placed in series to handle high voltages, connected in bridge circuits to provide full-wave rectification, or used in groups of three or six to convert a three-phase input to direct current.

3-22 MAGNETIC AMPLIFIERS

A *magnetic amplifier* is a device in which a control signal determines the inductive reactance of an output winding that is in series with an external load and thus controls the portion of the supply voltage that appears across the load. The reactance of the output winding is altered by varying the saturation of the iron core on which both the output and control windings are wound. Because of its principle of operation, this magnetic device is often referred to as a *saturable reactor*. The basic arrangement is shown in Fig. 3-56. With zero control input, the inductive reactance is a maximum. As a direct current is applied to the control winding, the magnetic core becomes saturated, the inductive

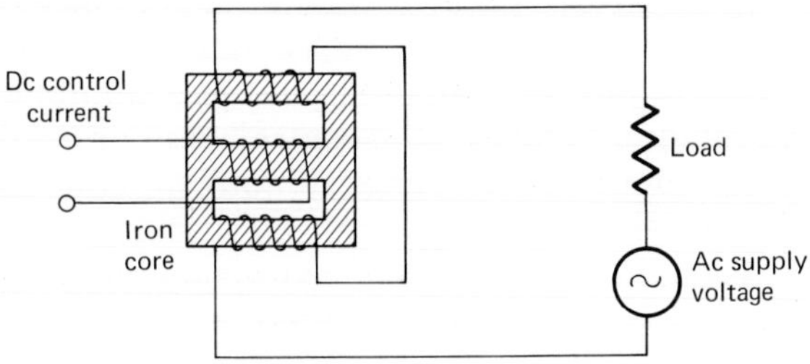

Figure 3-56 Magnetic amplifier (saturable reactor) circuit

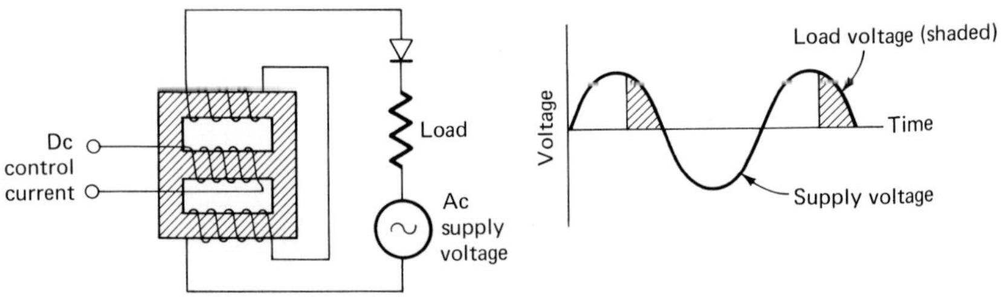

Figure 3-57 Self-saturating magnetic amplifier circuit and voltage waveform

reactance becomes virtually 0, and a nearly full supply voltage appears across the load. It should be noted that the load voltage is a sine wave whose amplitude varies with the input signal.

Since the energy in the control winding is small compared to the energy being controlled, power amplification is obtained. Power gains of several hundred are possible using the saturable reactor.

The power gain can be increased substantially by the addition of a series rectifier in the output winding as is indicated in Fig. 3-57. In this case current is permitted to flow through the load in a direction that aids the control current in saturating the core. Assuming that a square-loop core material is employed, the value of the control current has the effect of determining the time during the supply-voltage cycle at which the core will abruptly saturate. This behavior is shown by the waveform in Fig. 3-57. Note that the characteristic is similar to that of a thyratron or SCR device. This self-saturating type of amplifier is what is most generally regarded as a magnetic amplifier, and power gains of over 10,000 are possible.

Since current flows through the load only during half of each cycle of the ac supply, the output is half-wave direct current. Additional output windings and rectifiers can be used to provide either full-wave direct current or alternating current to the load. Other winding arrangements are also available for either push-pull or three-phase operation. A feedback winding is sometimes added to provide positive feedback and increase the gain of the amplifier. It is common practice to provide several isolated control windings; this permits several signals to be combined in a single amplifier without incurring circuit complications.

Magnetic amplifiers are extremely rugged, reliable, efficient, and long-lived and have been widely used in medium-power applications. Their main disadvantages as compared to electronic amplifiers are their size, weight, and slow response time. In general, magnetic amplifiers are not operated at response times of less than three or four cycles of the supply frequency. Special circuits may be used which fundamentally give half-cycle response time, but usually the simplest approach is to use a higher-frequency supply voltage. An additional response limitation is due to the inductive time constant of the control winding, which is generally of the order of several milliseconds.

3-23 ROTATING POWER AMPLIFIERS

Dc machines are sometimes used in the power output stage of electromechanical servomechanisms requiring between 50 W and 50 hp. A widely used arrangement is the *Ward Leonard system* shown in Fig. 3-58, in which a dc generator drives a dc motor. The generator armature is driven at a constant speed by a prime mover. The field of the motor is excited by a constant voltage or, for some smaller motors, by a permanent magnet. The generator output voltage, which is determined by the magnitude of its field current, controls the motor speed. Reversal of the direction of the generator field current causes a polarity reversal of the generator output and, in turn, a reversal of the direction of motor rotation. Power gains of the order of 10 to 20 are typical for a system of this type.

When larger power gains are required, two or more generators may be cascaded; that is, the output of one generator excites the field of a second generator, etc.

A problem in cascading dc generators is the accumulation of time delays from input to output due to the inductance of the field and armature circuits. Since the resistance of the armature must be small to achieve reasonable power amplification, the inductive time constant of the armature is a principal source of delay.

A form of cascaded generator that provides improved dynamic performance is the crossed-field type of machine known by the trade name Amplidyne. This is essentially a two-stage generator; however, the two stages are combined using a single yoke and armature, and the amplification takes place in the magnetic circuit of the machine. In the Amplidyne a small control field excites a short-circuited arma-

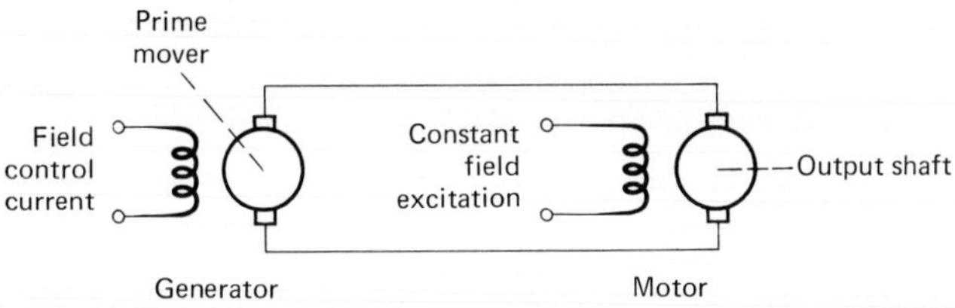

Figure 3-58 Ward Leonard system

ture winding which then produces a larger quadrature field to excite the main armature winding that is connected to the load. The Amplidyne can provide power amplification as high as 10,000. In addition to smaller delays, it occupies less space and weighs less than would two equivalent cascaded generators.

3-24 MODULATORS AND DEMODULATORS

In this chapter we have described certain transducers that produce dc signals and others that yield ac outputs. Likewise some amplifiers operate with dc inputs while others function best with ac signals. Finally power actuators, such as electric motors, are available in both dc and ac forms.

To obtain optimum performance and maximum efficiency, it is sometimes desirable to use a number of seemingly incompatible dc and ac devices in one system. For example, an ac motor may be used in what is otherwise an all dc control system. Or a dc amplifier may be needed to provide compensation in an ac system. In such cases it is necessary to convert dc signals to those that are ac in form, and vice versa. These conversions are performed by devices known as *modulators* and *demodulators,* respectively. In concluding our discussion of control system components, we will briefly describe several means of modulation and demodulation.

A modulator is a device with a dc input and an ac output that is proportional to the magnitude of the input. In addition, the phase of the output with respect to an ac reference reverses when the polarity of the input is reversed.

Many modulators function as devices which "chop" the dc signal into a square wave of some fixed frequency. Hence they are frequently referred to as *choppers,* a name that is descriptive of their action.

Modulators (or choppers) are available in many forms. Electromechanical, transistor, and photoelectric modulators are the most common.

The *electromechanical chopper* was the first modulator to find wide acceptance, and it is still popular because of its excellent performance at microvolt signal levels. The construction of a mechanical chopper is similar to that of a high-speed, light-duty relay. It consists of a contact-carrying metal reed positioned between fixed contacts and mechanically actuated by an ac voltage applied to a drive coil. The excitation causes the reed to vibrate in synchronism and the contacts open and close on each half-cycle of the excitation. The magnetic circuit is designed so that switching occurs close to the zero point of

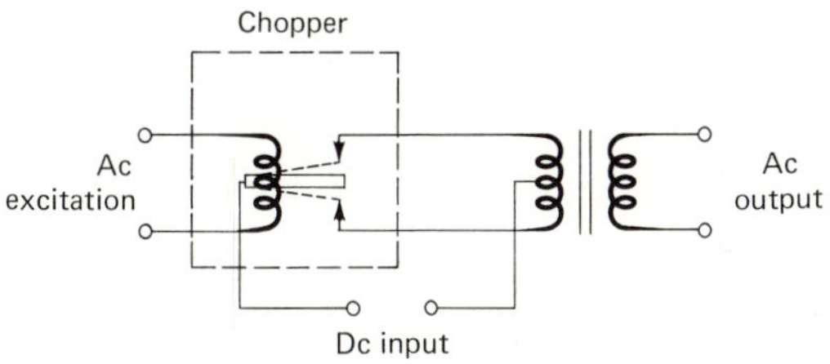

Figure 3-59 Electromechanical modulator circuit

the exciting waveform. The 60-Hz ac power line is commonly used for excitation; however, some electromechanical choppers are designed to perform at higher frequencies in the range of 400 to 1,000 Hz.

Figure 3-59 shows a modulator circuit using an electromechanical chopper. The contacts alternately connect the dc signal to the top and bottom of the transformer primary. This produces an alternating output from the secondary of the transformer. The frequency of this output is the same as the excitation, and the magnitude is proportional to the magnitude of the dc input. If the polarity of the dc input is reversed, the direction of transformer current is reversed for any given excitation. Thus the phase of the ac output is reversed by reversing the input polarity.

Operating life is an important consideration in applying electromechanical choppers. With proper use, an operating life from 5,000 to 10,000 hours can generally be expected.

Despite improvements in electromechanical choppers, the trend is toward the use of electronic modulators. Solid-state modulators overcome many of the limitations of electromechanical choppers including limited life, low chopping frequency, and sensitivity to shock and vibration. Although the low-level performance of mechanical choppers is unequaled, the characteristics of some solid-state choppers now approach those of the best mechanical units.

The *transistor chopper* is shown in its basic form in Fig. 3-60. The

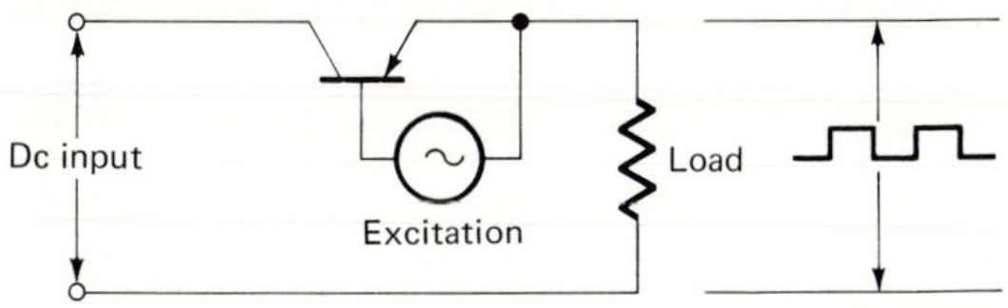

Figure 3-60 Basic transistor modulator circuit

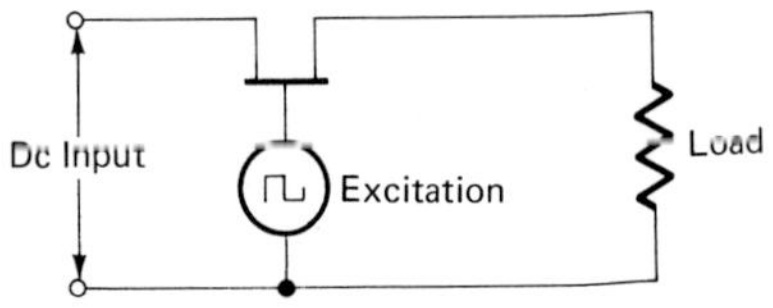

Figure 3-61 Basic FET modulator circuit

transistor is alternately biased to saturation and cutoff by the excitation voltage. This causes the current flow to the load to be interrupted at the frequency of the excitation, and the dc signal is thereby converted to an ac voltage across the load. Most practical transistor chopper circuits use two transistors in such a way that internal voltage and current effects cancel each other at the output.

Field-effect transistors have proven to be especially suitable as electronic modulators. Two characteristics of the FET make it an excellent device for low-level switching: it does not have an inherent voltage drop between the input and output terminals, and the input impedance is extremely high.

The simple modulator circuit shown in Fig. 3-61 offers most of the advantages of both the mechanical and transistor choppers and few of the disadvantages. With no bias applied from gate to source, current flows freely through the channel. When a reverse bias is applied, a very high value of resistance will be seen from drain to source. Since the gate-to-source junction is reverse-biased, a very high impedance is seen at the gate and no transformer is needed to isolate the drive.

As noted earlier, a *demodulator* is a device that converts an ac input signal to a dc output. The magnitude of the dc output is proportional to the amplitude of the ac input, and its polarity will change with the phase of the input. Demodulators of this type are sometimes referred to as *phase-sensitive detectors* or *rectifiers*.

Figure 3-62 shows the electromechanical chopper used as a demodulator. This circuit is identical with that of Fig. 3-59 except that the positions of the input and output signals are interchanged.

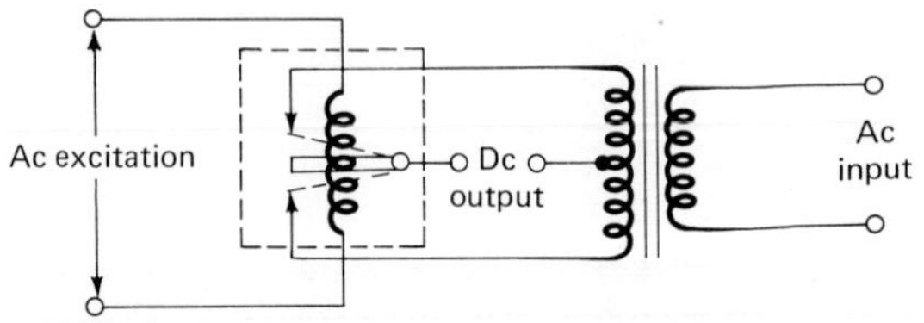

Figure 3-62 Electromechanical demodulator circuit

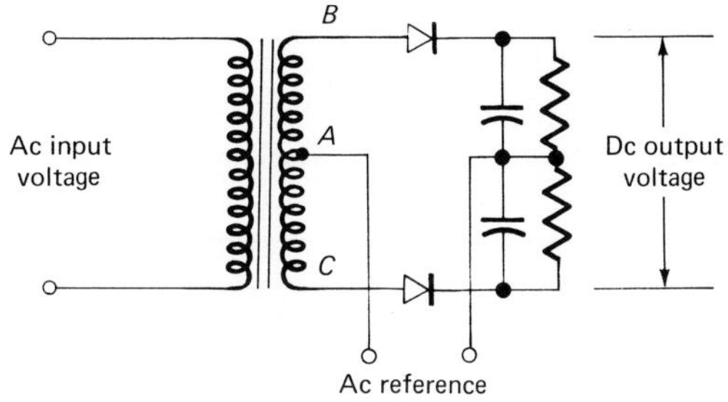

Figure 3-63 Diode demodulator circuit

Since demodulators are frequently used at relatively high signal levels, the diode circuit shown in Fig. 3-63 has found wide acceptance. The reference voltage is applied symmetrically to both diodes. With zero input signal, the diodes conduct when points *B* and *C* are positive, and equal currents then flow through the load resistors in opposite directions. Hence the output voltage is 0. During the negative half-cycle the diodes do not conduct and the output is also 0. With an ac signal applied the voltage on the secondary of the input transformer adds to the reference voltage on one side of the center tap and subtracts on the other side. During the positive half-cycle of the reference, one diode conducts more than the other so that a net output voltage is produced. The output voltage is half-wave rectified direct current, and a suitable filter is required to smooth out the pulsating waveform. Other forms of the circuit can provide a full-wave demodulator for improved ripple characteristics.

In describing demodulator action, we have assumed that the ac input signal is exactly in phase (or 180° out of phase) with the reference voltage. This is rarely the case in practice since slight shifts in the phase of the signal are introduced by various system components. To determine the effect of phase mismatch, let us consider the case when the phase difference is exactly 90°. An input so related to the reference is called a *quadrature input*. During any given half-cycle of the reference, the quadrature signal will be positive half the time and negative half the time resulting in an average value of 0. Since no dc output is obtained, we see that quadrature rejection is a characteristic of the demodulator.

From ac circuit theory we can recall that a sinusoidal waveform of given phase angle can be resolved into two component waveforms that have a phase difference of 90°. If one component is the in-phase signal, then the other is quadrature. Since the demodulator responds only to the in-phase component, an exact phase relationship between input and reference is not necessary to demodulator performance. Thus system phase shifts are not critical.

Another potential difficulty in ac systems is extraneous ac voltages that appear as a result of stray inductive or capacitive coupling. Fortunately such coupling invariably introduces a near-90° phase shift. Since the stray voltages are now at quadrature with the signal, they will be rejected by the demodulator.

PROBLEMS

A. *Transducers*

3-1. What is the output voltage of the 1-kΩ potentiometer shown below if the excitation is 2 V and the setting ratio is 0.4?

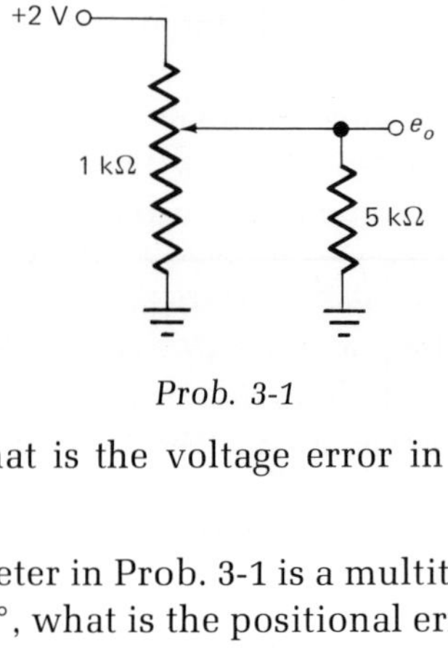

Prob. 3-1

3-2. In Prob. 3-1, what is the voltage error in the output due to the loading?

3-3. If the potentiometer in Prob. 3-1 is a multiturn unit having a shaft rotation of 1,080°, what is the positional error associated with the loading?

3-4. A 10-kΩ potentiometer is operated with a grounded center tap as shown below. What is the maximum output-voltage error due to the loading?

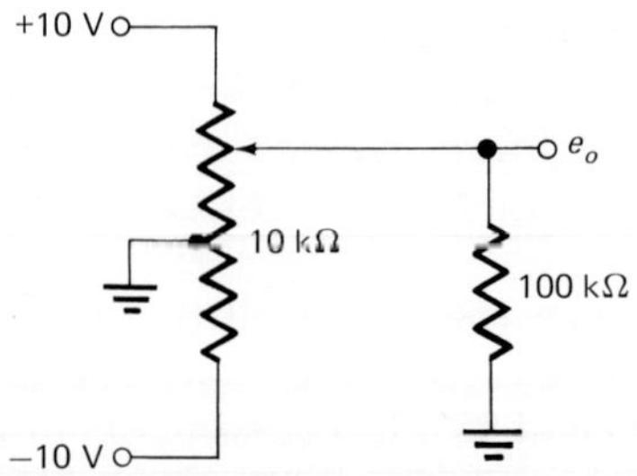

Prob. 3-4

3-5. If the potentiometer in Prob. 3-4 has a rotation angle of 350°, what is the gain constant expressed in units of volts per radian?

3-6. Determine the number of turns of wire needed to provide a potentiometer with a resolution of 0.05 percent.

3-7. A wire-wound potentiometer has the following specifications: resistance—2,500 Ω ± 10 percent; wire turns—3,000; and linearity—0.1 percent. The potentiometer is operated single-ended with a 100-kΩ load. If the various errors that affect the output combine in the same direction, what is the worst-case error in the output expressed as a percentage of the input voltage?

3-8. The voltage applied to the rotor of a synchro control transmitter is 28 V rms. The rotor shaft is moved 60° from the zero position. Determine the stator voltages with respect to the common stator connection if $K = 1$.

3-9. In Prob. 3-8 determine the voltages between terminals S_1 and S_2, S_2 and S_3, and S_3 and S_1.

3-10. Repeat Probs. 3-8 and 3-9 for a rotor angle of −45°.

3-11. A synchro transmitter is connected to a control transformer whose sensitivity is 0.1 V/°. If the transmitter shaft is advanced 5° with respect to the transformer shaft, what is the transformer output voltage?

3-12. What is the shaft angle of a synchro transmitter if the stator voltages are $S_{31} = 14$ V in phase, $S_{12} = 28$ V out of phase, and $S_{23} = 14$ V in phase?

3-13. Repeat Prob. 3-12 for $S_{31} = 21.8$ V in phase, $S_{12} = 0$ V, and $S_{23} = 21.8$ V out of phase.

3-14. A synchro transmitter, differential, and receiver are connected as shown in Fig. 3-13. If $\theta_i = 240°$ and $\theta_d = -20°$, at what angle will the receiver come to rest?

3-15. An induction resolver with $K = 1.0$ has stator input voltages $S_{13} = 10$ V and $S_{24} = 5$ V. Determine the two rotor output voltages if $\theta = 25°$.

3-16. Repeat Prob. 3-15 if voltage S_{13} is out of phase.

3-17. If a resolver has a functional accuracy of 0.1 percent, determine the angular error at an angle of 30°.

3-18. A tachometer has a sensitivity of 5 V/1,000 rpm. Express the gain constant of the tachometer in units of volts per radian per second.

3-19. A tachometer whose internal resistance is 50 Ω is connected to a 10-kΩ load. What will be the percentage change in the output voltage due to this loading?

3-20. A dc tachometer is constructed such that there are eight ripple cycles per revolution of the armature. If the shaft speed is 500 rpm, determine the ripple frequency expressed in hertz.

3-21. An accelerometer contains a sensing mass whose weight is 0.5 oz. The mass is supported by springs having a constant of 1.1 lb/in. What displacement will occur if the unit undergoes an acceleration of 10 g?

3-22. The strain gage in the bridge circuit below has a gage factor of 3. Determine the bridge output voltage e_o if an applied force causes a 2 percent stretch in the active gage.

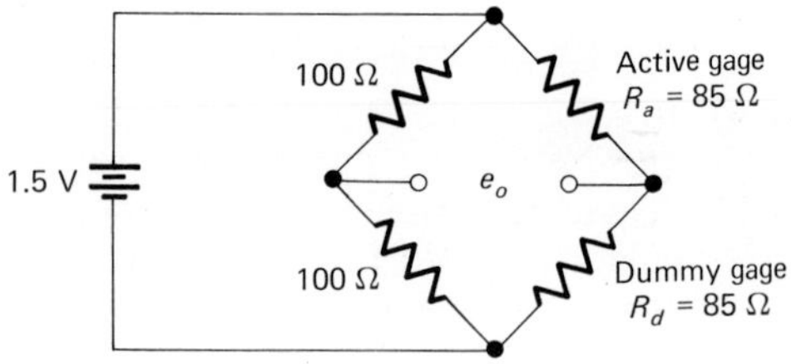

Prob. 3-22

3-23. A copper wire is used as a resistance thermometer for measuring temperatures. If its resistance is 40 Ω at 20°C, how much will its resistance increase for a temperature rise of 1°C?

3-24. A platinum-wire resistance thermometer R_t is connected as an arm of the Wheatstone bridge shown below. At 20°C, the bridge balances with the slide-wire resistance R_s at 272 Ω. With the thermometer placed in a freezing mixture, a new balance is obtained at 240 Ω. Find the temperature of the mixture.

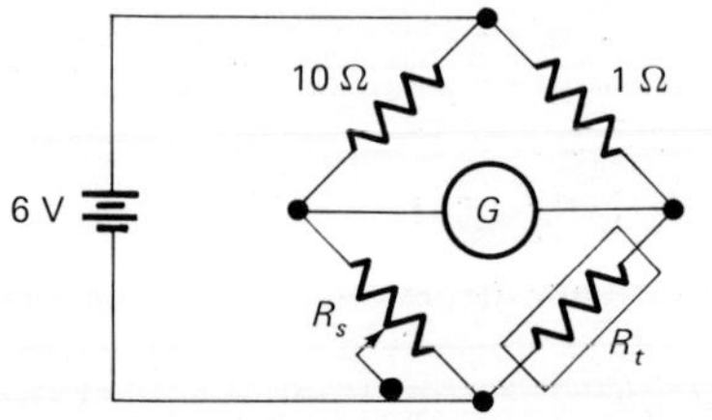

Prob. 3-24

3-25. Compute the emf generated by a copper-constantan thermocouple, the hot and cold junctions of which are at 200 and 20°C, respectively.

3-26. To measure the temperature at an internal point in an electric motor, one junction of an iron-constantan thermocouple is placed at this point and the other junction is maintained at 0°C. If the thermocouple produces an emf of 1.92 mV, compute the motor temperature being measured.

B. Power Actuators

3-27. Plot a family of linear speed-torque curves for a dc motor which has a stall torque of 12 oz-in. and a no-load speed of 6,000 rpm at a rated armature voltage of 30 V. Determine the torque and damping constants of the motor.

3-28. Under rated conditions a motor can deliver a torque of 2 lb-ft at a speed of 3,000 rpm. What is the rated horsepower of the motor? (Recall 1 hp = 550 ft-lb/s.)

3-29. The speed-torque curves of a dc motor with field control are shown in Prob. 3-29. Determine the torque and damping constants. The field resistance is 450 Ω.

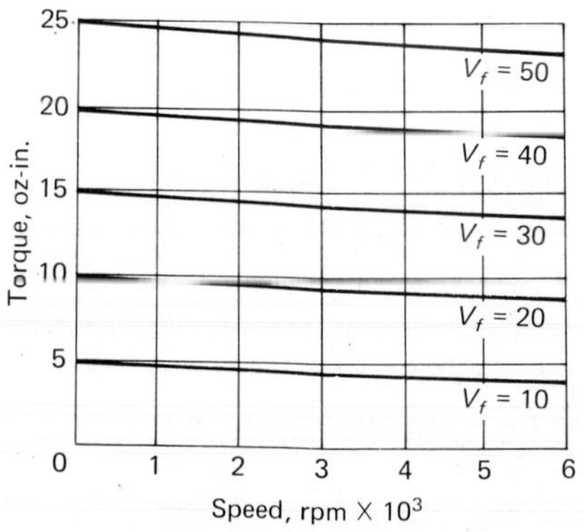

Prob. 3-29

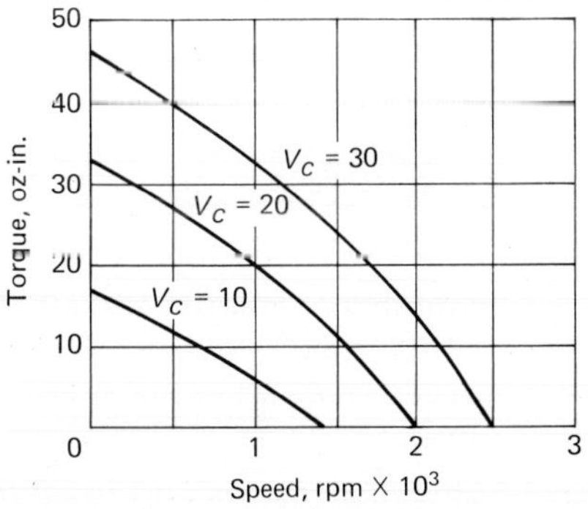

Prob. 3-30

3-30. The speed-torque curves of a two-phase ac motor are shown in Prob. 3-30. Determine the torque and damping constants assuming operation at low values of the control voltage.

3-31. A 400-Hz motor has a control-field-winding resistance of 200 Ω and a reactance of 150 Ω. What value of parallel capacitance is required to make the winding appear purely resistive? What is the impedance of the control-winding circuit with this parallel capacitance added?

3-32. A mechanical load is coupled to a motor through a gear train. If the rated motor speed is 2,000 rpm, what gear ratio is required to provide a maximum load speed of 50 rpm? What torque is then delivered to the load for a motor torque of 0.2 lb-ft?

3-33. A motor pinion that has 10 teeth drives a gear with 80 teeth. Determine the gear ratio and load speed for a motor speed of 10,000 rpm.

3-34. A motor that develops 60 oz-in. of torque at a speed of 3,600 rpm is coupled to a load by means of the gear train shown below. Determine the torque and speed at the load shaft.

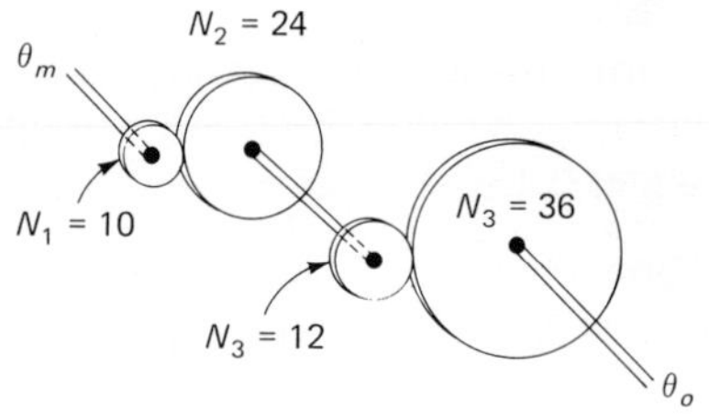

Prob. 3-34

3-35. The armature inertia of a motor is 5×10^{-4} oz-in.-s^2 and the damping constant is 3×10^{-3} oz-in./(rad/s). Determine the inertia and damping as they appear at the output of a gear train whose ratio is 64.

3-36. A mechanical load is coupled to a motor by a gear train. The motor gear has 18 teeth and the load gear has 108 teeth. The inertia of the load is 1.0 oz-in.-s^2 and the inertia of the rotor is 0.01 oz-in.-s^2. Determine the gear ratio to provide maximum acceleration to the load. Diagram a gear train to provide this ratio using the specified motor and load gears.

3-37. An inertia load of 0.017 oz-in.-s^2 must be given an acceleration from stall of 400 rad/s^2. Assuming a motor that is capable of a maximum acceleration of 17,600 rad/s^2, determine a suitable gear ratio. What must be the stall torque of the motor?

3-38. Liquid glycerin (density $= 78\ \text{lb/ft}^3$) flows through a valve at a maximum rate of 25 gal/min. If the pressure drop across the valve is limited to 10 psi, what size valve is required? (Recall that water density $= 62.4\ \text{lb/ft}^3$.)

C. *Amplifiers*

3-39. Determine the power gain of the relay circuit below. The coil resistance is 150 Ω and the signal voltage to actuate the relay is 28 V.

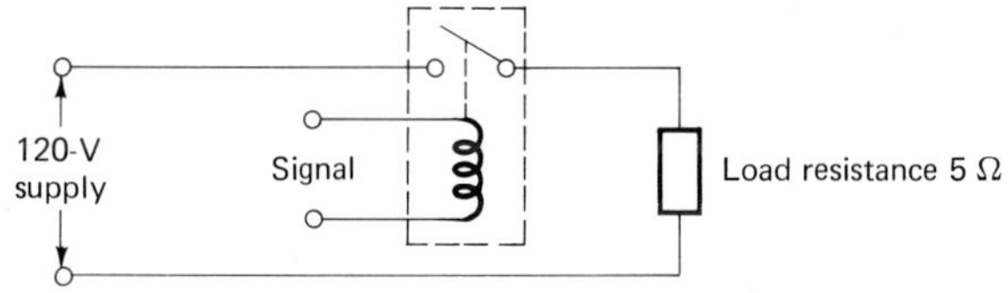

Prob. 3-39

3-40. The operational amplifier shown below functions as an error detector and voltage amplifier in a control system. Determine the equation for the amplifier output voltage e_o as a function of the input voltage e_i and system feedback voltage $-e_f$.

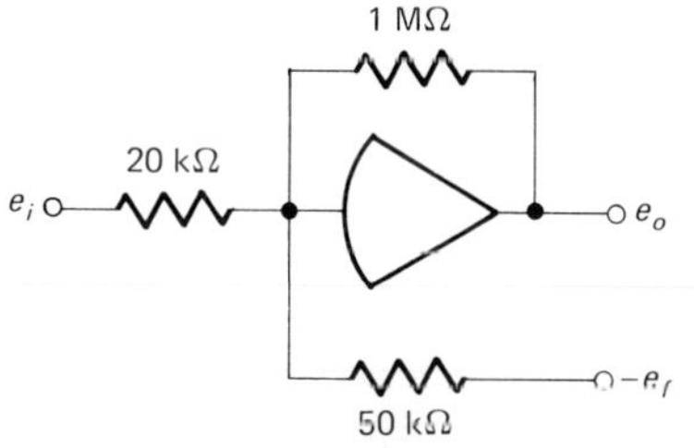

Prob. 3-40

3-41. A dc input voltage of magnitude 0.1 V is connected to the circuit below. Sketch the amplifier output voltage as a function of time for $0 \le t \le 100$ s.

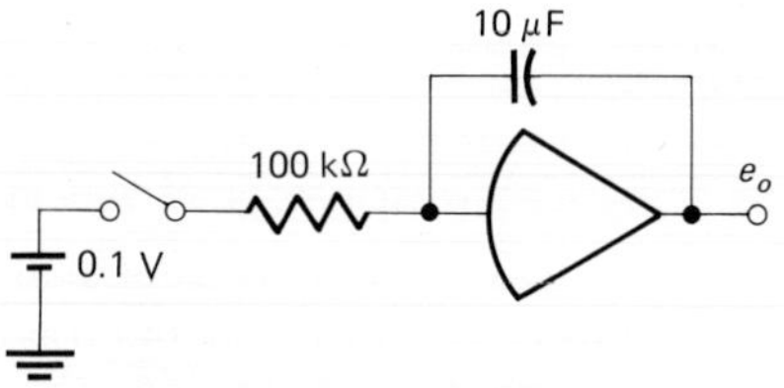

Prob. 3-41

3-42. Determine the voltage output e_o.

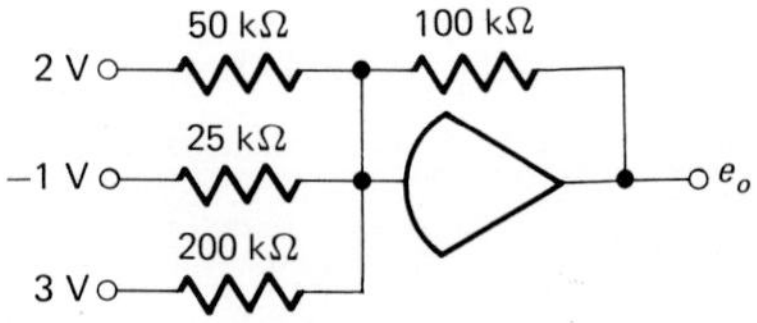

Prob. 3-42

3-43. Show that the operational-amplifier circuit below functions as a voltage follower if the amplifier gain A is very high.

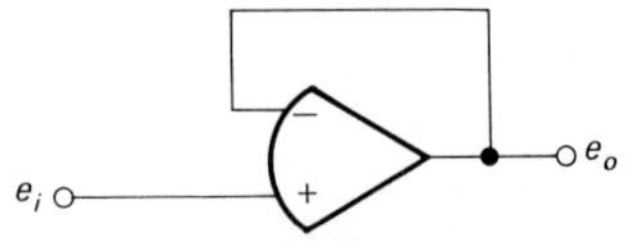

Prob. 3-43

3-44. A noninverting amplifier having high input impedance is obtained with the following operational-amplifier circuit. If the gain A is very high, show that $e_o/e_i = (R_1 + R_2)/R_1$.

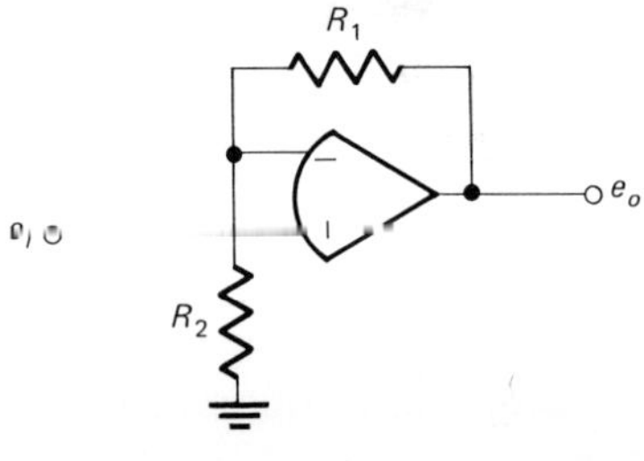

Prob. 3-44

3-45. A positional servomechanism includes a servomotor with a torque constant of 0.2 oz-in./V. The error detector is a synchro control transformer with a sensitivity of 0.5 V/°. The gear ratio from the motor to output is 160. If the output shaft is to provide a torque gradient of 10,000 oz-in./°, what voltage gain is required of the servo amplifier?

3-46. Sketch the output voltage waveform for the diode demodulator circuit of Fig. 3-63 for the following combinations of input and reference voltages (assume filter capacitors are removed).

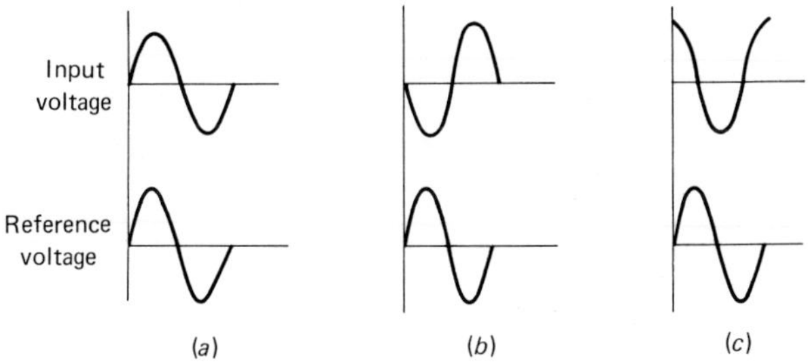

Prob. 3-46

Chapter Four

transient response of physical systems

4-1 INTRODUCTION

It was stated in Chap. 2 that the dynamic behavior of physical systems is extremely important in control work. A control system continually receives inputs, either as commands or disturbances, and its response to these inputs must be analyzed in order to achieve satisfactory system performance. The purpose of this chapter is to provide an introduction to dynamic behavior by describing the transient response of several basic physical systems, including simple feedback control systems.

In Chap. 2 we have seen how differential equations may be used to describe a physical system. Predicting the transient response of the system and achieving a satisfactory design involve writing and solving the differential equations for the system.

This chapter discusses the transient behavior of basic first- and second-order systems (those whose behavior is defined by first- and second-order differential equations). This is not to say that most control systems are simple enough to be represented by lower-order equations. Frequently system equations are quite complex and contain nonlinear relationships between the variables. However, by judicious approximations and restriction of the region of operation (as will be described in Chap. 9), we can frequently use first- and second-order equations as a convenient starting point in analysis and design. In addition, the behavior of many of the components that make up control systems can be adequately defined by lower-order equations.

In this chapter we will examine the transient response of first- and second-order systems to a widely used input, the step function. The actual process of solving the system differential equations to obtain the response to this and other inputs will be deferred until the next chapter, where the Laplace transform method will be introduced.

Since electric circuits are a very basic kind of physical system, the contents of this chapter are presented by first considering the transient response of several familiar circuits. It should be apparent though that a control system is seldom purely electrical in nature; it may contain mechanical, hydraulic, thermal, or other elements. However, the equations of these other elements are similar in form to those of electric circuits. The difference is in the coefficients and the physical significance of the variables. Thus the forms of response for familiar electric circuits may be applied to equations for other dynamic systems, including systems for automatic control.

4-2 INPUT FUNCTIONS

Various input functions are commonly used in evaluating the dynamic performance of physical systems. The use of certain test inputs has two advantages in control systems work. First, if the system can be described mathematically, the response to these inputs can be predicted using well-defined analytical techniques. Second, these inputs can be used as a basis for experimentally verifying the theoretical results. The choice of test input depends upon the form of the inputs to the system in actual operations, the physical makeup of the control system, and the specifications for its performance.

The *step function* is perhaps the most widely used input function in dynamic analysis. A plot of this function is shown in Fig. 4-1. The step function assumes that all previous transients have completely died out prior to time $t = 0$. At $t = 0$, the amplitude r of the function instantaneously rises to some level A and remains at that value until all transients in the system response caused by the step have died out.

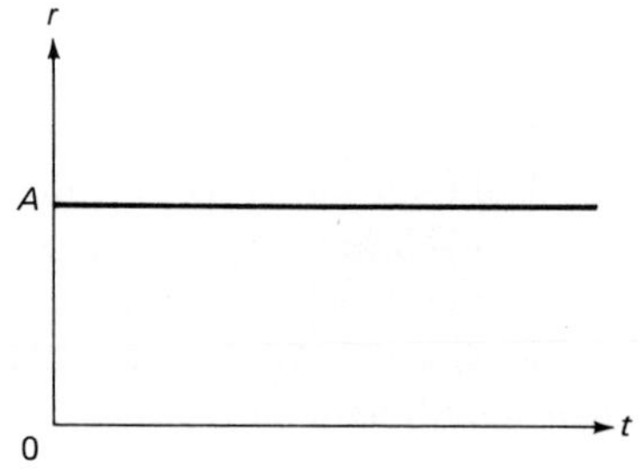

Figure 4-1 Step function

Stated mathematically:

$$r = 0 \text{ for } t < 0$$
$$r = A \text{ for } t \geq 0 \qquad (4\text{-}1)$$

In an actual physical system it is obviously not possible to change instantly from one level to another. However, an instant change can be approximated with little error, and so the step function is widely used in the study of control systems.

The step function is frequently used where the amplitude of the step is assigned a value of 1. The function is then called the *unit step function*. The unit step function eliminates the necessity of specifying a value for A.

The *rectangular pulse function* shown in Fig. 4-2 is an extension of the step function. The pulse function assumes that all previous transients have completely died out prior to time $t = 0$. At $t = 0$, the function instantaneously rises to level A and remains at that level until time $t = t_1$. Time t_1 is chosen such that all measurable transients caused by the change in amplitude have died out. At time $t = t_1$, the function instantaneously drops back to 0 and remains there. Stated mathematically:

$$r = 0 \text{ for } t < 0$$
$$r = A \text{ for } 0 \leq t < t_1 \qquad (4\text{-}2)$$
$$r = 0 \text{ for } t \geq t_1$$

Pulse functions find application in experimental work primarily because of their ease in generation. For example, in an electrical system,

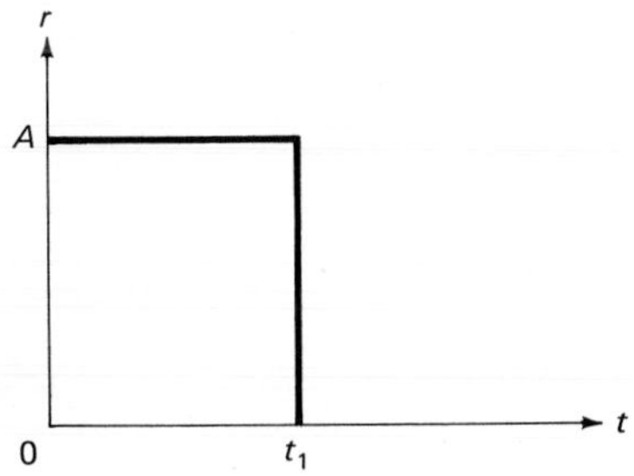

Figure 4-2 Rectangular pulse function

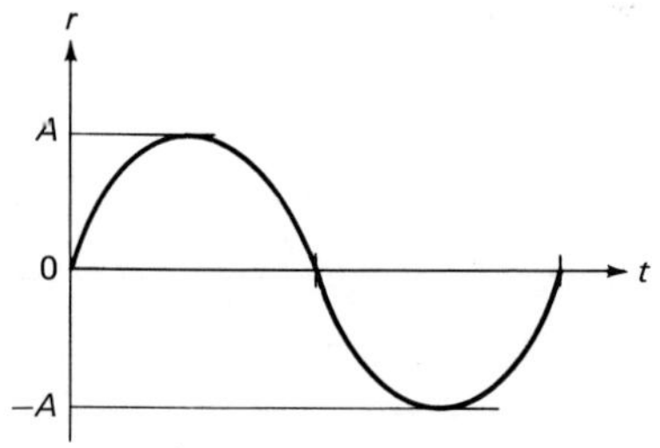

Figure 4-3 Sinusoidal function

pulse functions can be readily obtained from a simple square-wave generator.

Another input function that finds wide use is the *sinusoidal function* shown in Fig. 4-3. Although the sinusoidal function can be viewed as a time function, it is primarily used with the frequency ω taken as the independent variable. Stated mathematically:

$$r = A \sin \omega t \tag{4-3}$$

where A is the peak amplitude and ω is the angular frequency expressed in radians per second.

In the *ramp function* shown in Fig. 4-4, the input is changed at a constant rate. Prior to time $t = 0$, the function assumes that all previous transients have died out. At $t = 0$, the function begins to increase at a constant rate designated by the slope K. Stated mathematically:

$$\begin{aligned} r &= 0 \text{ for } t < 0 \\ r &= Kt \text{ for } t \geq 0 \end{aligned} \tag{4-4}$$

Although this function has important theoretical considerations, it

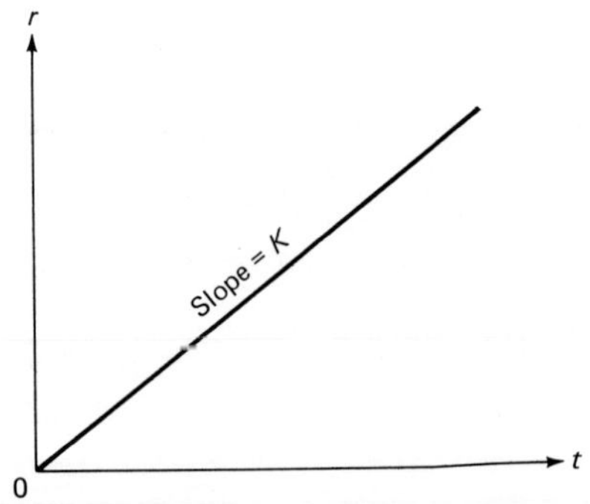

Figure 4-4 Ramp function

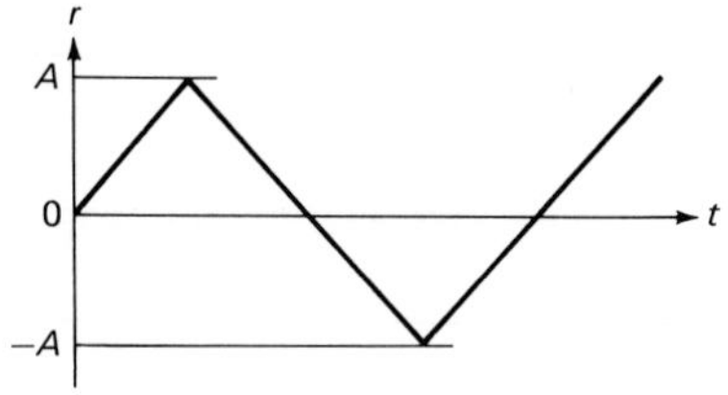

Figure 4-5 Triangular function

cannot be used conveniently in experimental work since there is a limit to the value that can be reached. The ramp function is often approximated by a *triangular wave* as shown in Fig. 4-5. This approximation allows the amplitude to be limited to a value A that is consistent with the system under test. The period of the waveform is chosen so that any transients die out before the ramp changes slope.

Another function to be noted is a *random function* as shown in Fig. 4-6. This is a specialized type of input based on statistical considerations and is commonly termed *white noise* since it randomly contains all frequencies. The advantage of the random input function is that the amplitude and time duration of the input signal can be kept small. Thus, random signals can be applied experimentally to a control system with a minimum effect on system operation. This may be important in control systems that can be disturbed only by a small amount or can be tested only in actual operation. For example, the position of the control rods in a nuclear reactor could only be changed slightly in the experimental design of a positioning control system. Since random input analysis requires statistical considerations and mathematical facility beyond the scope of this text, it will not be considered further.

4-3 FIRST-ORDER SYSTEMS

A physical system that is described by a first-order differential equation is identified as a *first-order system*. A characteristic of a first-order

Figure 4-6 Random function

system is that it contains one element in which energy may be stored in some form. Thus, an example of a first-order system is a moving object possessing kinetic energy.

A first-order system is defined by a first-order equation of the general form

$$a\frac{dx}{dt} + bx = f(t) \tag{4-5}$$

where x is the dependent variable, $f(t)$ is an input function, and a and b are constants.

In Sec. 2-9 it was pointed out that the solution of a differential equation where time is the independent variable gives the response of the system to an input function. This response consists of two parts: a *steady-state part* and a *transient part*. The total response is the sum of these two solutions. Thus

Total response = steady-state response + transient response

Both solutions must be found to completely define the response.

In the steady state the input has been present long enough so that the effects of any previous change have died out. In general the steady-state response is of the same form as the input function. Thus, with a step input applied to the system described by Eq. (4-5), the condition for steady-state operation is x unchanging or

$$\frac{dx}{dt} = 0$$

from which

$$x_{ss} = \frac{f(t)}{b} = \frac{R}{b} \tag{4-6}$$

where x_{ss} = steady-state value of x
R = magnitude of step input

The transient response shows the manner in which the variable changes from its initial value to the steady-state value as the form of the input is changed. For a step input to a stable first-order system it can be shown that the transient response x_t is an exponential decay having the form

$$x_t = Ae^{-bt/a} \tag{4-7}$$

where a and b are the coefficients from the differential equation and A is a constant related to the initial or starting conditions in the system. Note that the transient response approaches 0 as time approaches ∞.

The total response of a first-order system is then

$$
\begin{aligned}
x &= x_{ss} + x_t \\
&= \frac{R}{b} + Ae^{-bt/a}
\end{aligned}
\tag{4-8}
$$

The exponential term $e^{-bt/a}$ reveals the decay of the transient part of the solution. It will be noted that the coefficient of t determines the rate of decay. The reciprocal of this coefficient, which has the dimensions of time, is called a *time constant* and is a parameter that is used in describing the response of first-order systems. The time constant is denoted by the symbol τ and is defined as the time for the transient term to be reduced to $e^{-1} = 0.368$ (approximately 37 percent) of its initial value. It should be noted that the time constant is a characteristic of the system and is independent of the input function. With respect to the general first-order system given by Eq. (4-5) the time constant is $\tau = a/b$, and the exponential factor may then be written as $e^{-t/\tau}$.

For a step-function input to a first-order system where the initial condition is 0, the total response is

$$
x = x_{ss}(1 - e^{-t/\tau}) \tag{4-9}
$$

Figure 4-7 shows the step-response of a first-order system plotted on a normalized basis. On the normalized curve, the horizontal axis is dimensionless since it is scaled in units of time divided by the time constant. The normalized response plotted on the vertical axis has a unity steady-state value. Thus the vertical scale shows the ratio of the output at any instant to the final value. A knowledge of the normalized curve allows one to readily plot the transient response of a first-order system by multiplying the normalized vertical scale by the value of the steady state response and the horizontal scale by the time constant. It can be seen that the response transient essentially disappears after

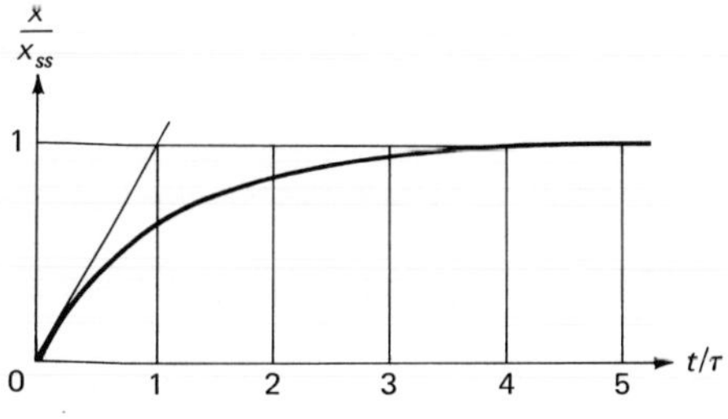

t/τ	Normalized response
1	0.632
2	0.865
3	0.951
4	0.981
5	0.993

Figure 4-7 Normalized response of first-order system to step input

five time constants have elapsed. Further, it can be shown that the response of the system would reach its final value at $t = \tau$ if the initial rate of response was maintained.

4-4 SERIES RESISTANCE-INDUCTANCE NETWORK

Simple electric circuits consisting of either a series-connected resistance and inductance or resistance and capacitance are examples of first-order systems. A voltage applied to the circuit will cause a change in both the current and voltage drops. When the network contains an energy-storage element in the form of an inductor or capacitor, these changes are not instantaneous. Rather, some period of time is required for the variable quantities to change from their initial to final volume. The behavior of the variables during this transition time may be determined by solving the circuit differential equation.

Figure 4-8 shows a series resistance-inductance circuit to which a battery of voltage E is connected by closing the switch at a time $t = 0$ s. Let us examine the response of the current i to this step input of voltage. Referring to Sec. 2-5, the differential equation of the circuit is

$$L\frac{di}{dt} + Ri = E \tag{4-10}$$

The steady-state current i_{ss} is dependent only upon the applied voltage E and has the same form as E. Since E is a constant, i_{ss} does not change with time. Using Eq. (4-6),

$$i_{ss} = \frac{E}{R} \tag{4-11}$$

The transient current i_t is of the form given by Eq. (4-7).

$$i_t = Ae^{-Rt/L} \tag{4-12}$$

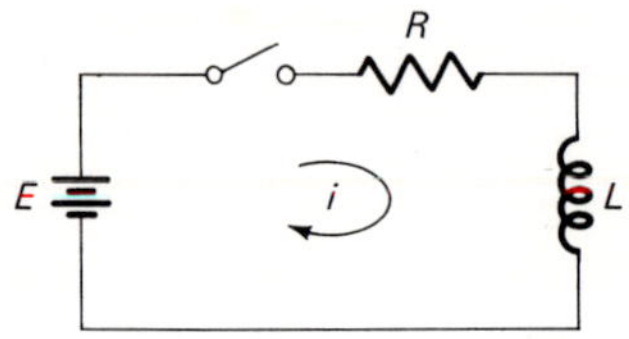

Figure 4-8 Series resistance-inductance network

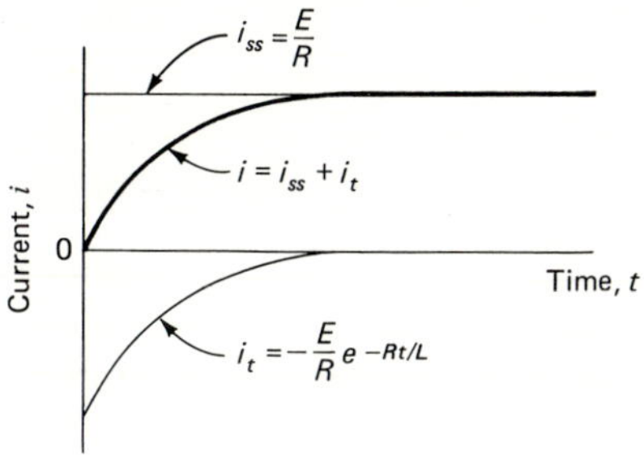

Figure 4-9 Current through series RL circuit after closing switch

At any instant the total current i is the sum of the steady-state and transient solutions. Thus

$$i = \frac{E}{R} + Ae^{-Rt/L} \tag{4-13}$$

To evaluate the magnitude of A, the initial condition at $t = 0$ may be applied. With the switch open, the current is 0. Since there cannot be an instantaneous change in energy when the switch is closed, it follows that at $t = 0$, $i = 0$. Substituting this condition in Eq. (4-13) gives

$$A = -\frac{E}{R}$$

and so

$$i = \frac{E}{R} - \frac{E}{R} e^{-Rt/L} = \frac{E}{R}(1 - e^{-Rt/L}) = \frac{E}{R}(1 - e^{-t/\tau}) \tag{4-14}$$

This result corresponds to the form of response given by Eq. (4-9). The time constant is $\tau = L/R$ seconds, where R has units of ohms and L is measured in henrys.

Figure 4-9 is a plot of the current, including both steady-state and transient components, as a function of time. The exponential nature of this response is a characteristic of first-order systems.

EXAMPLE 4-1

Determine the current equation in the series RL circuit of Fig. 4-8 when $R = 10\,\Omega$, $L = 1$ H, and $E = 1$ V.

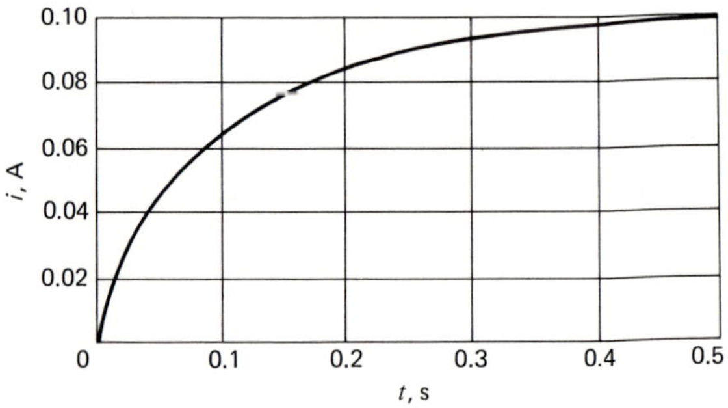

Figure 4-10 Current rise for RL circuit of Example 4-1

Solution

From Eq. (4-14), the current is

$$i = \frac{E}{R}(1 - e^{-Rt/L})$$

Substituting circuit values, we have

$$i = 0.1(1 - e^{-10t})\ \text{A}$$

Note that the circuit time constant $\tau = L/R = 0.1$ s.

The current in Fig. 4-10 exhibits the exponential characteristic of first-order systems that was shown in Fig. 4-7.

In the above example we have selected a familiar electrical circuit to illustrate the response of a first-order system. It is important to note that the response of a first-order system is characterized by variables which increase or decrease in an exponential fashion. Other examples of first-order response include the heating or cooling of a body, motion with resistance proportional to velocity, mixing of solutions, and radioactive decay. As we shall see in subsequent sections, some types of closed-loop control systems can also be analyzed as first-order systems.

4-5 SERVOMOTOR

Other dynamic systems may be analyzed in a manner similar to the treatment of first-order electrical circuits. An example pertinent to control systems is a servomotor where the armature voltage is varied and the shaft velocity is the controlled variable.

Let us assume that the motor characteristic can be approximated by linear speed-torque curves such as were shown in Fig. 3-30. With constant field excitation, motor torque is proportional to armature current. Referring to Eq. (3-34), this relationship is expressed as

$$T_m = K_i i_a \tag{4-15}$$

As the motor armature rotates, a back electromotive-force voltage e_b is generated proportional to motor speed. Thus we have

$$e_b = K_b \frac{d\theta}{dt} = K_b \dot{\theta} \tag{4-16}$$

where $\frac{d\theta}{dt} = \dot{\theta}$ = shaft velocity, rad/s

K_b = voltage constant, V/(rad/s)

If the armature winding inductance is assumed negligible, the armature current is

$$i_a = \frac{v_a - e_b}{R_a} = \frac{v_a - K_b\dot{\theta}}{R_a} \tag{4-17}$$

where R_a is the resistance of the armature winding.

The torque supplied to the mechanical load (motor armature plus coupled load) is used to accelerate the rotating inertia as well as overcome the viscous friction. Thus

$$T_L = J\ddot{\theta} + F\dot{\theta} \tag{4-18}$$

where T_L = load torque, oz-in.
$\ddot{\theta}$ = acceleration, rad/s^2
J = inertia, oz-in.-s^2
F = friction coefficient, oz-in./(rad/s)

The load torque must equal the torque delivered by the motor. Substituting Eq. (4-17) into Eq. (4-15) and equating the result to Eq. (4-18) gives

$$\frac{K_i}{R_a}(v_a - K_b\dot{\theta}) = J\ddot{\theta} + F\dot{\theta} \tag{4-19}$$

Solving Eq. (4-19) for the armature voltage yields

$$v_a = \frac{JR_a}{K_i}\ddot{\theta} + \frac{R_a}{K_i}\left(F + \frac{K_bK_i}{R_a}\right)\dot{\theta} \tag{4-20}$$

Recall from Eq. (3-36) that the slope of the motor speed-torque curve represents the motor damping constant B_m. For a given armature voltage V_a, the slope is $T_s/\dot{\theta}_{NL}$, where T_s and $\dot{\theta}_{NL}$ are the stall torque and no-load speed corresponding to V_a. Expressing the slope in terms of the motor constants, we have

$$B_m = \frac{T_s}{\dot{\theta}_{NL}} = \frac{K_i V_a/R_a}{V_a/K_b} = \frac{K_b K_i}{R_a} \tag{4-21}$$

The term B_m represents the damping that is inherent in the motor characteristic. The total damping B is the sum of the viscous friction and the motor damping. Thus $B = F + B_m$, and Eq. (4-20) may be written as

$$v_a = \frac{JR_a}{K_i}\ddot{\theta} + \frac{BR_a}{K_i}\dot{\theta} \tag{4-22}$$

Equation (4-22) is a first-order differential equation of the dependent variable $\dot{\theta}$. Although the equation does contain a second derivative $\ddot{\theta}$, this should be interpreted as the *first* derivative of the variable $\dot{\theta}$. Note that Eq. (4-22) is similar in form to the differential equation given in Eq. (4-5).

The motor torque constant is more commonly expressed in terms of armature voltage rather than current. Referring to Eq. (3-35),

$$K_t = \frac{T_s}{V_a} = \frac{K_i I_a}{V_a} = \frac{K_i}{R_a} \tag{4-23}$$

Rewriting Eq. (4-22) to include K_t, we have

$$v_a = \frac{J}{K_t}\ddot{\theta} + \frac{B}{K_t}\dot{\theta} \tag{4-24}$$

If a step input of voltage of magnitude V_a is applied to the armature, the velocity will change exponentially in the manner characteristic of first-order systems. Setting the acceleration $\ddot{\theta}$ to 0 in Eq. (4-24), the steady-state velocity may be determined as

$$\dot{\theta}_{ss} = \frac{K_t V_a}{B} = K_m V_a \tag{4-25}$$

The constant K_m relates steady-state motor speed to armature voltage; hence K_m may be interpreted as a velocity gain constant numerically equal to K_t/B, which has units of (rad/s)/V.

Taking the ratio of the coefficients of $\ddot{\theta}$ and $\dot{\theta}$ in Eq. (4-24), the motor time constant τ_m is found to be

$$\tau_m = \frac{J/K_t}{B/K_t} = \frac{J}{B} \tag{4-26}$$

If the motor is accelerated from rest ($\dot{\theta} = 0$ at $t = 0$), the response to a step input will be of the form given by Eq. (4-9). Thus

$$\dot{\theta} = \dot{\theta}_{ss}(1 - e^{-t/\tau_m}) \tag{4-27}$$

where $\dot{\theta}_{ss}$ and τ_m are defined by Eqs. (4-25) and (4-26), respectively.

It should be noted that Eq. (4-27) is similar in form to the response of the series resistance-inductance network given by Eq. (4-14). This is to be expected since the differential equations describing these two systems are alike in form. Recall that Table 2-1 showed that inductance and resistance in an electric circuit are analogous to inertia and friction (or damping) in a mechanical system. This analogy is evident in comparing Eqs. (4-14) and (4-26).

EXAMPLE 4-2

A dc servomotor of the permanent-magnet type has the following characteristics: an armature inertia of 5×10^{-5} oz-in.-s^2, a speed of 5,000 rpm, and a stall torque of 1.5 oz-in. with 115 V applied to the armature. Determine the response of the motor when 25 V is applied to the armature terminals.

Solution

Using Eq. (4-21), the damping as given by the slope of the speed-torque curve is

$$B = \frac{T_s}{\dot{\theta}_{NL}} = \frac{1.5}{5{,}000 \times 2\pi/60} = 2.87 \times 10^{-3} \text{ oz-in./(rad/s)}$$

From Eq. (4-23), the torque constant is

$$K_t = \frac{T_s}{V_a} = \frac{1.5}{115} = 0.013 \text{ oz-in./V}$$

From Eq. (4-25), the steady-state speed is

$$\dot{\theta}_{ss} = \frac{0.013 \times 25}{2.87 \times 10^{-3}} = 113 \text{ rad/s}$$

From Eq. (4-26), the time constant is

$$\tau_m = \frac{5 \times 10^{-5}}{2.87 \times 10^{-3}} = 0.0174 \text{ s}$$

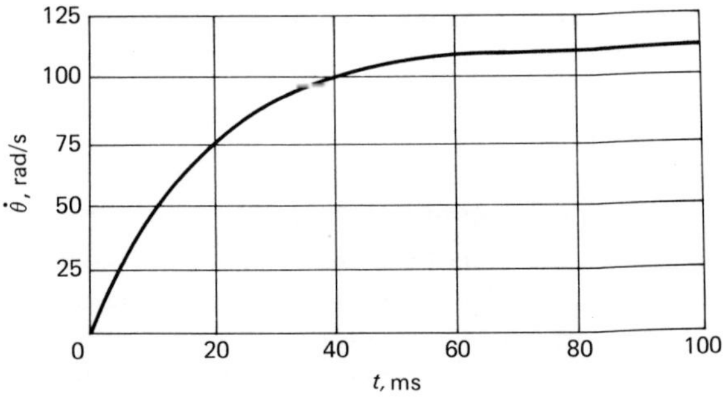

Figure 4-11 Speed response of motor in Example 4-2

Using Eq. (4-27), the motor response is then

$$\dot{\theta} = 113(1 - e^{-t/0.0174}) \text{ rad/s}$$

Figure 4-11 shows the response curve for motor speed as a function of time.

4-6 SPEED-REGULATING SYSTEM (RATE SERVOMECHANISM)

Many control applications in industry are concerned with maintaining a constant value of a variable such as temperature, pressure, flow, liquid level, or another physical quantity. An automatic system that maintains a fixed level of the controlled variable is identified as a *regulating system*. Such a system exhibits the basic property that a constant actuating signal results in a constant value of the output. Control action is dependent on deviation from the desired value, and the control system acts to keep the deviation to a minimum despite load fluctuations, system disturbances, or changes in the characteristics of the control elements. For example, we have seen that an increase in load torque causes motor speed to decrease if the motor control voltage remains constant. The obvious way to maintain a constant motor speed is to increase the control voltage as the speed decreases. For this purpose, a closed-loop regulating system incorporating negative feedback may be used.

A speed-regulating system is shown in basic form in Fig. 4-12. The system is composed of a motor whose shaft speed $\dot{\theta}$ is to be held constant, a mechanical load driven by the motor shaft, a power amplifier that controls the motor voltage, a tachometer coupled to the

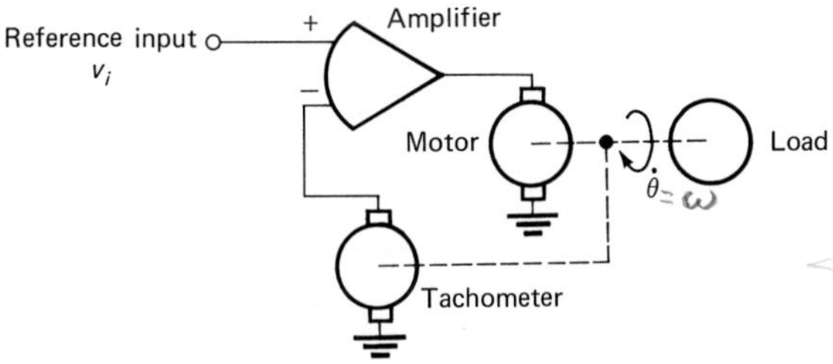

Figure 4-12 Speed-regulating system

motor shaft, and a circuit that detects the deviation or error between the reference input and the feedback signals. Because the velocity of the output is the controlled variable, this system is sometimes called a *velocity* or *rate servomechanism*. Figure 4-13 is a block-diagram representation of the system.

In analyzing the speed-control system, we will first determine the input-output relationship of each of the components. If the amplifier has a voltage gain K_a,

$$v_a = K_a v_e \tag{4-28}$$

The tachometer generates a voltage v_f proportional to the speed $\dot{\theta}$, and so

$$v_f = K_g \dot{\theta} \tag{4-29}$$

where K_g is the tachometer gain constant.

If the amplifier provides a differential input, the reference and feedback voltages may be combined so that the net input to the amplifier is the error voltage v_e. This may be expressed as

$$v_e = v_i - v_f = v_i - K_g \dot{\theta} \tag{4-30}$$

At the motor output, the torques resulting from inertia and damping must be equal to that developed by the motor. Thus, beginning with

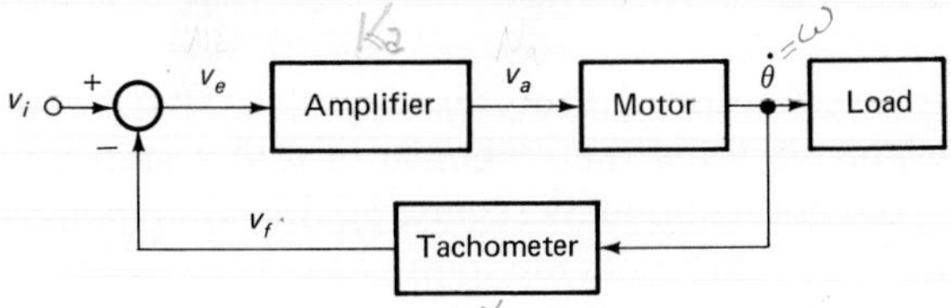

Figure 4-13 Block diagram of speed-regulating system

Eq. (4-24), we have

$$\frac{J}{K_t}\ddot{\theta} + \frac{B}{K_t}\dot{\theta} = v_a = K_a v_e = K_a(v_i - K_g\dot{\theta})$$

Solving for v_i gives

$$v_i = \left(\frac{B}{K_a K_t} + K_g\right)\dot{\theta} + \frac{J}{K_a K_t}\ddot{\theta} \tag{4-31}$$

Equation (4-31) is a first-order equation in $\dot{\theta}$, and the solution will give $\dot{\theta}$ as a function of time for a given input command. If the input is a constant V_i, then in the steady state $\ddot{\theta} = 0$, and the steady-state output is

$$\begin{aligned}\dot{\theta}_{ss} &= \left(\frac{K_a K_t}{B + K_a K_g K_t}\right)V_i \\ &= \left(\frac{K_a K_t/B}{1 + K_a K_g K_t/B}\right)V_i \\ &= \left(\frac{K_a K_m}{1 + K_a K_m K_g}\right)V_i\end{aligned} \tag{4-32}$$

where K_m is the velocity gain defined by Eq. (4-25). Since a regulating system is designed primarily to maintain the controlled variable at a constant, the steady-state output given by Eq. (4-32) is of foremost interest.

Regulating systems are generally not intended for transient operation. However, we will examine the response of the speed control system to a step input. Taking the ratio of the coefficients of $\ddot{\theta}$ and $\dot{\theta}$ in Eq. (4-31), the time constant τ_c of the closed-loop system is

$$\tau_c = \frac{J}{B + K_a K_g K_t} = \frac{J/B}{1 + K_a K_m K_g} \tag{4-33}$$

The response of the system to a step input is then

$$\dot{\theta} = \dot{\theta}_{ss}(1 - e^{-t/\tau_c}) \tag{4-34}$$

where $\dot{\theta}_{ss}$ and τ_c are given by Eqs. (4-32) and (4-33). Figure 4-14 is a graph showing the response of $\dot{\theta}$ to a step input of voltage.

Several comments regarding this system are significant. The product $K_a K_m K_g$ represents the loop gain of the system. If the loop gain is large so that $K_a K_m K_g \gg 1$, then in the steady state

$$\dot{\theta}_{ss} \cong \frac{V_i}{K_g} \tag{4-35}$$

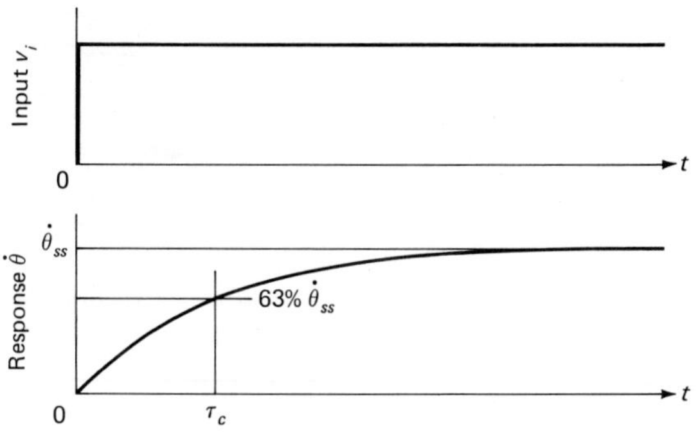

Figure 4-14 Response of speed control system to step input

Thus, the controlled velocity depends only on the magnitude of the input and the characteristic of the feedback element, which in this case is the tachometer. The velocity does not depend upon the actual characteristics of the amplifier or motor nor upon the load at the motor output.

A comparison of Eqs. (4-26) and (4-33) shows that the time constant of the closed-loop system is reduced as compared to the motor itself by virtue of the loop gain. The resulting decrease in response time is an inherent feature of closed loop control. However, the response time cannot decrease indefinitely as the loop gain is increased. At some point the response will be limited by the maximum velocity and/or acceleration of which the motor is capable. The system then enters a nonlinear region of operation, and the equations based on linear operating conditions are no longer valid. The matter of limiting is discussed in more detail in Chap. 9.

EXAMPLE 4-3

The motor in Example 4-2 is used in the velocity servomechanism shown in Fig. 4-13. The gain of the servo amplifier is 100 V/V and the tachometer constant is 10 mV/(rad/s). Determine the response of the system to a 1-V step input.

Solution

Using Eq. (4-32), the steady-state output is

$$\dot{\theta}_{ss} = \frac{K_a K_t V_i}{B + K_a K_g K_t} = \frac{100 \times 0.013 \times 1.0}{2.87 \times 10^{-3} + 100 \times 0.01 \times 0.013} = 82 \text{ rad/s}$$

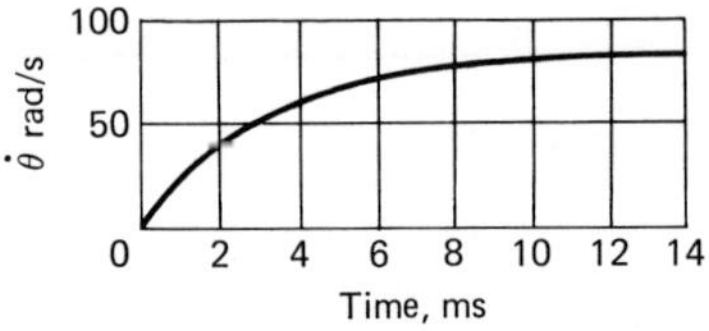

Figure 4-15 Response of servomechanism in Example 4-3 to 1-V step input

Using Eq. (4-33), the time constant is

$$\tau_c = \frac{J}{B + K_a K_g K_t} = \frac{5 \times 10^{-5}}{2.87 \times 10^{-3} + 100 \times 0.01 \times 0.013} = 0.00315 \text{ s}$$

From Eq. (4-34) the response is

$$\dot{\theta} = 82(1 - e^{-t/0.00315}) \text{ rad/s}$$

Figure 4-15 shows the response curve for the servomechanism.

4-7 SECOND-ORDER SYSTEMS

Certain elementary systems can be described by second-order equations, and some systems of greater complexity can be adequately approximated by second-order equations. Systems of this type are characterized by elements that can store energy in two different forms and are known as *second-order systems.* Examples are mechanical systems possessing both potential and kinetic energies and electric circuits with energy contained in both magnetic and electric fields. In addition to energy-storage elements, second-order systems invariably include some form of energy absorption or damping. This might be viscous friction in a mechanical system or resistance in an electric circuit.

A second-order system is defined by a second-order equation of the general form

$$a\frac{d^2x}{dt^2} + b\frac{dx}{dt} + cx = f(t) \tag{4-36}$$

If we refer to Eqs. (2-10), (2-15), and (2-25), we see that they are of this form; hence the systems described by these equations are second-order.

A second-order system can exhibit various forms of response to a given input depending upon the relative values of its parameters. However, the typical response is an oscillation which decays or dies out. We are familiar with the fact that many objects such as bells, springs, and electrical circuits "ring" or vibrate for a brief period when shocked. This is the kind of behavior that is commonly associated with second-order systems.

Two parameters are convenient in describing the response of a second-order system: the frequency of the oscillation and the rate at which the oscillation damps out. These parameters are represented by the symbols

ω_n = undamped natural frequency, rad/s (frequency of oscillation if no damping is present)

ζ = damping ratio (dimensionless)

In terms of these two parameters, a second-order linear differential equation with constant coefficients can be written as

$$\frac{1}{\omega_n^2}\frac{d^2x}{dt^2} + \frac{2\zeta}{\omega_n}\frac{dx}{dt} + x = f(t) \tag{4-37}$$

The solution of Eq. (4-37) when $f(t)$ is a step input and the damping ratio ζ is less than unity will be

$$x = x_{ss}\left[1 - \frac{e^{-\zeta\omega_n t}}{\sqrt{1-\zeta^2}}\sin\left(\sqrt{1-\zeta^2}\,\omega_n t + \cos^{-1}\zeta\right)\right] \tag{4-38}$$

We note that this response is oscillatory with a *damped frequency* ω_d equal to $\omega_n\sqrt{1-\zeta^2}$ rad/s. At the same time, the amplitude of the oscillation decreases with time because of the exponential term $e^{-\zeta\omega_n t}$. The response given by Eq. (4-38) is plausible from a physical point of view; the damping element dissipates energy from the system with the result that the oscillation eventually ceases. This form of response is referred to as *underdamped*.

A convenient way of evaluating the response of a second-order system to a step input is by the normalized family of curves shown in Fig. 4-16. The abscissa is a dimensionless variable, which can be translated to units of time if ω_n is known. The vertical axis is the normalized response x/x_{ss}. From a knowledge of the system's natural frequency and damping, the transient response can readily be found by referring to these curves.

As the damping ratio increases, the response becomes less oscillatory. At $\zeta = 1$, the overshoot in the response disappears. This is known as *critical damping*. For values of ζ greater than unity, the response becomes even slower and the system is described as *overdamped*.

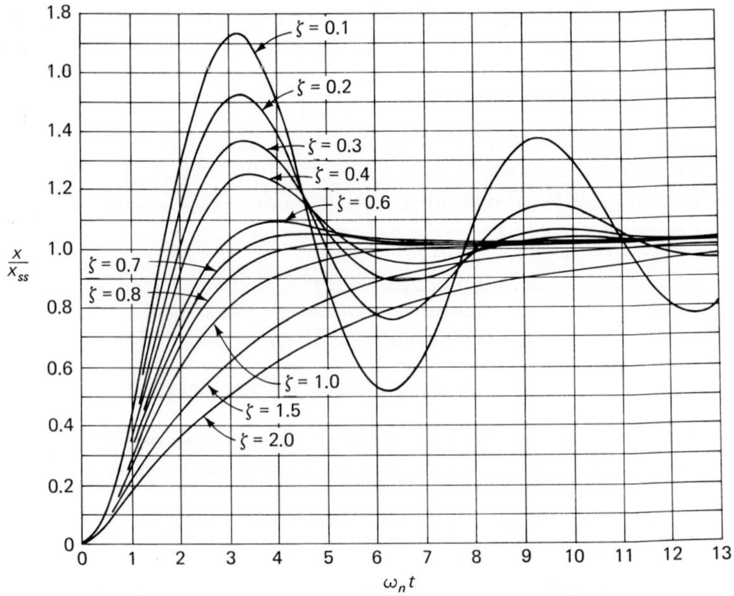

Figure 4-16 Universal transient response curves for second-order systems subjected to a step input

4-8 SERIES RESISTANCE-INDUCTANCE-CAPACITANCE NETWORK

To illustrate the response of a second-order system, consider a circuit containing a resistor, an inductor, and a capacitor connected in series to a dc voltage as shown in Fig. 4-17. We are interested in the behavior of the circuit when the switch is closed at a time $t = 0$. Applying Kirchhoff's voltage law, the voltages around the circuit are related by the equation

$$L\frac{di}{dt} + Ri + \frac{1}{C}\int i\,dt = E \tag{4-39}$$

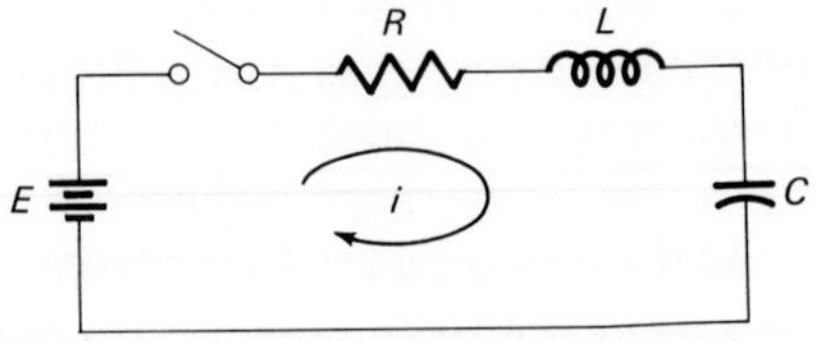

Figure 4-17 Series resistance-inductance-capacitance network

where i is the current in the circuit. Since $i = dq/dt$, the integral may be removed from Eq. (4-39) by using charge q as the dependent variable. Thus we obtain the second-order differential equation

$$LC\frac{d^2q}{dt^2} + RC\frac{dq}{dt} + q = EC \tag{4-40}$$

If we compare Eq. (4-40) with Eq. (4-37), we note that they are similar in form. The coefficients of like derivatives may then be equated giving the relations

$$\omega_n{}^2 = \frac{1}{LC} \quad \text{and} \quad \frac{2\zeta}{\omega_n} = RC \tag{4-41}$$

Solving these identities for ω_n and ζ yields

$$\omega_n = \frac{1}{\sqrt{LC}} \quad \text{and} \quad \zeta = \frac{R}{2}\sqrt{\frac{C}{L}} \tag{4-42}$$

Thus the natural frequency and damping ratio are defined in terms of the circuit constants.

By setting the derivatives in Eq. (4-40) equal to 0, the steady-state charge is found to be

$$q_{ss} = EC \tag{4-43}$$

Referring to Eq. (4-38), the charge as a function of time after closing the switch is

$$q = EC\left[1 - \frac{e^{-\zeta\omega_n t}}{\sqrt{1-\zeta^2}}\sin\left(\sqrt{1-\zeta^2}\,\omega_n t + \cos^{-1}\zeta\right)\right] \tag{4-44}$$

where ω_n and ζ are defined by Eq. (4-42).

4-9 POSITIONAL SERVOMECHANISM

Probably the most common application of control systems is mechanically positioning a load; that is, for a given value of the reference input the load is to assume a particular position. Positioning systems find use in a wide variety of ways, including aircraft control surfaces, machine tools, communication antennas, reactor control rods, and indicating instruments.

In position control, one is usually concerned with both steady-state and transient operating conditions. The accuracy of the final position is obviously important. However, the response as the system moves from one position to another is also of importance.

A load could be positioned by allowing a motor to run at a given

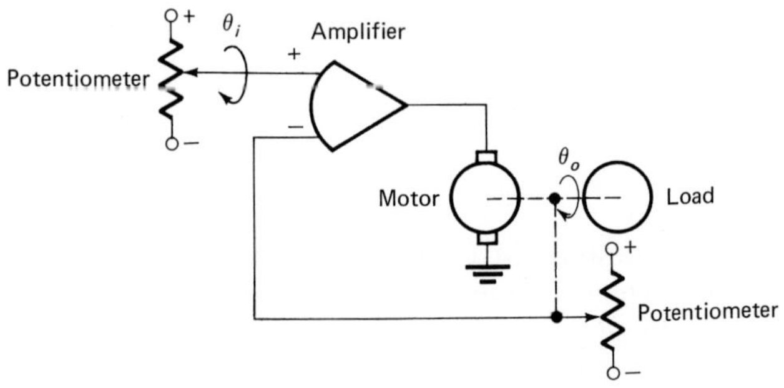

Figure 4-18 Positional servomechanism

speed for a specified interval of time. However, such a velocity system would not be too accurate for positioning since motor speed would vary (even though slightly) with load. But if the feedback signal is proportional to the *position* of the load, the system can operate closed-loop with high accuracy. Such a system is known as a *positional servomechanism*.

A simplified positional servomechanism is shown in Fig. 4-18. The motor and amplifier may be the same components used in the speed-regulating system of Fig. 4-13. However, changing the feedback transfer function completely alters the operation of the system.

Figure 4-19 is a block diagram of this positional servomechanism. The input-output characteristics for the motor and amplifier are given by Eqs. (4-24) and (4-28), respectively. The input and feedback elements are potentiometers that produce a voltage proportional to shaft position. If these units are identical,

$$v_f = K_p\theta_o \qquad \text{and} \qquad v_i = K_p\theta_i \tag{4-45}$$

where K_p is the potentiometer gain constant. The error voltage v_e is

$$v_e = v_i - v_f = K_p(\theta_i - \theta_o) \tag{4-46}$$

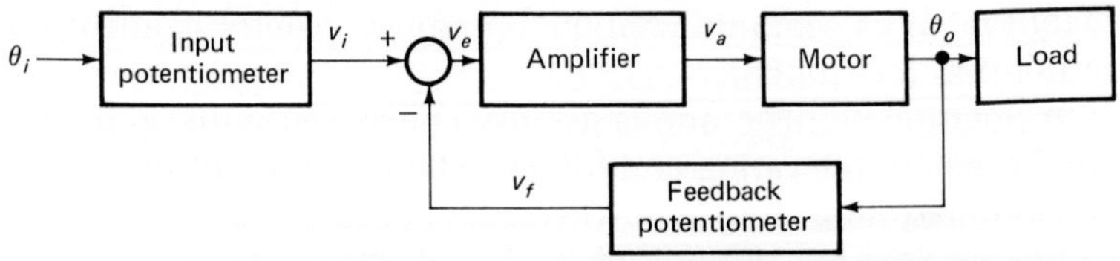

Figure 4-19 Block diagram of positional servomechanism

At the output shaft the torque delivered by the motor must equal the torque from the load. Therefore

$$J\ddot{\theta}_o + B\dot{\theta}_o = K_t v_a = K_a K_t v_e = K_p K_a K_t(\theta_i - \theta_o)$$

or

$$\theta_i = \frac{J}{K}\ddot{\theta}_o + \frac{B}{K}\dot{\theta}_o + \theta_o \tag{4-47}$$

where K is the loop gain equal to $K_p K_a K_t$.

Equation (4-47) is a second-order differential equation in the dependent variable θ_o. For a given input command, the solution will yield an equation for θ_o which will consist of both steady-state and transient terms. As noted in Sec. 4-7, the response can take on various forms depending upon the relative magnitudes of J, B, and K. For example, a system in which the friction coefficient B is large will have a slow response and tend to creep toward its final position. On the other hand, the output of a system with little friction will oscillate back and forth for some time before settling down.

When Eq. (4-47) is compared with Eq. (4-37), we see that

$$\frac{2\zeta}{\omega_n} = \frac{B}{K} \qquad \text{and} \qquad \omega_n{}^2 = \frac{K}{J} \tag{4-48}$$

from which

$$\omega_n = \sqrt{\frac{K}{J}} \qquad \text{and} \qquad \zeta = \frac{B}{2\sqrt{KJ}} \tag{4-49}$$

If the input θ_i is a step function, the steady-state position of the output may be evaluated by taking the derivatives $\ddot{\theta}_o$ and $\dot{\theta}_o$ as 0. This gives the result

$$\theta_o = \theta_i \tag{4-50}$$

or the input and output positions are at correspondence.[1]

The total response (transient plus steady-state) is the solution of the system differential equation. Assuming a step-input function and using the general form for an underdamped second-order system as given by Eq. (4-38), the response is

$$\theta_o = \theta_i\left[1 - \frac{e^{-\zeta\omega_n t}}{\sqrt{1-\zeta^2}}\sin\left(\omega_n\sqrt{1-\zeta^2}\,t + \cos^{-1}\zeta\right)\right] \tag{4-51}$$

1 *Strictly speaking this is true in theory for a completely linear system. However, in practice static friction in the system will cause a steady-state position error that is not included in this linear analysis. Static friction, as well as other nonlinear effects, will be discussed in Chap. 9.*

Although critical damping in a second-order system produces a response with no overshoot, the curves of Fig. 4-16 show that a system with a lower value of damping, for instance 0.6 to 1.0, approaches correspondence more quickly. Furthermore the time for the response to come and remain within the vicinity of correspondence is shorter. However, a further reduction in damping produces an oscillatory response that can be quite objectionable. Thus, although the choice of damping is within the discretion of the designer, values of $\zeta = 0.6$ or thereabouts are commonly used in control-system work.

EXAMPLE 4-4

The motor in Example 4-2 is used in a servomechanism as shown in Fig. 4-18 to position an inertia load to 10^{-4} oz-in.-s^2. The gain of the servo amplifier is 100 V/V and the input and feedback potentiometers are 10-turn units excited by ± 10 V. Determine the response of the servo to a step change in input angle of 30°.

Solution

From Eq. (3-8) the potentiometer gain is

$$K_p = \frac{E_i}{\theta_T} = \frac{20}{10 \times 2\pi} = 0.318 \text{ V/rad}$$

The total inertia is

$$J_T = J_M + J_L = 5 \times 10^{-5} + 10^{-4} = 1.5 \times 10^{-4} \text{ oz-in.-s}^2$$

The loop gain is

$$K = K_p K_a K_t = 0.318 \times 100 \times 0.013 = 0.414 \text{ oz-in./rad}$$

Using Eq. (4-49),

$$\zeta = \frac{B}{2\sqrt{KJ}} = \frac{2.87 \times 10^{-3}}{2\sqrt{0.414 \times 1.5 \times 10^{-4}}} = 0.182$$

$$\omega_n = \sqrt{\frac{K}{J}} = \sqrt{\frac{0.414}{1.5 \times 10^{-4}}} = 52.5 \text{ rad/s}$$

Substituting values in Eq. (4-51) gives the response

$$\theta_o = 30[1 - 1.02e^{-9.55t} \sin(52.5t + 1.39)] \text{ degrees}$$

The response of the servo can be graphed with the aid of the universal response curves of Fig. 4-16. Figure 4-20 shows the response using $\zeta = 0.182$ and $\omega_n = 52.5$ rad/s.

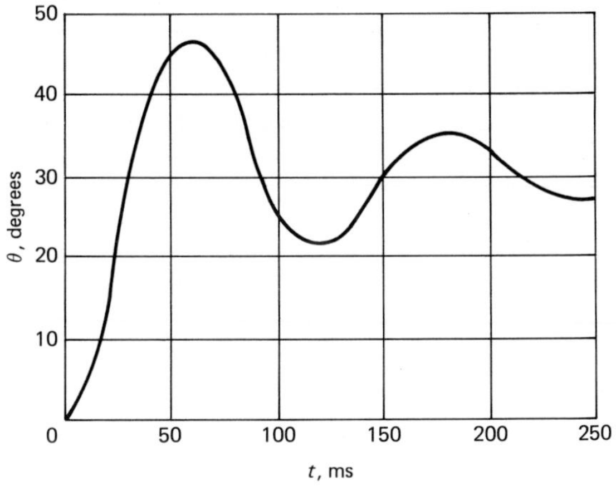

Figure 4-20 Response of servomechanism in Example 4-4 to step input of 30°

4-10 TRANSIENT-RESPONSE CHARACTERISTICS

Control systems are designed according to some set of specifications, and the response to a step-function input is commonly used as a means of specifying and evaluating the transient response. We have seen that the damping ratio and resonant frequency are parameters used in describing the response of a second-order system. In this section we will describe several other parameters associated with the transient response of second order systems.

The transient response of a second-order system exhibits several characteristics of interest. As indicated by the curves of Fig. 4-16, the overdamped system is relatively slow in responding and does not oscillate about the final value. In some applications this absence of oscillation may be important. For example, it would be undesirable for an elevator control to permit oscillation at each stop.

The underdamped system reaches a final value faster than the overdamped case, but the response does oscillate about the final value. Where this oscillation can be tolerated, the underdamped system provides faster action. For example, an underdamped system would generally be preferred in positioning a machine tool in response to a series of input commands.

The step-function response can be characterized in terms of three quantities that are shown in Fig. 4-21: rise time, overshoot, and settling time.

The *rise time* t_r is defined as the time required for the step-function

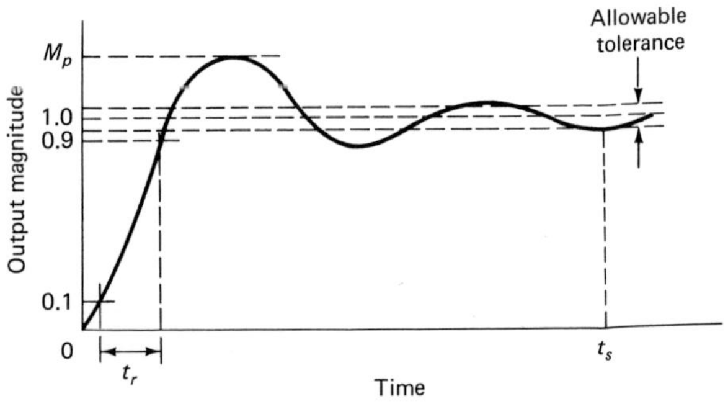

Figure 4-21 Transient-response specifications

response to rise from 10 to 90 percent of the final or steady-state value.

The *overshoot* M_p is the magnitude of the first overshoot which may be expressed as a percent of final value according to the equation

Percent overshoot =

$$\frac{\text{maximum value in overshoot} - \text{steady-state value}}{\text{steady-state value}} \times 100 \qquad \text{(4-52)}$$

The *settling time* t_s is taken as the time required for the output to decrease to a specified percent of the final value and thereafter remain within that value. An error of 5 percent is commonly used to specify settling time.

The settling time based on the 5 percent–error criterion can be taken as three time constants, where the time constant is contained in the exponential factor $e^{-\zeta\omega_n t}$ in Eq. (4-38) as

$$\tau = \frac{1}{\zeta\omega_n} \qquad \text{(4-53)}$$

As indicated by the curves in Fig. 4-16 the overshoot is dependent only on the system damping ratio ζ. Figure 4-22 shows the percent overshoot as a function of the damping ratio.

The damping ratio of a second-order system is a function of several system parameters, and it can be varied by changing one or more of these quantities. For example, the damping of a closed-loop control system may be altered by changing the gain. Referring to Eq. (4-49), we note that the damping ratio of a positional servomechanism can be increased by decreasing the loop gain. Unfortunately the gain adjustment in itself is generally inadequate since a satisfactory damping ratio may not provide sufficient amplification to achieve the desired steady-state accuracy.

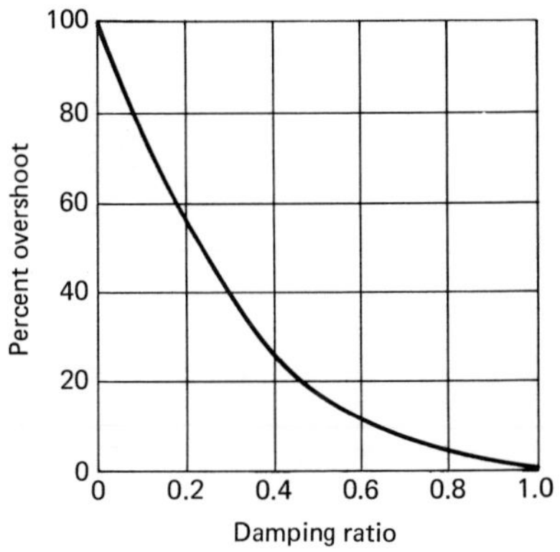

Figure 4-22 Overshoot versus damping ratio for a second-order system

If the system gain is increased beyond a certain limit, the system will become unstable and will oscillate with increasing amplitude. As pointed out in Sec. 1-8, the tendency toward instability is due to the delays that are inherent in a practical system. Fortunately various means are available to the system designer to provide high gain together with stable operation. Although this subject will be discussed in detail in subsequent chapters, we will conclude this chapter by describing one method of obtaining satisfactory transient response without sacrificing loop gain.

EXAMPLE 4-5

Determine the rise time, overshoot, and settling time of the step-function response of the servomechanism in Example 4-4.

Solution

The rise time may be determined from Fig. 4-16. For $\zeta = 0.182$, the rise time is

$$t_r = \frac{(1.6 - 0.5)}{\omega_n} = \frac{1.1}{52.5} = 0.021 \text{ s}$$

Using Eq. (4-52), the percent overshoot is

$$\frac{1.55 - 1.0}{1.0} \times 100 = 55 \text{ percent}$$

Applying Eq. (4-53), the settling time for a 5 percent error is

$$t_s = \frac{3}{\zeta\omega_n} = \frac{3}{0.182 \times 52.5} = 0.314 \text{ s}$$

4-11 DAMPING BY DERIVATIVE FEEDBACK

Damping is introduced into a mechanical system by viscous friction. However, this is not a desirable form of damping since the friction is not constant and requires an expenditure of energy. Recall that friction damping is characterized by increasing load torque as the velocity increases. The same effect could be achieved if the driving torque were decreased with increasing speed. This form of damping may be introduced in a closed-loop positioning system by introducing negative feedback that is proportional to speed. This is known as *velocity feedback*. It is also described by the more general term *derivative feedback* since velocity represents the derivative of the controlled output, which is position.

In the positional servomechanism, a tachometer may be added to provide a measure of output velocity as shown in Fig. 4-23. The tachometer voltage is fed back with a polarity such that it also opposes

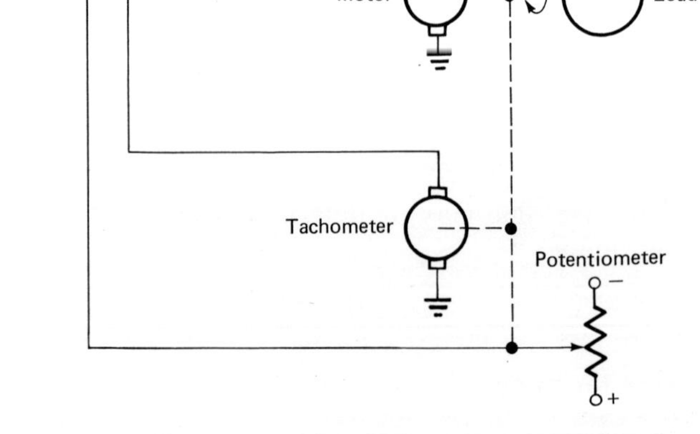

Figure 4-23 Positional servomechanism including velocity feedback

the input signal. The error voltage V_e is now

$$v_e = K_p(\theta_i - \theta_o) - K_g\dot{\theta}_o \tag{4-54}$$

where K_p and K_g are the potentiometer and tachometer gain constants, respectively.

Equating the load torque to that developed by the motor gives

$$J\ddot{\theta}_o + B\dot{\theta}_o = K_pK_aK_t(\theta_i - \theta_o) - K_gK_dK_t\dot{\theta}_o \tag{4-55}$$

where K_d is the amplifier gain associated with the voltage from the tachometer. Solving Eq. (4-55) for θ_i yields the second-order equation

$$\theta_i = \frac{J}{K}\ddot{\theta}_o + \left(\frac{B + K_gK_dK_t}{K}\right)\dot{\theta}_o + \theta_o \tag{4-56}$$

where K is again the loop gain $K_pK_aK_t$. The damping ratio associated with Eq. (4-56) is

$$\zeta = \frac{B + K_gK_dK_t}{2\sqrt{KJ}} \tag{4-57}$$

Comparing Eq. (4-57) with Eq. (4-49), we note that the damping ratio has been increased by the addition of the derivative feedback. The magnitude of the damping can be varied independently of the loop gain K by changing the tachometer input gain K_d. Thus it is possible to adjust the loop gain to achieve a desired steady-state accuracy and then establish K_d for a particular damping.

The way in which velocity feedback can prevent overshoot and subsequent oscillation can be described with the aid of Fig. 4-24. When

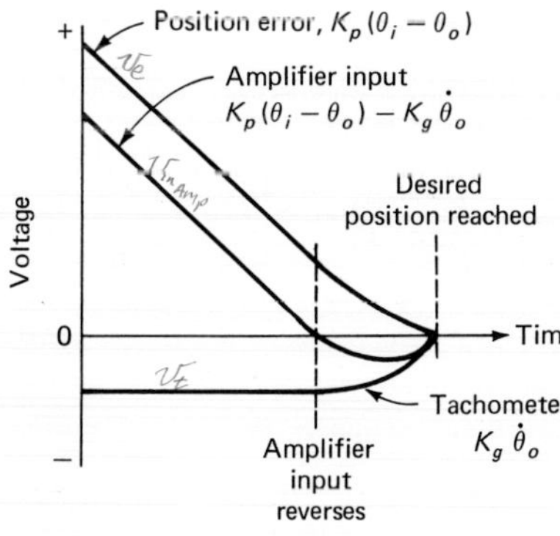

Figure 4-24 Error-voltage characteristic

moving the load from one position to another the motor will run at a constant sluing speed due to a large error that saturates the amplifier. The tachometer voltage reduces the amplifier input so that it becomes 0 before the correct position is actually reached. As the motor continues to run because of inertia, the amplifier input reverses and creates a reversed torque. This torque produces a braking effect that will bring the output to rest at the desired position with little or no overshoot.

EXAMPLE 4-6

It is required that the positional servomechanism in Example 4-4 be critically damped ($\zeta = 1$) while maintaining the same loop gain. A tachometer having a sensitivity of 1 V/1,000 rpm is added to the system to provide velocity feedback. Determine the needed amplifier gain for the tachometer input.

Solution

The tachometer gain constant is

$$K_g = \frac{1}{1{,}000 \times 2\pi/60} = 9.55 \times 10^{-3}\ \mathrm{V/(rad/s)}$$

Using Eq. (4-57),

$$\zeta = \frac{B + K_g K_d K_t}{2\sqrt{KJ}} = 1.0 = \frac{2.87 \times 10^{-3} + 9.55 \times 10^{-3} \times 0.013 \times K_d}{2\sqrt{0.414 \times 1.5 \times 10^{-4}}}$$

Solving for the desired gain gives

$$K_d = 105$$

PROBLEMS

4-1. The mass M shown below is accelerated by a constant force F. Motion of the mass is opposed by friction having a coefficient B.

$$M = 32\ \text{slugs}$$
$$B = 0.1\ \text{lb/(ft/s)}$$
$$F = 1\ \text{lb}$$

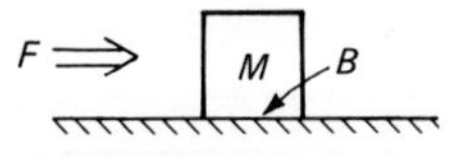

Prob. 4-1

a. Write a first-order differential equation describing the velocity v of the mass.
b. Determine the steady-state velocity.
c. Determine the equation for the response assuming the mass starts from rest ($v = 0$ at $t = 0$).
d. Sketch the curve showing velocity as a function of time.

4-2. A capacitor is charged through a resistance as shown below.

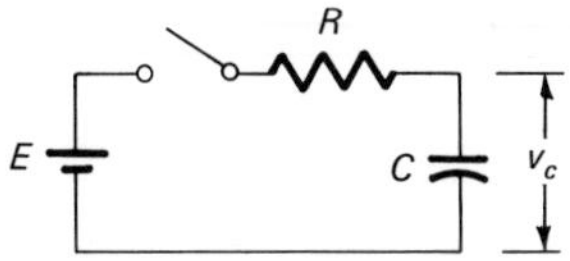

Prob. 4-2

a. Write a differential equation describing the voltage across the capacitor after the switch is closed.
b. Determine the steady-state capacitor voltage.
c. Determine the equation for the capacitor voltage assuming the capacitor is initially uncharged.

4-3. A small servomotor has the following characteristics:

Rated armature voltage	115 V
Stall torque at rated voltage	5 oz-in.
No-load speed at rated voltage	4,500 rpm
Rotor inertia	5×10^{-4} oz-in.-s^2

Determine the equation and sketch the curve for the speed response of the motor when a step input of 50 V is applied to the armature.

4-4. The motor in Prob. 4-3 is coupled to a mechanical load whose inertia is 2×10^{-3} oz-in.-s^2 and viscous friction is 0.05 oz-in./(rad/s). Sketch the speed response of the loaded motor to a step input of 50 V. Compare the results with the unloaded case.

4-5. The speed of the motor in Prob. 4-3 is controlled by the regulating system shown below in block-diagram form. Determine the equation for the speed response of the motor when a step input of 1.0 V is applied to the system.

$K_a = 200$ V/V
$K_g = 0.05$ V/(rad/s)

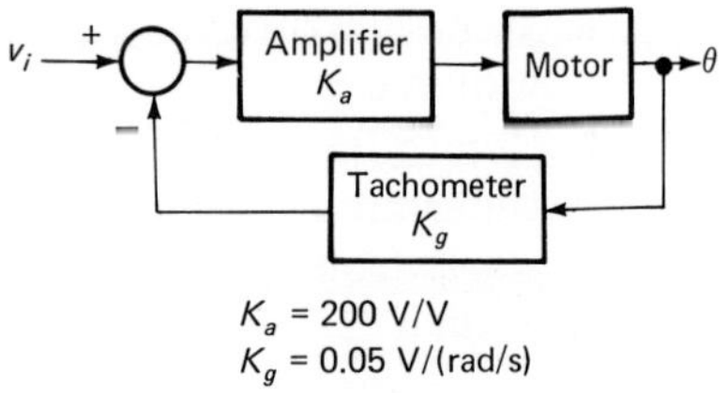

Prob. 4-5

4-6. The load in Prob. 4-4 is coupled to the regulating system in Prob. 4-5. Determine the equation for the speed response. Compare the loading effect of the motor alone with the speed control system.

4-7. A dc motor has an armature resistance and inductance of 14 Ω and 3.5 mH, respectively. The rotor inertia is 0.15×10^{-3} oz-in.-s^2, and the damping coefficient is 6×10^{-3} oz-in./(rad/s). Compare the electrical and mechanical time constants of the motor.

4-8. Speed-torque curves for a small ac servomotor are shown below. The rotor inertia is 0.1 oz-in.^{2} (note units). Sketch the speed response of the motor when the control winding voltage is suddenly increased from 10 to 20 V.

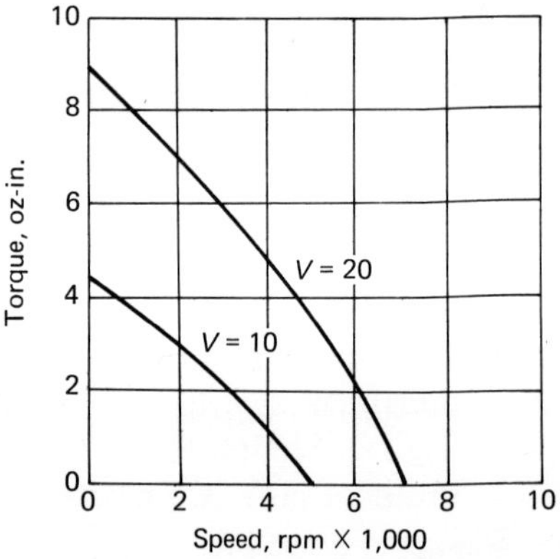

Prob. 4-8

4-9. Assume a servomotor is running at some constant speed. Derive an equation for the speed response when the control voltage to the motor is suddenly reduced to 0.

4-10. The cylindrical tank shown below has a cross-sectional area of 5 ft^2 and the hydraulic resistance R of the outlet is 10 s/ft^2.

a. Write a differential equation for the liquid level h with the inflow q_i as the input function.

b. If the tank is initially empty and an inflow of 0.1 ft^3/s is suddenly applied, determine the equation for the liquid level h as a function of time. What will be the steady-state level? What time (approximate) is required to reach this level? Sketch the response curve.

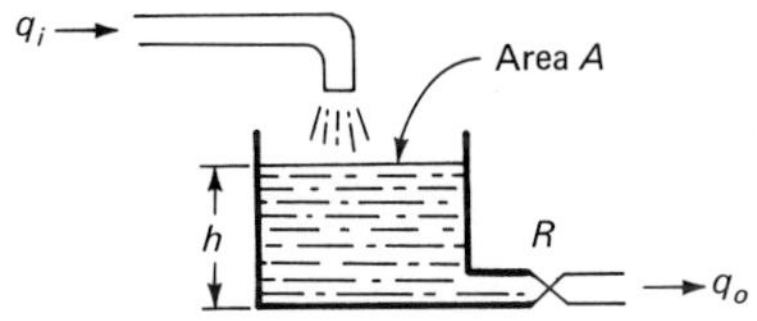

Prob. 4-10

4-11. Repeat Prob. 4-10 if a second outlet identical to the first is added to the tank.

4-12. A simple float-type level control is added to the tank in Prob. 4-10. The inflow is now proportional to the difference between the desired level h_d and the actual level h; that is, $q_i = K(h_d - h)$, where K is a proportionality constant.

a. Draw a block diagram of the level control system.

b. Write a differential equation for the level with the desired level as the input function.

c. For $K = 2$, determine the equation for the level h as a function of time when h_d is suddenly changed.

4-13. The differential equation of the spring-mass system in Example 2-1 was found to be

$$\frac{d^2y}{dt^2} + 1.6\frac{dy}{dt} + 128y = 0$$

Determine the undamped natural frequency and the damping ratio of the system.

4-14. Determine the equation for the rotation speed $\dot{\theta}$ if a step input torque T is applied to the system below. Assume that the system is initially at rest.

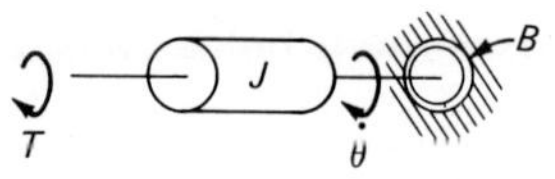

Prob. 4-14

4-15. Determine the equation for the angular position θ_o if a step-input position θ_i is applied to the system shown below. Assume an underdamped system that is initially at rest.

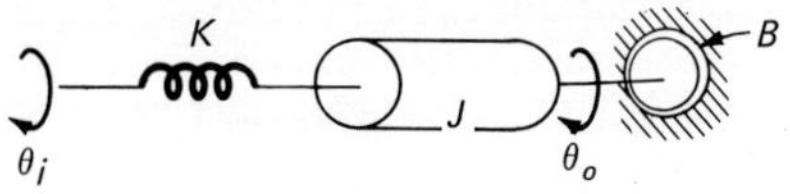

Prob. 4-15

4-16. A series resistance-inductance-capacitance network is shown below. Determine the undamped natural frequency and the damping ratio of the circuit. Write an equation for the charge q as a function of time after the switch is closed.

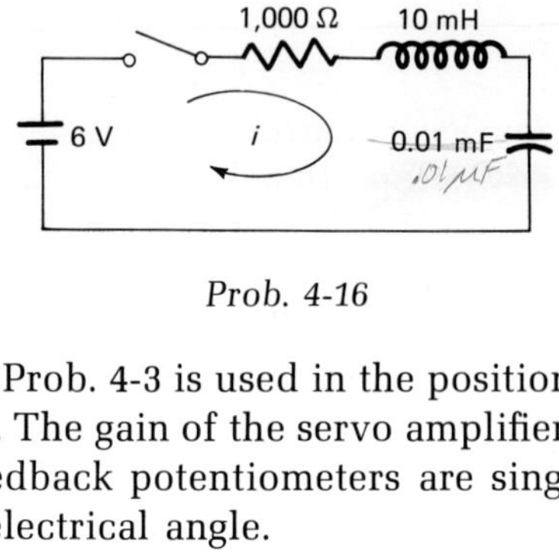

Prob. 4-16

4-17. The motor in Prob. 4-3 is used in the positional servomechanism shown below. The gain of the servo amplifier is 300 V/V, and the input and feedback potentiometers are single-turn units which have a 350° electrical angle.

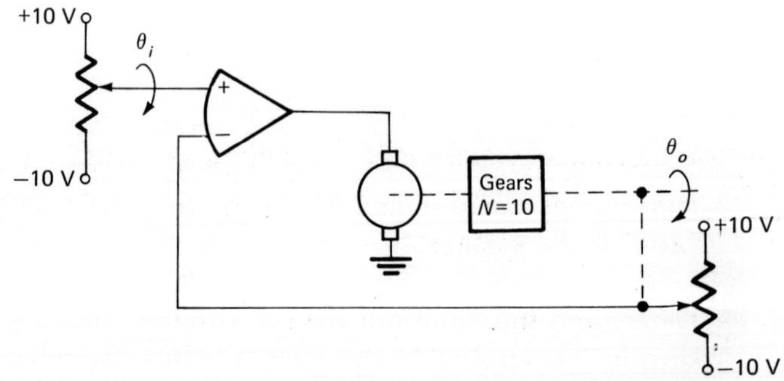

Prob. 4-17

a. Determine the undamped natural frequency and the damping ratio of the servo.
b. Write the equation for the response of θ_o for a step change of $10°$ in θ_i.
c. Sketch the curve showing θ_o versus t for part b.

4-18. Repeat Prob. 4-17 with an inertia load of 0.15 oz-in.-s^2 coupled to the servo output.

4-19. Referring to the servomechanism described in Example 4-4, determine the amplifier gain so that the transient-response overshoot is reduced to 10 percent. Determine the rise time and settling time of the step-function response.

4-20. In Prob. 4-17 a tachometer having a gain constant of 0.05 V/rad is *directly* coupled to the motor shaft. The tachometer output is connected to an amplifier input. Determine the amplifier gain from this input to provide a system damping ratio of 0.7.

Chapter Five

transient analysis using the Laplace transform

5-1 INTRODUCTION

In Chap. 4 we saw that the solution of a differential equation in which time is the independent variable will yield the transient response of the system described by the equation. The effort involved in using classical mathematical approaches to the solution of linear differential equations becomes exceedingly complex as the order of the equation increases. Fortunately the work may be considerably simplified by a mathematical method known as the *Laplace transform*. The Laplace transform method has found wide acceptance in the analysis of automatic control systems, and much of the progress in automatic control during the past several decades can be attributed to its application.

This chapter will introduce the Laplace transform from the standpoint of a mathematical tool useful in control-system design. Emphasis is therefore on the application of the method rather than on its underlying theory and derivation. In addition to its relative simplicity, the Laplace transform method has the advantage that both the steady-state and transient forms of the solution of the differential equation are obtained simultaneously. Another distinct advantage is that the method permits system initial conditions to be handled directly as part of the solution.

The concept of a mathematical transformation can perhaps be better understood by recalling the use of logarithms. It frequently proves convenient in performing the arithmetic operations of multiplication, division, and raising a number to a power to convert or transform the numbers to their logarithms, generally through the use of tables of logarithms. The original operations are then replaced by a more simple additive process, and the desired answer is then obtained by an inverse transform, or antilogarithmic process, again using tables.

EXAMPLE 5-1

Using the method of logarithms, determine the value of x for the product

$$x = (1.7364)(3.9432)$$

Solution

Transforming the equation to logarithmic form gives

$$\log x = \log (1.7364) + \log (3.9432)$$

Substituting the values of the logarithms as obtained from a table gives

$$\log x = 0.23965 + 0.59585 = 0.83550$$

The antilogarithm (or inverse transformation) of both sides gives the product

$$x = \log^{-1} (0.83550) = 6.8470$$

Referring to the above example, we note that the method of solution can be represented in block-diagram form as shown in Fig. 5-1*a*.

The Laplace transform method provides an analogous approach to solving differential equations. An equation that contains derivatives with respect to time is converted to an algebraic equation in which a new quantity, the Laplace variable s, replaces the derivatives. Thus we may think of the equation as transformed from the time domain to a new domain in which the Laplace variable s rather than time *t* is the independent variable. The new equation is then algebraically manipulated to obtain the desired dependent variable as a function of s. This expression is then inversely transformed to yield the desired solution in the time domain. As with logarithms the use of Laplace transforms is facilitated by tables of transform pairs.

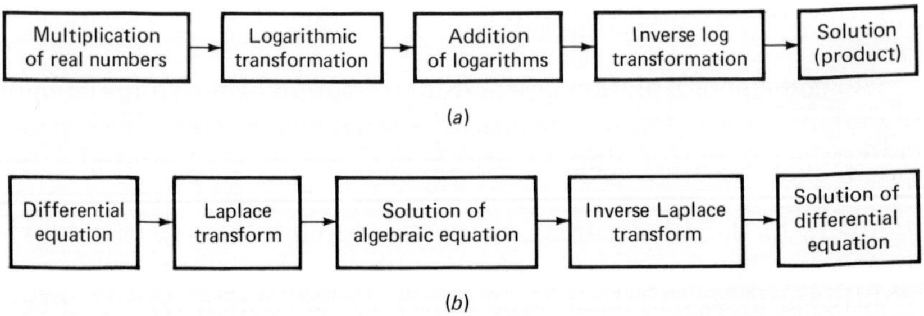

Figure 5-1 (a) Block diagram of multiplication by logarithms; (b) block diagram of differential-equation solution using Laplace transforms

A block diagram of the Laplace transform method is shown in Fig. 5-1*b*. This diagram should be compared with that of Fig. 5-1*a*.

Once the physics of a dynamic system has been established and the appropriate differential equation formulated, the Laplace transform may be applied to obtain the solution in a somewhat routine fashion. However, before concluding this section, a word of caution should be given. Although the Laplace transform is a powerful tool in solving differential equations, it is *not* applicable to all equations. It can only be used with linear differential equations containing constant coefficients. Fortunately most equations used in automatic-control-systems work are of this type.

5-2 NATURE OF THE LAPLACE TRANSFORMATION

In performing the Laplace transformation, an expression is changed from a function of time $f(t)$ to a new function designated as $F(s)$. To illustrate this statement, assume that the position of a mass as a function of time is defined as $x(t)$, or simply x. After the differential equation containing x is transformed to the Laplace domain, the position will be identified as $X(s)$. In this transformation s is a complex quantity similar to the familiar complex number $a + jb$. However, s is defined by $s = \sigma + j\omega$, where σ (sigma) and ω (omega) are real numbers and $j - \sqrt{-1}$.

In order for the Laplace transform to be used, a function must be real and continuous over the time interval of interest. Mathematically the direct Laplace transform is given by the integral

$$\mathcal{L}[f(t)] = \int_0^{\infty} f(t)e^{-st}\,dt = F(s) \tag{5-1}$$

In this equation $\mathcal{L}[f(t)]$ designates the Laplace transform of the function $f(t)$, which then becomes identified as the function $F(s)$. In performing the transformation according to Eq. (5-1), $f(t)$ is multiplied by the complex quantity e^{-st} and the product is then integrated with respect to time over an interval from $t = 0$ (beginning of system response) to $t = \infty$.

The inverse Laplace transform by which the time-domain response is obtained from the Laplace function is given by the definition

$$f(t) = \mathcal{L}^{-1}[F(s)] \tag{5-2}$$

where $F(s)$ is defined by Eq. (5-1) and $\mathcal{L}^{-1}$ designates the inverse-transform operation.

Common practice involves using a table of transform pairs in which the time function $f(t)$ and the corresponding Laplace function $F(s)$ are

TABLE 5-1 SHORT TABLE OF LAPLACE TRANSFORM PAIRS

Time Function: $f(t)$	Laplace Function: $F(s)$
1. Unit impulse	1
2. Constant value of unity, or the unit step function	$\frac{1}{s}$
3. t	$\frac{1}{s^2}$
4. t^n	$\frac{n!}{s^{n+1}}$
5. e^{-at}	$\frac{1}{s+a}$
6. $\frac{t^{n-1}e^{-at}}{(n-1)!}$	$\frac{1}{(s+a)^n}$
7. $\frac{1}{a}(1-e^{-at})$	$\frac{1}{s(s+a)}$
8. $\frac{1}{a^2}(at-1+e^{-at})$	$\frac{1}{s^2(s+a)}$
9. $\frac{1}{b-a}(e^{-at}-e^{-bt})$	$\frac{1}{(s+a)(s+b)}$
10. $\frac{1}{ab}\left[1+\frac{1}{a-b}(be^{-at}-ae^{-bt})\right]$	$\frac{1}{s(s+a)(s+b)}$
11. $\frac{1}{b-a}(be^{-bt}-ae^{-at})$	$\frac{s}{(s+a)(s+b)}$
12. $\sin \omega t$	$\frac{\omega}{s^2+\omega^2}$
13. $\cos \omega t$	$\frac{s}{s^2+\omega^2}$
14. $\frac{1}{\omega_n^2}-\frac{e^{-\zeta\omega_n t}}{\omega_n^2\sqrt{1-\zeta^2}}\sin(\omega_n\sqrt{1-\zeta^2}\,t+\phi)$, $\phi=\cos^{-1}\zeta$	$\frac{1}{s(s^2+2\zeta\omega_n s+\omega_n^2)}$
15. $e^{-at}\sin \omega t$	$\frac{\omega}{(s+a)^2+\omega^2}$
16. $e^{-at}\cos \omega t$	$\frac{s+a}{(s+a)^2+\omega^2}$

grouped. The table may be used for both the direct and inverse Laplace transform, that is, either to obtain $F(s)$ from $f(t)$ or vice versa. A brief table of transform pairs for the most frequently encountered time functions is given in Table 5-1. A more extensive table of transforms is provided in Appendix A.

The table in Appendix A includes many if not most of the Laplace transforms encountered in control systems practice, especially in transforming a function from the time to the Laplace domain. However, inverse-transform requirements are less specific because of the diversity of transformed system equations, and a table will on occasion prove inadequate. In this case the Laplace function may be expanded in a series of partial fractions which may then be transformed on an individual basis. A discussion of partial-fraction expansion is given in Appendix B.

It was mentioned in Sec. 4-2 that the step and sine functions are frequently used in control system analysis. To further clarify the nature of the Laplace transform, we will derive the transform of these functions in the following examples.

EXAMPLE 5-2

Find the Laplace transform of a step function of magnitude A defined by

$$f(t) = 0 \text{ for } t < 0$$

$$f(t) = A \text{ for } t \geq 0$$

Solution

Since we are concerned with $f(t)$ for $t \geq 0$, we will substitute A for $f(t)$ in Eq. (5-1), giving

$$\mathcal{L}[A] = \int_0^\infty Ae^{-st}\,dt$$

Performing the indicated integration gives

$$\mathcal{L}[A] = -\frac{A}{s}e^{-st}\Big|_0^\infty$$

Applying the specified limits yields

$$\mathcal{L}[A] = \frac{A}{s} = F(s)$$

Thus the Laplace transform of the step function is simply $F(s) = A/s$.

EXAMPLE 5-3

Find the Laplace transform of the sine function

$$f(t) = \sin \omega t$$

Solution

From Eq. (5-1)

$$\mathcal{L}[\sin \omega t] = F(s) = \int_0^\infty (\sin \omega t)e^{-st}\,dt$$

The integration may be accomplished most easily by a table of integrals. Alternatively it can be performed by a substitution based on Euler's identities. Using the identity

$$\sin \omega t = \frac{e^{j\omega t} - e^{-j\omega t}}{2j}$$

we have

$$F(s) = \int_0^\infty \left(\frac{e^{j\omega t} - e^{-j\omega t}}{2j}\right) e^{-st}\,dt$$

$$= \frac{1}{2j}\left[-\frac{e^{-(s-j\omega)t}}{s - j\omega} + \frac{e^{-(s+j\omega)t}}{s + j\omega}\right]_0^\infty$$

$$= \frac{1}{2j}\left(\frac{1}{s - j\omega} - \frac{1}{s + j\omega}\right) = \frac{\omega}{s^2 + \omega^2}$$

5-3 LAPLACE TRANSFORM THEOREMS

Several theorems that are useful in applying Laplace transforms to the solution of differential equations are presented in this section. Detailed proofs of these theorems will not be given but may be found in a number of higher-mathematics texts.

A. *Linearity* *If a function is multiplied by a constant, the Laplace transform of the product is the constant times the transform of the function. Thus*

$$\mathcal{L}[Af(t)] = AF(s) \tag{5-3}$$

where A is an arbitrary constant.

B. *Superposition* *The Laplace transform of the sum of two functions is the sum of the Laplace transforms of the functions.*

$$\mathcal{L}[f_1(t) \pm f_2(t)] = F_1(s) \pm F_2(s) \tag{5-4}$$

C. Differentiation The first derivative of a function that can be treated by the Laplace transform method can also be treated by this method, and the value of its transform is

$$\mathcal{L}\left[\frac{df(t)}{dt}\right] = sF(S) - f(0) \tag{5-5}$$

where $F(s)$ is the transform of $f(t)$ and where the term $f(0)$ is the value of the function $f(t)$ at time $t = 0$.

The transform of the second derivative is

$$\mathcal{L}\left[\frac{d^2f(t)}{dt^2}\right] = s^2F(S) - sf(0) - \frac{df(0)}{dt} \tag{5-6}$$

where $df(0)/dt$ is the value of the first derivative at $t = 0$.

Note that the transform includes the initial conditions whereas in the classical method of solution the initial conditions are handled separately to obtain a particular solution from the general solution.

In general the transform of the nth derivative of $f(t)$ is

$$\mathcal{L}\left[\frac{d^nf(t)}{dt^n}\right] = s^nF(s) - s^{n-1}f(0) - \cdots - \frac{d^{n-1}f(0)}{dt^{n-1}} \tag{5-7}$$

D. Integration The integral of a function that can be treated by the Laplace transform method can also be treated by this method, and the value of its transform is

$$\mathcal{L}\left[\int f(t)\,dt\right] = \frac{F(s)}{s} + \frac{\int f(0)\,dt}{s} \tag{5-8}$$

The term $\int f(0)\,dt$ is the initial value of the integral, where $f(0)$ is the value of the function at $t = 0$.

Equation (5-8) can be extended to cover the higher-order n^{th} integral as

$$\mathcal{L}\left[\int^n f(t)\,dt\right] = \frac{F(s)}{s^n} + \frac{\int f(0)\,dt}{s^n} + \cdots + \frac{\int^n f(0)\,dt}{s} \tag{5-9}$$

E. Final value The final-value theorem expresses the value of $f(t)$ as t approaches ∞ according to

$$\lim_{t\to\infty} f(t) = \lim_{s\to 0} sF(s) \tag{5-10}$$

This theorem is useful in predicting steady-state behavior in the time domain without performing the entire inverse transformation of $F(s)$.

F. Initial value *The initial-value theorem expresses the value of $f(t)$ as t approaches 0 according to*

$$\lim_{t \to 0} f(t) = \lim_{s \to \infty} sF(s) \tag{5-11}$$

This theorem is useful in establishing the initial value of $f(t)$ without carrying out the complete inverse transformation.

5-4 APPLICATION OF LAPLACE TRANSFORMS TO SOLVING DIFFERENTIAL EQUATIONS

The following is a summary of the steps to be followed in applying the Laplace transform method to solving a differential equation in which time is the independent variable.

1. Formulate the differential equation that describes the behavior of the physical system under study.
2. Transform the differential equation on a term-by-term basis using a table of transform pairs together with the theorems presented in Sec. 5-3. Initial conditions should be included in the transformation according to Theorem C.
3. Solve the transformed equation for the variable of interest expressed as a function of s.
4. Inverse transform the equation from step 3 to obtain the variable of interest as a function of time.

The above steps can be extended to cover the case of simultaneously solving two or more differential equations. Each equation is individually transformed, and the resulting algebraic equations are solved simultaneously for the variables of interest. These variables are then inverse transformed to obtain the desired solution.

This procedure will be illustrated by means of several examples.

EXAMPLE 5-4

Using the Laplace transform method, determine the equation for the current i in the circuit shown in Fig. 5-2. Assume the switch is closed at time $t = 0$. The circuit in this example is the first-order system described in Sec. 4-4; hence this example will verify the response presented earlier.

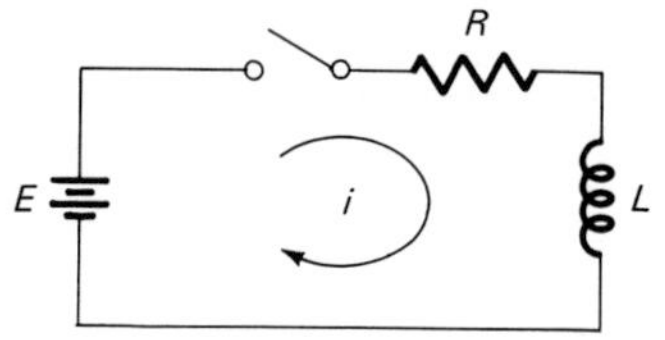

Figure 5-2 Circuit for Example 5-4

Solution

As given in Eq. (4-10) the differential equation of the current in the circuit is

$$Ri + L\frac{di}{dt} = E$$

As a result of the superposition property given in Eq. (5-4), the Laplace transform of the left side of the given equation is the sum of the Laplace transforms of the individual terms. Also the transform of the left side is equal to the transform of the right side. Using Table 5-1, we may proceed to transform each term on both sides of the equation. Thus

$$\mathcal{L}\left[Ri + L\frac{di}{dt}\right] = RI(s) + L[sI(s) - i(0)] = \mathcal{L}[E] = \frac{E}{s}$$

where $I(s)$ is the current as a function of the Laplace variable s. Note that Eq. (5-5) has been applied in transforming the term containing the derivative di/dt and that Transform Pair 2 of Table 5-1 is used in transforming the dc (constant) voltage E.

Applying the initial condition $i(0) = 0$ at $t = 0$, the transformed equation becomes

$$(R + Ls)I(s) = \frac{E}{s}$$

Solving for the variable of interest,

$$I(s) = \frac{E}{s(R + Ls)} = \frac{E}{L}\left[\frac{1}{s(s + R/L)}\right]$$

We may take the inverse transform of the right side using Transform Pair 7 of Table 5-1. This gives the solution as

$$i = \frac{E}{R}(1 - e^{-Rt/L})$$

which is identical to the solution given in Eq. (4-14).

EXAMPLE 5-5

Apply the final-value and initial-value theorems to the circuit in Example 5-4.

Solution

Applying Eq. (5-10) to the transformed equation, we have

$$\lim_{t\to\infty} i(t) = \lim_{s\to 0} s\left[\frac{E}{s(R + Ls)}\right] = \frac{E}{R}$$

This result may be verified by allowing $t \to \infty$ in the solution for i in Example 5-4. Applying Eq. (5-11) gives

$$\lim_{t\to 0} i(t) = \lim_{s\to\infty} s\left[\frac{E}{s(R + Ls)}\right] = 0$$

This result of course agrees with the specified initial condition for the current.

EXAMPLE 5-6

Using the Laplace transform method, determine the current in a series *RLC* circuit with the capacitor initially charged to a voltage as indicated in Fig. 5-3. This example pertains to the second-order system described in Sec. 4-8.

Solution

Applying Kirchhoff's voltage law, the differential equation for the current i is

$$L\frac{di}{dt} + Ri + \frac{1}{C}\int i\,dt = 0$$

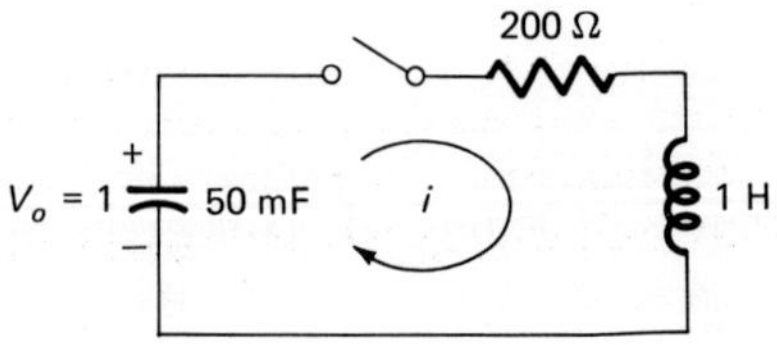

Figure 5-3 Circuit for Example 5-6

The corresponding transform equation is

$$L[sI(s) - i(0)] + RI(s) + \frac{1}{Cs}\left[I(s) - \int i(0)\,dt\right] = 0$$

The integral $\int i(0)\,dt$ represents the initial charge $q(0)$ on the capacitor. Since capacitor voltage = charge/capacitance ($V_c = Q/C$), the last term in the equation can be written as

$$\frac{1}{Cs}\int i(0)\,dt = \frac{q(0)}{Cs} = \frac{V_o}{s}$$

Because the polarity of V_o is opposite the voltage drops due to i, the term $\int i(0)\,dt$ has a minus sign associated with it. Substituting the circuit parameters, the transformed equation becomes

$$sI(s) + 200I(s) + \frac{I(s)}{50 \times 10^{-6}s} - \frac{1}{s} = 0$$

Solving for $I(s)$, the transform variable of interest, gives

$$I(s) = \frac{1}{s^2 + 200s + 2 \times 10^4}$$

or, on completing the square,

$$I(s) = \frac{1}{(s + 100)^2 + (100)^2}$$

This result may be put in the form of Transform Pair 15 of Table 5-1. Thus

$$I(s) = \frac{1}{100}\left[\frac{100}{(s + 100)^2 + 100^2}\right]$$

Taking the inverse transform then gives the solution

$$i = \frac{1}{100}\left(e^{-100t} \sin 100t\right) \quad \text{A}$$

EXAMPLE 5-7

Determine the current delivered by the battery in the circuit shown in Fig. 5-4 after closure of the switch.

Solution

In order to illustrate the solution of simultaneous differential equations, the mesh-current method will be applied to this problem. Sum-

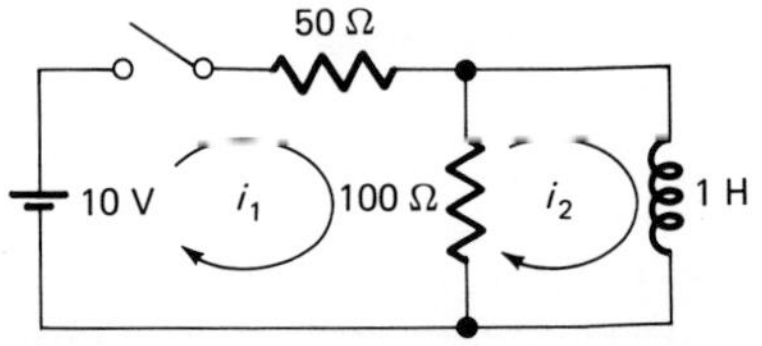

Figure 5-4 Circuit for Example 5-7

ming the voltage drops around each loop, the mesh equations are

$$10 = 50i_1 + 100i_1 - 100i_2$$

$$0 = -100i_1 + 100i_2 + \frac{di_2}{dt}$$

The corresponding transform equations are

$$\frac{10}{s} = 150I_1(s) - 100I_2(s)$$

$$0 = -100I_1(s) + (100 + s)I_2(s)$$

where the conditions at $t = 0$ are $i_1 = 0$ and $i_2 = 0$.

The battery current $I_1(s)$ may be found by solving the second equation for $I_2(s)$ and substituting the result in the first equation. Thus

$$I_2(s) = \frac{100}{100 + s} I_1(s)$$

Substituting,

$$\frac{10}{s} = 150I_1(s) - \frac{(100)(100)}{100 + s} I_1(s)$$

Solving for $I_1(s)$,

$$I_1(s) = \frac{10(100 + s)}{s[150(100 + s) - (100)(100)]}$$

$$= \frac{100 + s}{s(500 + 15s)}$$

$$= \frac{\frac{100}{15}}{s(s + \frac{500}{15})} + \frac{\frac{1}{15}}{s + \frac{500}{15}}$$

Taking the inverse transform of both terms gives

$$i_1 = \tfrac{1}{5}(1 - e^{-500t/15}) + \tfrac{1}{15}e^{-500t/15}$$

$$= \tfrac{1}{5} - \tfrac{2}{15}e^{-500t/15} \text{ A}$$

5-5 TRANSFER FUNCTIONS IN THE LAPLACE DOMAIN

In Sec. 1-5 the concept of a transfer function as a means of characterizing a linear element or system was introduced. However, our consideration at that time was limited to a simple proportional relationship between the input and output of the element. We have subsequently seen that many system elements have a response that varies as a function of time. In the time domain this dynamic characteristic is described by a differential equation, which is not directly applicable as a transfer function. However, the dynamic characteristic can be represented by a transfer function incorporating the Laplace variable s. Through the use of such a transfer function, a relatively simple method of time-domain analysis is available for predicting system response to a defined input signal.

The Laplace-domain transfer function for a linear element or system can be defined as the ratio of the Laplace transform of the output to the Laplace transform of the input with all initial conditions assumed to be 0. The transfer function for a particular element can be obtained as follows:

1. Write the differential equation which defines the behavior of the element.
2. Transform the equation into the Laplace domain assuming all initial conditions to be 0.
3. Form the ratio of the output to the input as

$$\frac{O(s)}{I(s)} = G(s) \tag{5-12}$$

where $O(s)$ is the Laplace transform of the output
$I(s)$ is the Laplace transform of the input
$G(s)$ is the transfer function in the Laplace domain

The element can be represented in block-diagram form as shown in Fig. 5-5. The block is interpreted to mean that the transformed output may be obtained by multiplying the transformed input by the transfer function. Since a transformed equation can be treated as an algebraic

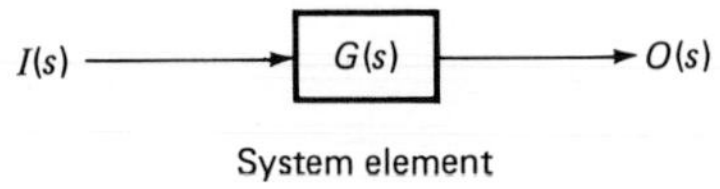

Figure 5-5 Block-diagram representation of system element in Laplace domain

equation, the transformed system block diagram can be manipulated according to the block-diagram techniques described in Chap. 1. It should be emphasized that the transfer function is a property of the element only, and is independent of the input and the initial conditions, if any. Because the transfer function characterizes the element in determining the form of its transient response, it is also known as the *characteristic function* of the element.

EXAMPLE 5-8

Determine the transfer function $E_o(s)/E_i(s)$ for the circuit shown in Fig. 5-6.

Solution

The transfer function may be found by deriving an appropriate differential equation based on Kirchhoff's voltage law. Thus

$$e_i = e_o + iR$$

By definition

$$i = C\frac{de_o}{dt}$$

and so

$$e_i = e_o + RC\frac{de_o}{dt}$$

Transforming both sides of the equation gives

$$E_i(s) = E_o(s) + RCsE_o(s)$$

Solving for the desired output-input ratio yields the result

$$\frac{E_o(s)}{E_i(s)} = \frac{1}{1 + RCs} = \frac{1}{1 + \tau s}$$

where $\tau = RC$, the circuit time constant.

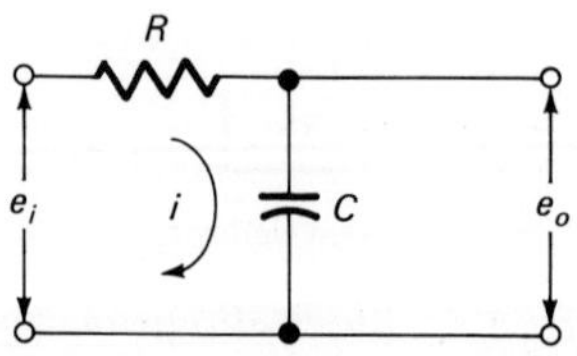

Figure 5-6 Circuit for Example 5-8

EXAMPLE 5-9

Determine the transfer function of the spring-mass system shown in Fig. 5-7, where the input displacement is x and the output displacement is *y*.

Solution

Referring to Sec. 2-4, we may write the system differential equation as

$$M\frac{d^2y}{dt^2} + B\frac{dy}{dt} + K(y - x) = 0$$

Transforming the equation gives

$$Ms^2Y(s) + BsY(s) + KY(s) = KX(s)$$

The transfer function is then found to be

$$\frac{Y(s)}{X(s)} = \frac{1}{(M/K)s^2 + (B/K)s + 1}$$

This result can be rewritten as

$$\frac{Y(s)}{X(s)} = \frac{1}{s^2/\omega_n{}^2 + (2\zeta/\omega_n)s + 1}$$

where $\omega_n = \sqrt{\dfrac{K}{M}}$

$$\zeta = \frac{B}{2\sqrt{MK}}$$

It should be noted that the transfer functions derived in Examples 5-8 and 5-9 are for first- and second-order systems, respectively. In both

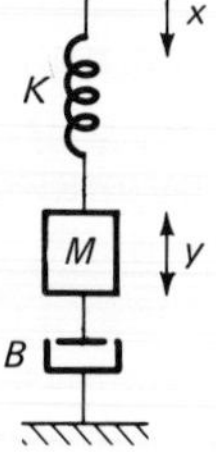

Figure 5-7 Spring-mass system for Example 5-9

cases the transfer functions have been rewritten in a general form using the parameters presented in Chap. 4.

5-6 SERVOMOTOR TRANSFER FUNCTION

In Sec. 4-5 a differential equation was derived for a servomotor using armature control. In this section we will consider the transfer function of such a motor. Additionally the transfer function of a motor with field control will be presented.

Beginning with the motor differential equation that was given in Eq. (4-24), we have

$$v_a = \frac{J}{K_t}\ddot{\theta} + \frac{B}{K_t}\dot{\theta}$$

Treating this equation by the Laplace transform method gives

$$V_a(s) = \left(\frac{J}{K_t}s + \frac{B}{K_t}\right)\dot{\theta}(s) \tag{5-13}$$

which when written in transfer-function form gives

$$\frac{\dot{\theta}(s)}{V_a(s)} = \frac{1}{(J/K_t)s + B/K_t} = \frac{K_t/B}{(J/B)s + 1} = \frac{K_m}{\tau_m s + 1} \tag{5-14}$$

where K_m = velocity gain constant defined by Eq. (4-25)
τ_m = mechanical time constant defined by Eq. (4-26)

As a next step, let us assume that the position θ of the output shaft is the variable of interest. In this case the transformed equation becomes

$$V_a(s) = \left(\frac{J}{K_t}s^2 + \frac{B}{K_t}s\right)\theta(s) \tag{5-15}$$

from which the resulting transfer function may be written as

$$\frac{\theta(s)}{V_a(s)} = \frac{K_t/B}{s(Js/B + 1)} = \frac{K_m}{s(\tau_m s + 1)} \tag{5-16}$$

Note that Eq. (5-16) differs from Eq. (5-14) by the additional s factor in the denominator. This is to be expected when we recall that (1) a Laplace function is integrated by multiplying the function by 1/s and (2) position is the integral of velocity with respect to time.

In Sec. 3-12 it was pointed out that a dc motor may also be controlled by varying its field current. In this arrangement the back emf does not affect the control (field) current. However, the inductance of the field winding is generally large enough so that its effect must be considered in determining the transient response of the motor.

The field current i_f may be expressed by the differential equation

$$v_f = i_f R_f + L_f \frac{di_f}{dt} \tag{5-17}$$

where v_f = control voltage applied to field winding
R_f = resistance of field winding
L_f = inductance of field winding

By transforming Eq. (5-17) we may obtain a transfer function relating field current to input voltage. Thus

$$\frac{I_f(s)}{V_f(s)} = \frac{1/R_f}{(L_f/R_f)s + 1} \tag{5-18}$$

The torque developed by the motor is proportional to the field current so that

$$T_m = K_i i_f \tag{5-19}$$

where K_i is the motor torque constant. Equating the motor torque to the load torque gives

$$K_i i_f = J\ddot{\theta} + B\dot{\theta} \tag{5-20}$$

Equation (5-20) may be transformed and written as

$$\frac{\theta(s)}{I_f(s)} = \frac{K_i/B}{s[(J/B)s + 1]} \tag{5-21}$$

Substituting for $I_f(s)$ from Eq. (5-18) into Eq. (5-21) gives the desired transfer function

$$\frac{\theta(s)}{V_f(s)} = \frac{K_i/BR_f}{s[(J/B)s + 1][(L/R_f)s + 1]} = \frac{K_i/BR_f}{s(\tau_m s + 1)(\tau_f s + 1)} \tag{5-22}$$

where τ_f is the time constant of the field winding.

EXAMPLE 5-10

A dc motor has the following specifications:

Rated armature voltage	120 V
Stall torque at rated voltage	6 oz-in.
No-load speed at rated voltage	6,000 rpm
Rotor inertia	5×10^{-4} oz-in.-s^2

1. Determine the position transfer function.
2. Determine the transient response of thc motor to a step input of 50 V.

Solution

1. The various motor parameters are found to be

$$K_t = \frac{T_s}{V_a} = \frac{6}{120} = 0.05 \text{ oz-in./V}$$

$$B = \frac{T_s}{\dot{\theta}_{NL}} = \frac{6}{6{,}000 \times 2\pi/60} = 0.00955 \text{ oz-in./(rad/s)}$$

$$K_m = \frac{K_t}{B} = 5.24\text{(rad/s)/V}$$

$$\tau_m = \frac{J}{B} = \frac{5 \times 10^{-4}}{95.5 \times 10^{-4}} = 0.0524 \text{ s}$$

Substituting these values in Eq. (5-16) gives the desired transfer function

$$\frac{\theta(s)}{V_a(s)} = \frac{5.24}{s(0.0524s + 1)}$$

2. The transient response may be determined by applying a 50-V step input to the above transfer function. Thus

$$\theta(s) = \left[\frac{5.24}{s(0.0524s + 1)}\right] V_a(s) = \left[\frac{5.24}{s(0.0524s + 1)}\right] \frac{50}{s}$$

$$= \frac{262}{s^2(0.0524s + 1)}$$

Note that 50/s is the Laplace transform of the 50-V step input. To obtain the inverse transform with the aid of Transform Pair 8 in Table 5-1, the equation must be rewritten as

$$\theta(s) = \frac{262/0.0524}{s^2(s + 1/0.0524)} = \frac{5{,}000}{s^2(s + 19.1)}$$

Then

$$\theta = \frac{5{,}000}{19.1^2}(19.1t - 1 + e^{-19.1t})$$

$$= 262t - 13.7(1 - e^{-19.1t}) \text{ rad}$$

Figure 5-8 is a graph of this response. We note that the position increases linearly with time after an initial lag caused by the inertia of the system.

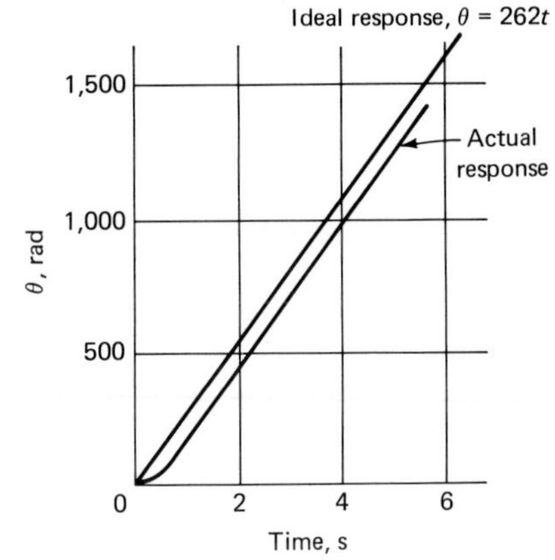

Figure 5-8 Response of motor in Example 5-10

5-7 TRANSIENT ANALYSIS OF A POSITIONAL SERVOMECHANISM

The differential equation for a simple positional servomechanism was derived in Sec. 4-9. The transient response to a step input was then obtained by simply referring to the general response curves of an underdamped second-order system. In this section we will apply the Laplace transform method to determine the response of this positional servomechanism. This application will serve to introduce Laplace transforms as a means of analyzing the behavior of closed-loop systems.

A closed-loop transfer function for the positional servomechanism can be derived following the steps outlined in Sec. 5-5. Thus, referring to Eq. (4-47), we have

$$\theta_i(s) = \frac{J}{K}s^2\theta_o(s) + \frac{B}{K}s\theta_o(s) + \theta_o(s)$$

$$= \left(\frac{J}{K}s^2 + \frac{B}{K}s + 1\right)\theta_o(s)$$

which can be written as

$$\frac{\theta_o(s)}{\theta_i(s)} = \frac{1}{(J/K)s^2 + (B/K)s + 1} \tag{5-23}$$

The transfer function is that of a second-order system. For the underdamped case it can be written using the damping ratio and undamped natural frequency. Substituting these constants as defined by Eq. (4-49) gives

$$\frac{\theta_o(s)}{\theta_i(s)} = \frac{1}{s^2/\omega_n^2 + (2\zeta/\omega_n)s + 1} = \frac{\omega_n^2}{s^2 + 2\zeta\omega_n s + \omega_n^2} \tag{5-24}$$

If the input $\theta_i(s)$ is a step function of magnitude θ_i,

$$\theta_i(s) = \frac{\theta_i}{s} \tag{5-25}$$

and the response is

$$\theta_o(s) = \frac{\theta_i \omega_n{}^2}{s(s^2 + 2\zeta\omega_n s + \omega_n{}^2)} \tag{5-26}$$

Equation (5-26) may be inverse transformed using Transform Pair 14 of Table 5-1 to obtain the transient response. Thus

$$\theta_o = \theta_i \omega_n{}^2 \left[\frac{1}{\omega_n{}^2} - \frac{e^{-\zeta\omega_n t}}{\omega_n{}^2\sqrt{1-\zeta^2}} \sin\left(\omega_n \sqrt{1-\zeta^2}\, t + \phi\right) \right] \tag{5-27}$$

Since $\phi = \cos^{-1}\zeta$, this may be written as

$$\theta_o = \theta_i \left[1 - \frac{e^{-\zeta\omega_n t}}{\sqrt{1-\zeta^2}} \sin\left(\omega_n \sqrt{1-\zeta^2}\, t + \cos^{-1}\zeta\right) \right] \tag{5-28}$$

which is identical with Eq. (4-51).

Wc should note that the above analysis of a positional servomechanism was based on applying Laplace transforms to the differential equation that describes the *closed-loop* behavior of the system. Unfortunately it becomes extremely difficult to write a differential equation for a closed-loop system as the complexity of the system increases. Therefore some alternative approach in applying Laplace transforms is needed if the method is to be at all universal.

The method of analysis that is widely used in control systems work involves a block diagram of the system wherein the variables and block transfer functions are expressed in the Laplace domain. The transformed block diagram can then be analyzed using the methods described in Chap. 1 to obtain an expression for the closed-loop transfer function.

To illustrate the block-diagram method let us again consider the positional servomechanism. Figure 5-9 shows a block diagram of the servo. This diagram is similar to that of Fig. 4-19 except that the variables and some transfer functions are shown as functions of s. Note that the transfer function for the motor is that given in Eq. (5-16).

Recall from Chap. 1 that the transfer function of a closed-loop system may be stated as

$$\frac{C}{R} = \frac{G}{1 + GH}$$

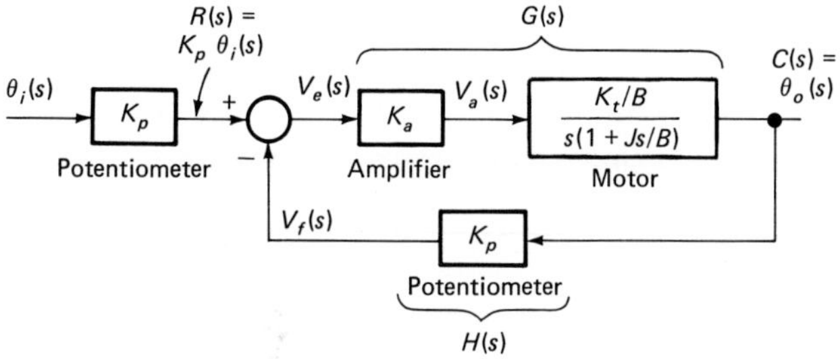

Figure 5-9 Block diagram of positional servomechanism

which in the Laplace domain becomes

$$\frac{C(s)}{R(s)} = \frac{G(s)}{1 + G(s)H(s)} \tag{5-29}$$

If this relationship is applied to the servo system in Fig. 5-9, we have

$$\frac{\theta_o(s)}{K_p\theta_i(s)} = \frac{(K_aK_t/B)/[s(1 + Js/B)]}{1 + (K_aK_tK_p/B)/[s(1 + Js/B)]}$$

or

$$\frac{\theta_o(s)}{\theta_i(s)} = \frac{K_pK_aK_t/B}{(J/B)s^2 + s + K_aK_tK_p/B} = \frac{K/B}{(J/B)s^2 + s + K/B} \tag{5-30}$$

where K is again the gain product $K_pK_aK_t$. Equation (5-30) may be rewritten to become

$$\frac{\theta_o(s)}{\theta_i(s)} = \frac{1}{(J/K)s^2 + (B/K)s + 1} \tag{5-31}$$

We observe that the closed loop transfer function obtained from the system block diagram is the same as the one previously obtained by transforming the differential equation [Eq. (5-23)].

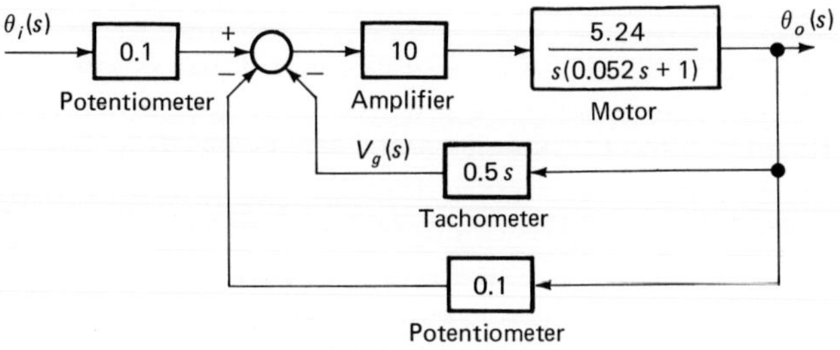

Figure 5-10 Positional servomechanism for Example 5-11

EXAMPLE 5-11

The servomotor of Example 5-10 is used in a positional servomechanism that incorporates tachometer feedback for improved performance. A block diagram of the servo including gain constants is shown in Fig. 5-10. Derive the closed-loop system transfer function.

Solution

The tachometer output voltage v_g is proportional to the derivative of the controlled variable, so that the transfer function is

$$\frac{V_g(s)}{\theta_o(s)} = K_g s = 0.5s$$

Because the two feedback paths are in parallel, the overall feedback function $H(s)$ is

$$H(s) = K_p + K_g s = 0.1 + 0.5s$$

The closed-loop transfer function can be derived as

$$\frac{\theta_o(s)}{\theta_i(s)} = \frac{K_p K_a K_m / s(\tau_m s + 1)}{1 + (K_p + K_g s) K_a K_m / s(\tau_m s + 1)}$$

Substituting the gain constants gives

$$\frac{\theta_o(s)}{\theta_i(s)} = \frac{(0.1)(10)(5.24)/s(0.0524s + 1)}{1 + (0.1 + 0.5s)(10)(5.24)/s(0.0524s + 1)}$$

$$= \frac{5.24}{0.0524s^2 + 27.2s + 5.24}$$

5-8 STEADY-STATE ERRORS

It was pointed out in Chap. 4 that the response of a physical system can be divided into two parts: the transient and steady-state components. Since one of the fundamental reasons for using negative feedback in an automatic control system is to improve steady-state accuracy, the steady-state solution of the system differential equation is of interest. In particular the steady-state error resulting from the difference between the steady-state values of the input and the output variables is one of the measures of control system performance. In this section we will investigate the steady-state error of several forms of control systems when excited by common types of input functions. The method to be discussed has been widely applied in evaluating servomechanism performance.

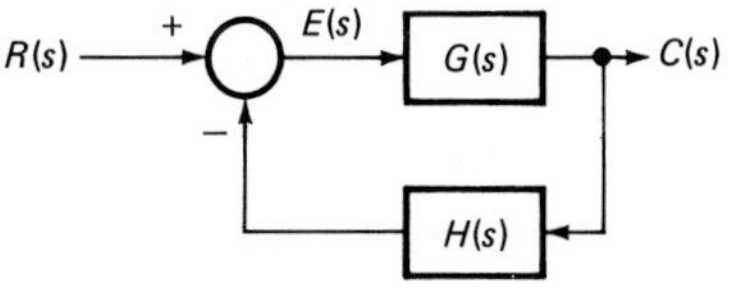

Figure 5-11 Generalized block diagram of feedback system

For the generalized feedback system shown in Laplace form in Fig. 5-11 the system actuating signal E(s) is a measure of the system error. The error may be expressed as

$$E(s) = R(s) - H(s)C(s) = R(s) - \left[\frac{G(s)H(s)}{1 + G(s)H(s)}\right]R(s)$$

$$= \left[\frac{1}{1 + G(s)H(s)}\right]R(s) \qquad (5\text{-}32)$$

For a unity-feedback system, $H(s) = 1$. Equation (5-32) then becomes

$$E(s) = \frac{R(s)}{1 + G(s)} \qquad (5\text{-}33)$$

The system steady-state error e_{ss} in the time domain can be determined by applying the Laplace transform final-value theorem to Eq. (5-33). Thus

$$e_{ss} = \lim_{t\to\infty} e(t) = \lim_{s\to 0} sE(s) = \lim_{s\to 0} \frac{sR(s)}{1 + G(s)} \qquad (5\text{-}34)$$

From Eq. (5-34) we see that the steady-state error depends on both the input function $R(s)$ and the open-loop transfer function $G(s)$. Before proceeding further let us examine the nature of $G(s)$ for various systems. Since $G(s)$ is the product of the various transfer functions that are cascaded in the loop, it can be expressed as the ratio of two factored polynomials. As an example consider

$$G(s) = \frac{K(\tau_1 s + 1)(\tau_2 s + 1)}{s^n(\tau_3 s + 1)(\tau_4 s + 1)} \qquad (5\text{-}35)$$

where K is the product of the gain constants of the various components in the system and the values of τ are the time constants of the various system elements. When $G(s)$ is expressed in the above form, the coefficient n in the denominator serves to identify a *system type number*. In terms of the physical system, n represents the number of integrations contained in the transfer function [recall Eq. (5-9)]. Thus a Type 0 system (for which $n = 0$) would not contain an integrating element, a

Type 1 system would include a single integrating element, etc. In practical systems the number of integrations is rarely greater than two.

If this method of classification is applied to the servomechanisms described in Chap. 4, we note that the velocity servo is a Type 0 system and the positional servo is a Type 1 system.

Let us consider the steady-state error when a step input of magnitude A is applied to a closed-loop system which has unity feedback. In this case the steady-state value of the controlled variable will be constant. For the step input $r(t) = A$, the transformed input is $R(s) = A/s$. Applying Eq. (5-34),

$$e_{ss} = \lim_{s \to 0} s\left(\frac{A}{s}\right)\left[\frac{1}{1 + G(s)}\right] = \frac{A}{1 + G(0)} \tag{5-36}$$

where $G(0)$ is the value of $G(s)$ with $s = 0$.

If the step input is applied to a Type 0 system, Eq. (5-35) becomes

$$G(0) = \lim_{s \to 0} \frac{K(\tau_1 s + 1)(\tau_2 s + 1)}{(\tau_3 s + 1)(\tau_4 s + 1)} = K$$

from which

$$e_{ss} = \frac{A}{1 + K} \tag{5-37}$$

Figure 5-12 is an example of the response of a Type 0 system to a step input showing the existence of a steady-state error.

If a step input is applied to a Type 1 system, we have

$$G(0) = \lim_{s \to 0} \frac{K(\tau_1 s + 1)(\tau_2 s + 1)}{s(\tau_3 s + 1)(\tau_4 s + 1)} = \frac{K}{0} = \infty$$

which gives

$$e_{ss} = \frac{A}{1 + \infty} = 0 \tag{5-38}$$

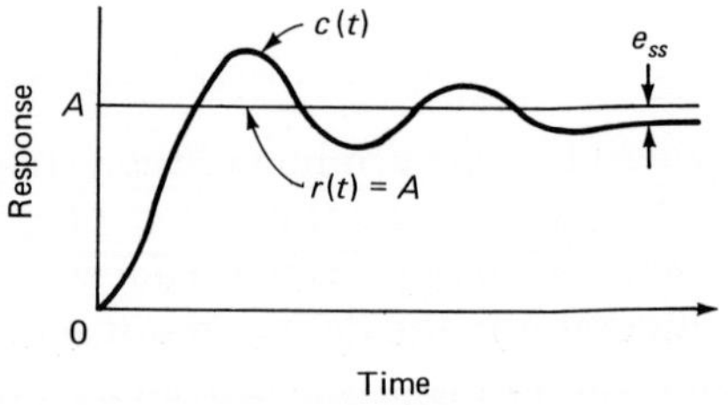

Figure 5-12 Step-function response for Type 0 system

TABLE 5-2 STEADY-STATE ERRORS FOR STABLE UNITY-FEEDBACK SYSTEMS

System Type No.	Input Function Step $r(t) = A$	 Ramp $r(t) = At$	 Parabolic $r(t) = \frac{1}{2}At^2$
0	$\dfrac{A}{1 + K_p}$	∞	∞
1	0	$\dfrac{A}{K_v}$	∞
2	0	0	$\dfrac{A}{K_a}$

and the output variable. For example, in many systems the feedback element is a transducer whose transfer function is a simple gain constant H which changes the output variable $C(s)$ to a quantity $HC(s)$ that can be compared with $R(s)$. Where $R(s)$ and $HC(s)$ have like dimensions, error constants can be derived by including H with $G(s)$ in the equations in this section. In the more general case where H is a function of s, the steady-state actuating signal may be found by applying the final-value theorem to Eq. (5-32).

An examination of Table 5-2 would appear to indicate that the higher the system type, the "better" the control system. If only steady-state accuracy is considered, this is indeed the case. However, the added integrations in higher systems cause a greater lag in the output so that these systems are increasingly difficult to stabilize. As with most engineering problems, the design of feedback control systems usually requires a compromise between the steady-state error and the stability requirements.

EXAMPLE 5-12

A velocity servomechanism is shown in block-diagram form in Fig. 5-15. Determine the steady-state error in the velocity $\dot{\theta}$ when the input voltage is 10 V.

Solution

To determine the error in $\dot{\theta}$, the servomechanism can be viewed as a unity-feedback system in which the tachometer voltage e_f is the controlled output (the block diagram has been drawn in this form). The

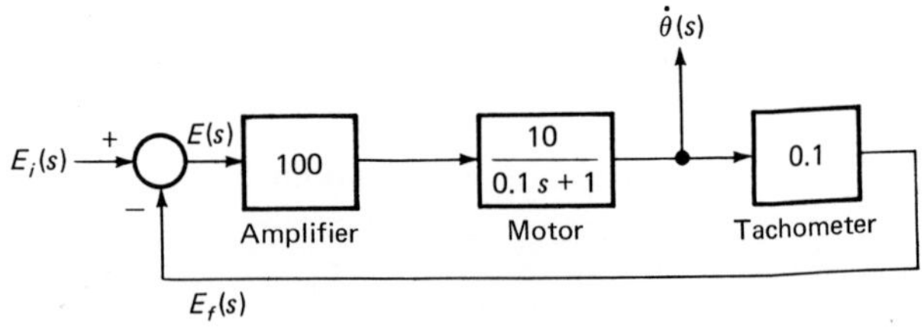

Figure 5-15 Velocity servomechanism for Example 5-12

open-loop transfer function is then

$$G(s) = \frac{E_f(s)}{E(s)} = \frac{(100)(10)(0.1)}{(0.1s + 1)} = \frac{100}{0.1s + 1}$$

which represents a Type 0 system. Assuming a constant input and using Eq. (5-43), $K_p = 100$.

Applying Eq. (5-37), the steady-state error e_{ss} in the tachometer voltage is

$$e_{ss} = \frac{A}{1 + K_p} = \frac{10}{101} = 0.099 \text{ V}$$

From the block diagram we see that $E_f(s) = 0.1\,\dot{\theta}(s)$. Thus the steady-state velocity error $\dot{\theta}_\varepsilon$ is

$$\dot{\theta}_\varepsilon = \frac{e_{ss}}{0.1} = \frac{0.099}{0.1} = 0.99 \text{ rad/s}$$

The analysis indicates that the actual output velocity will be less than the desired value by 0.99 rad/s.

EXAMPLE 5-13

A radar antenna is required to track a target with a maximum rate of 0.5 rad/s and a maximum tracking error of $0.5° = 0.0087$ rad. Determine the velocity constant K_v that is required for the antenna control system to meet this specification.

Solution

In tracking the target, the antenna is responding to a ramp input $r(t) = 0.5t$. Using Eq. (5-41),

$$K_v = \frac{A}{e_{ss}} = \frac{0.5}{0.0087} = 57.4 \text{ s}^{-1}$$

Thus a Type 1 system with a K_v of at least 57.4 is required.

5-9 SYSTEM STABILITY

In Sec. 1-8 it was pointed out that a system is said to be unstable if, in response to an input, it oscillates violently or drives to a limiting value. Conversely a stable system is one that will remain at rest unless excited by an input function and will return to rest if the excitation is removed. Instability is an unfortunate possibility with closed-loop control; an open-loop system cannot be unstable. Obviously to be of practical value a control system must be stable.

The system input has no effect on stability; a system that is stable for one input will be stable for all inputs. Instead, stability depends only upon the characteristics of the system itself.

The response of a stable system is characterized by transients that decay to 0 with increasing values of time. Thus, for stability, the coefficients of the exponential terms in the transient response must be either negative real numbers or complex numbers with negative real parts. For example, the first-order differential equation

$$\frac{dx}{dt} - x = 0 \tag{5-46}$$

can be shown to have the transient solution

$$x = Ae^{t} \tag{5-47}$$

Since the exponent is positive, the response grows with increasing time and the system is unstable. Actually in practice the response would not continue to infinity but would reach some limiting value determined by the nature of the physical system.

EXAMPLE 5-14

A closed-loop system is described by the differential equation

$$\frac{d^2c}{dt^2} + 2\frac{dc}{dt} + 5c = r(t)$$

Evaluate the stability of the system by determining the form of the transient response.

Solution

The transfer function of the system is found by transforming the system differential equation with the result

$$\frac{C(s)}{R(s)} = \frac{1}{s^2 + 2s + 5}$$

On completing the square, this may be written as

$$\frac{C(s)}{R(s)} = \frac{1}{(s+1)^2+4} = \frac{1}{2}\left[\frac{2}{(s+1)^2+4}\right]$$

Applying Transform Pair 15, the transient response of the system is found to be

$$y = \tfrac{1}{2}e^{-t}\sin 2t$$

This is an oscillatory response with an amplitude that decays with increasing time. Hence we conclude that the system is stable.

5-10 CHARACTERISTIC EQUATION

The transfer function of an element or system is also known as the *characteristic function* of the system, and this function determines the nature of the transient response. Hence it can provide information concerning the stability of the system.

The transfer function of the general closed-loop system shown in block-diagram form in Fig. 5-11 has been derived as

$$\frac{C(s)}{R(s)} = \frac{G(s)}{1+G(s)H(s)} \tag{5-48}$$

from which

$$C(s) = \left[\frac{1}{1+G(s)H(s)}\right][G(s)R(s)] \tag{5-49}$$

Equation (5-49) is typical of the transform solution of system behavior; that is,

$$\text{Response function} = (\text{system function})(\text{excitation function})$$

Since the excitation does not affect the form of the transient response, it has no bearing on whether the system is stable or not (the excitation will of course affect the steady-state response). Thus the excitation function, or numerator of Eq. (5-49), may be set equal to 0 without affecting the *form of the transient*. Thus

$$G(s)R(s) = C(s)[1+G(s)H(s)] = 0 \tag{5-50}$$

which, after dividing by C(s), becomes

$$1+G(s)H(s) = 0 \tag{5-51}$$

Equation (5-51) is called the *characteristic equation* of the closed-loop system, and from this equation system stability can be evaluated.

The open-loop transfer function $G(s)H(s)$ of a control system is typically a ratio of polynomials in s [recall Eq. (5-35)]. Thus we can write

$$G(s)H(s) = \frac{N(s)}{D(s)}$$

where $N(s)$ and $D(s)$ are the numerator and denominator polynomials, respectively. Then

$$1 + G(s)H(s) = 1 + \frac{N(s)}{D(s)} = \frac{D(s) + N(s)}{D(s)} \tag{5-52}$$

The characteristic equation derived from Eq. (5-51) can then be written as

$$\frac{D(s) + N(s)}{D(s)} = 0$$

$$D(s) + N(s) = [D(s)][0] = 0 \tag{5-53}$$

If the factors of $D(s)$ and $N(s)$ in Eq. (5-53) are multiplied together, the characteristic equation can be written in expanded form as

$$a_0 s^n + a_1 s^{n-1} + \cdots + a_{n-1} s + a_n = 0 \tag{5-54}$$

and this equation can then be factored to yield

$$D(s) + N(s) = (s + r_1)(s + r_2) \cdots (s + r_n) = 0 \tag{5-55}$$

The quantities $-r_1$, $-r_2$, etc., are the roots of the characteristic equation. For a stable system it is necessary that none of the roots of the characteristic equation have positive real parts. Thus by determining the roots of the characteristic equation of the system, we can evaluate the stability.

EXAMPLE 5-15

A control system has the following transfer functions:

$$G(s) = \frac{2}{s(s + 3)}$$

$$H(s) = 1$$

Determine the characteristic equation of the system, and evaluate the system's stability.

Solution

From Eq. (5-51) the characteristic equation is

$$1 + G(s)H(s) = 1 + \frac{2}{s(s+3)} = \frac{s^2 + 3s + 2}{s(s+3)} = 0$$

$$(s+2)(s+1) = 0$$

Since none of the factors has a positive root, the transient solution contains only negative exponents and the system is stable.

EXAMPLE 5-16

A control system has the transfer functions

$$G(s) = \frac{211}{s(0.01s+1)(0.02s+1)}$$

$$H(s) = 1$$

Determine the characteristic equation, and evaluate the stability.

Solution

The characteristic equation may be obtained by applying Eq. (5-53). Thus we have

$$D(s) + N(s) = s(0.01s+1)(0.02s+1) + 211$$

$$= 0.0002s^3 + 0.03s^2 + s + 211 = 0$$

Dividing both sides of the equation by 0.0002 and factoring yields

$$D(s) + N(s) = s^3 + 150s^2 + 5{,}000s + 1.056 \times 10^6$$

$$= (s+160)(s^2 - 10s + 6{,}594)$$

$$= (s+160)(s - 5 - j81)(s - 5 + j81) = 0$$

The roots of the characteristic equation are -160, $5 + j81$, and $5 - j81$. Since the two complex roots have positive real parts, we conclude that the system is unstable.

5-11 ROUTH CRITERION

As we have seen, an investigation of the roots of the characteristic equation indicates whether a system is stable or not. However, finding the roots is a laborious task, especially if the equation is higher than second or third order. It is usually not necessary to know the exact solution when a system is unstable; hence a simple procedure to determine the existence of roots with positive real parts is useful. One such technique is the Routh–Hurwitz criterion, usually called the *Routh criterion*.

The Routh criterion is an algebraic method of determining system stability in the Laplace domain. It indicates the presence and number of unstable roots but not their values or possible steps to prevent instability.

Let us consider the characteristic equation in the form given by Eq. (5-54). For stability all powers of s must be present in the equation. If any coefficients other than a_n are 0 or if any coefficient is negative, there are imaginary roots or roots with positive real parts. The system is therefore known to be unstable or marginally stable and need not be investigated further.

The Routh criterion is applied by first constructing a *Routh array*. This array is of the form

$$\begin{array}{l} a_0 a_2 a_4 \cdots \\ a_1 a_3 a_5 \cdots \\ b_1 b_3 \cdots \\ c_1 c_3 \cdots \\ d_1 \cdots \\ e_1 \cdots \\ \cdots \\ \cdots \end{array}$$

The first row of the array consists of the first, third, fifth, etc., coefficients of the characteristic equation, and the second row consists of the second, fourth, sixth, etc., coefficients. The entries in the third row are computed by cross multiplication according to the following pattern:

$$b_1 = \frac{\begin{vmatrix} a_0 & a_2 \\ a_1 & a_3 \end{vmatrix}}{a_1} = \frac{a_1 a_2 - a_0 a_3}{a_1}$$
$$b_3 = \frac{\begin{vmatrix} a_0 & a_4 \\ a_1 & a_5 \end{vmatrix}}{a_1} = \frac{a_1 a_4 - a_0 a_5}{a_1} \tag{5-56}$$

The pattern is continued until the remaining b coefficients are all equal to 0. Succeeding rows are found in a similar manner. The terms in the fourth row are

$$c_1 = \frac{\begin{vmatrix} a_1 & a_3 \\ b_1 & b_3 \end{vmatrix}}{b_1} = \frac{b_1 a_3 - a_1 b_3}{b_1}$$
$$c_3 = \frac{\begin{vmatrix} a_1 & a_5 \\ b_1 & b_5 \end{vmatrix}}{b_1} = \frac{b_1 a_5 - a_1 b_5}{b_1} \tag{5-57}$$

The complete array is triangular in form, and the number of rows will correspond to one more than the order of the characteristic equation.

Routh's criterion states that the system represented by the array is stable if there are no sign changes in the first column of the array. This is an indication that there are no roots in the characteristic equation with positive real parts. If there are one or more changes in sign, the number of changes equals the number of roots with positive real parts.

In constructing the array, special cases may occasionally arise. Should the first entry in a row be 0 while the other entries are not 0, the 0 may be replaced with a small positive number ε and the array continued. Sign changes in the first column can be ascertained by letting ε approach 0. If a row of all 0s should occur, the system is not stable.

Routh's criterion is of limited usefulness in automatic control because it gives no indication of the degree of instability and the means of avoiding it. However, it does provide a simple stability test and it is possible to explore the effect of one or two system parameters on stability. For example, it is frequently useful to determine the loop gain at which a system just becomes unstable. This can be done by examining the marginal stability condition (transition from stable to unstable operation) using the Routh array.

EXAMPLE 5-17

Apply Routh's criterion to the characteristic equation of Example 5-16.

Solution

The characteristic equation is

$$s^3 + 150s^2 + 5{,}000s + 1.056 \times 10^6 = 0$$

The first two rows of the array are the coefficients of the terms in the equation.

$$\begin{array}{ll} 1 & 5{,}000 \\ 150 & 1.056 \times 10^6 \end{array}$$

The coefficients b_1 and c_1 are obtained using Eqs. (5-56) and (5-57), respectively. Thus

$$b_1 = \frac{150 \times 5{,}000 - 1.056 \times 10^6 \times 1}{150} = -2{,}040$$

$$c_1 = \frac{-2{,}040 \times 1.056 \times 10^6 - 150 \times 0}{-2{,}040} = 1.056 \times 10^6$$

The complete array is then

1	5,000
150	1.056×10^6
−2,040	
1.056×10^6	

There are two changes of sign in the first column of the array indicating that two roots have positive real parts and the system is unstable. This conclusion agrees with the result found in Example 5-16.

EXAMPLE 5-18

A closed-loop system is given in block-diagram form in Fig. 5-16. Determine the value of the gain K in the system to just produce instability.

Solution

The open-loop transfer function of the system is

$$G(s)H(s) = \frac{K}{s(s^2 + s + 2)}$$

From Eq. (5-53) the system characteristic equation is

$$s^3 + s^2 + 2s + K = 0$$

The Routh array is found to be

1	2
1	K
2 − K	
K	

For stability K must be a positive number. A second necessary condition for stability is that the b_1 term $2 - K$ be also positive. If b_1 is negative, the system will be unstable. Thus the system just becomes unstable

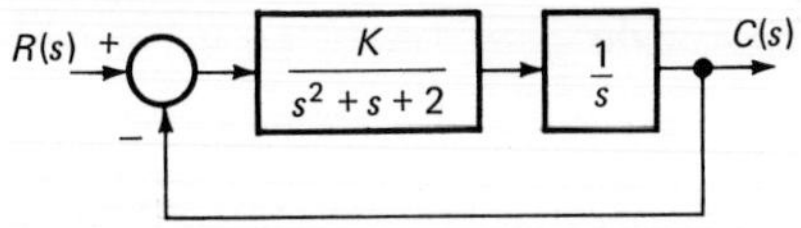

Figure 5-16 Closed-loop system for *Example 5-18*

at $b_1 = 0$. Thus

$$2 - K = 0$$

from which

$$K = 2$$

We conclude that the system is stable for values of $0 < K < 2$.

5-12 ROOT-LOCUS ANALYSIS

The preceding sections have pointed out that the transient response of a closed-loop system may be evaluated from a knowledge of the roots of the system characteristic equation. In particular, the roots must be either negative real numbers or complex numbers with negative real parts for the system to be stable.

The task of obtaining the roots of the characteristic equation becomes tedious as the complexity of the system increases. This is especially true in examining the transient response as some parameter of the system transfer function is changed. The usual parameter to be varied is the system gain, especially where the gain of an amplifier is easily changed. As the gain is increased, the response to a step input becomes more oscillatory and eventually becomes unstable. Hence gain is frequently adjusted to obtain best response.

A graphical method that provides means for determining all possible roots of the characteristic equation as the system gain is varied is the *root-locus method*. This method was devised by W. R. Evans in 1948.

The root-locus (locus of roots) diagram is the path of the roots on the complex or s plane as the gain is varied from 0 to ∞. A detailed study of the root-locus method is outside the scope of this text. Rather, we will attempt to introduce the concept underlying the method. The reader interested in techniques for constructing and analyzing root-locus diagrams is referred to a text of a more advanced nature.

Recall from Eq. (5-53) that the characteristic equation of a system may be expressed as $D(s) + N(s) = 0$, where $D(s)$ and $N(s)$ are the denominator and numerator of the open-loop transfer function $G(s)H(s)$. Thus a plot of the roots of Eq. (5-53) will provide a root-locus diagram. This concept will be illustrated by means of the following example.

EXAMPLE 5-19

Plot the root-locus diagram of the closed-loop system shown in Fig. 5-17.

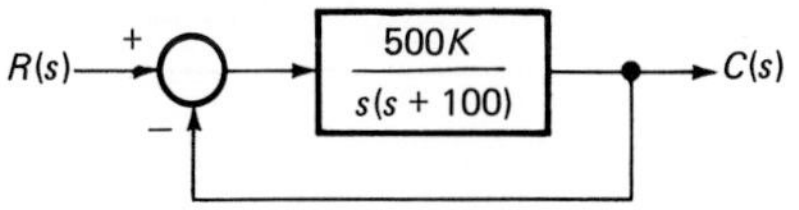

Figure 5-17 Closed-loop system for *Example 5-19*

Solution

The open-loop transfer function is

$$G(s)H(s) = \frac{500K}{s(s+100)} = \frac{N(s)}{D(s)}$$

from which the characteristic equation may be written as

$$D(s) + N(s) = 0$$

$$s(s + 100) + 500K = s^2 + 100s + 500K = 0$$

We now need to determine the roots of the above equation for $0 \leq K < \infty$. For $K = 0$, it is apparent that the roots are $s_1 = 0$ and $s_2 = -100$. These values are plotted on the complex plane shown in Fig. 5-18. The s_1 root is at the origin, and the s_2 root is on the negative real axis at -100. Since the characteristic equation is second-degree, roots for other values of K may be found by means of the quadratic equation. The roots for several selected values of loop gain are given in the table on page 220 and are also plotted in Fig. 5-18.

An examination of Fig. 5-18 shows that the roots move continuously along two branches of a locus as the gain is increased. Note that each branch originates at a value of s for which the denominator of $G(s)H(s)$

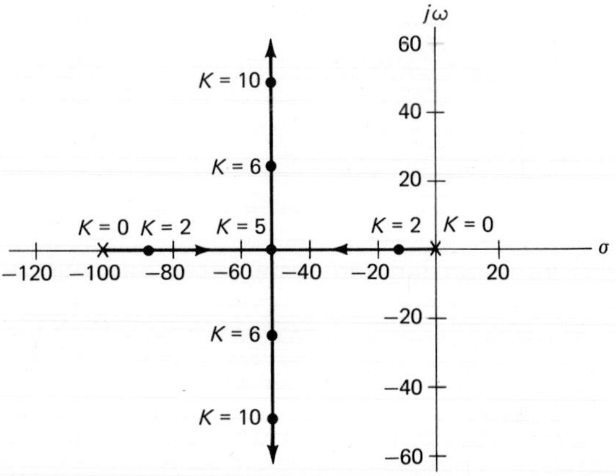

Figure 5-18 Root-locus diagram for Example 5-19

Gain	Roots of Characteristic Equation	
K	s_1	s_2
0	0	-100
2	-88.7	-11.3
5	-50	-50
6	$-50 + j22.4$	$-50 - j22.4$
10	$-50 + j50$	$-50 - j50$

is 0 (points designated by x). For values of $K < 5.0$, the roots are located in the negative real axis and the transient response will be that of an overdamped system. For $K > 5.0$, the roots are complex conjugates and the system response is underdamped. Critical damping occurs when the locus leaves the real axis since the roots then become complex. This occurs at $K = 5.0$ in the above example.

In Example 5-19 the root locations as the system gain is varied were calculated analytically only to demonstrate the nature of the root-locus method. In practice, the root-locus is rapidly obtained through application of a set of rules that aid in plotting. In addition a transparent protractor called the Spirule has been developed to facilitate constructing root-locus diagrams.

It has been stated that the location of the roots of the characteristic equation defines the transient behavior of the system. If all roots are in the left-half of the complex plane, the transient terms are decaying and the system is stable. If one or more roots exist in the right half of the plane, the system will be unstable. Values of gain for which roots exist on the imaginary axis represent the marginal stability condition.

For an underdamped second-order system, the damping ratio and damped frequency of oscillation can be found from the root locations associated with a particular gain. Referring to Eq. (5-24), the characteristic equation of a second-order system may be expressed as

$$s^2 + 2\zeta\omega_n s + \omega_n^2 = 0 \tag{5-58}$$

For an underdamped system the roots will be the complex conjugates

$$s_1, s_2 = -\zeta\omega_n \pm j\omega_n\sqrt{1 - \zeta^2} \tag{5-59}$$

For a given root the horizontal distance from the axis is $\zeta\omega_n$ [real part of Eq. (5-59)] and the vertical distance is the damped frequency $\omega_n\sqrt{1 - \zeta^2}$ [imaginary part of Eq. (5-59)]. This is shown in Fig. 5-19. The root location is also defined by the angle θ that is made with the

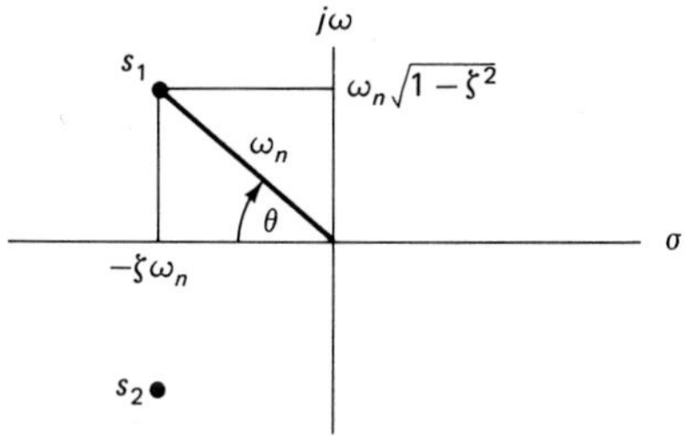

Figure 5-19 Location of complex roots on s plane

origin. This angle may be calculated from

$$\cos\theta = \frac{\zeta\omega_n}{\omega_n} = \zeta \tag{5-60}$$

Thus the damping ratio for a given gain may be easily found by computing the cosine of the angle θ. For example, if $K = 10$ in Example 5-19, the damped frequency of oscillation will be 50 rad/s. The angle θ is 45°, and the damping ratio is cos 45° = 0.707.

Figure 5-20 shows the form of the root-locus diagram for several common system transfer functions. The arrows on the loci indicate the direction of increasing gain.

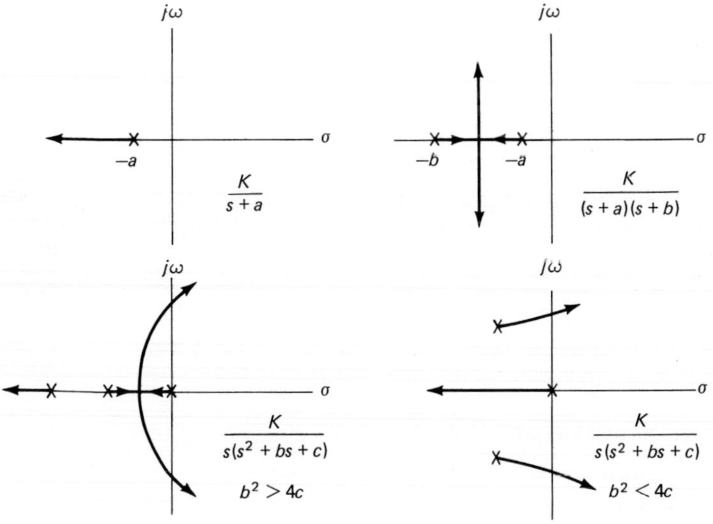

Figure 5-20 Common root-locus diagrams

PROBLEMS

5-1. Derive the Laplace transform of the exponential function.

$$f(t) = e^{-t}$$

5-2. Using Table 5-1, find the Laplace transform of the following functions of t.

a. $f(t) = t^4$ b. $f(t) = e^{-t} - e^{-2t}$

c. $f(t) = 4 \sin 2t$ d. $f(t) = 3e^{-t} \cos 10t$

5-3. Using Table 5-1, determine the inverse transform $f(t)$ of the following functions of s.

a. $F(s) = \dfrac{1}{s+3}$ b. $F(s) = \dfrac{s}{s^2+1}$

c. $F(s) = \dfrac{s-1}{s^2-2s+2}$ d. $F(s) = \dfrac{2}{s^2-6s+13}$

e. $F(s) = \dfrac{13}{s(s^2+4s+13)}$

5-4. Find the inverse transform of the following function.

$$F(s) = \frac{1}{s+1} - \frac{3}{s+10} + \frac{5}{s^2+2s+8}$$

5-5. Write the following differential equations in Laplace form. Solve the transformed equation for the independent variable as a function of s.

a. $\dfrac{dx}{dt} + 5x = 0$ with initial condition $x = 1$ at $t = 0[x(0) = 1]$

b. $\ddot{x} + 4\dot{x} + 3x = 10$ with $x(0) = -1$ and $\dot{x}(0) = 2$ (recall dot notation)

c. $\dfrac{d^2y}{dt^2} + 3\dfrac{dy}{dt} + y = t$ with $y = 0$ and $\dfrac{dy}{dt} = -2$ at $t = 0$

5-6. Apply the initial- and final-value theorems to the following functions of s.

a. $F(s) = \dfrac{1}{s(s+1)}$ b. $F(s) = \dfrac{1}{(s+1)(s+4)}$

c. $F(s) = \dfrac{5}{s(s^2 + 2s + 5)}$ d. $F(s) = \dfrac{10}{(s + 5)^3}$

e. $F(s) = \dfrac{10(s^2 + 2s + 1)}{5(0.1s^2 + s + 1)}$

5-7. Solve the following differential equations using the Laplace transform method.

a. $\dfrac{dy}{dt} + 10y = 2$ with $y = 0$ at $t = 0$

b. $\ddot{x} + 3\dot{x} + x = 5$ with $x(0) = 0$ and $\dot{x}(0) = 0$

c. $\dot{q} + 100q = 300$ with $\dot{q}(0) = 50$

Hint: As a first step determine $q(0)$.

5-8. Using the Laplace transform method, determine the equation for the current i in the circuit below. Assume that the capacitor is initially uncharged.

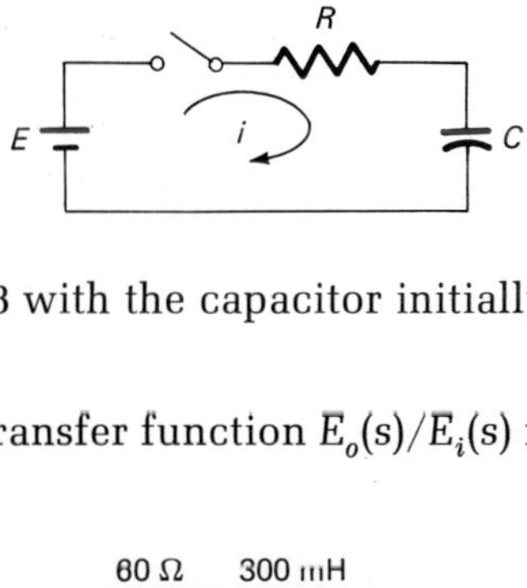

Prob. 5-8

5-9. Repeat Prob. 5-8 with the capacitor initially charged to a voltage V_o.

5-10. Determine the transfer function $E_o(s)/E_i(s)$ for each of the following circuits.

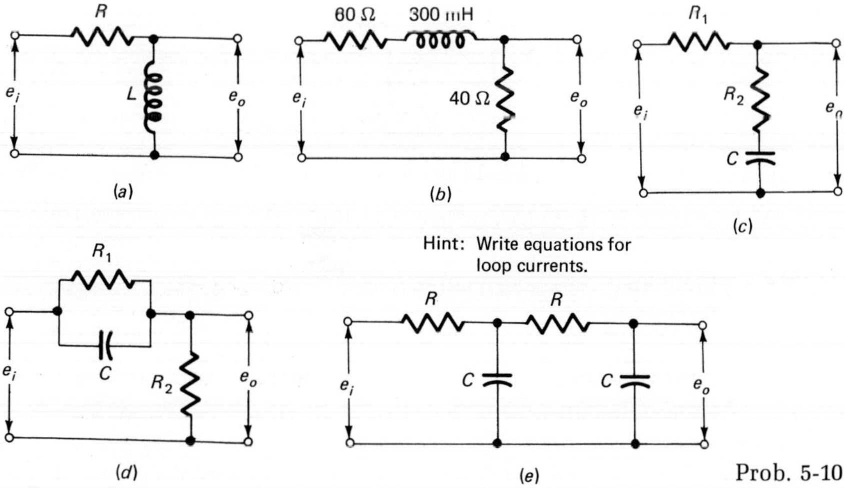

Hint: Write equations for loop currents.

Prob. 5-10

5-11. Write the differential equation in Laplace form for the motion of the mass in the system below, where the mass is given an initial displacement X_o from its equilibrium point and then released.

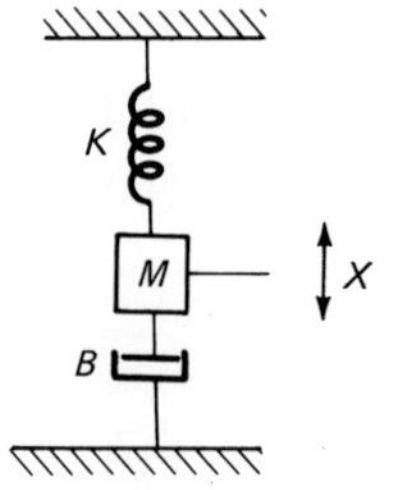

Prob. 5-11

5-12. Determine the transfer function $\theta_o(s)/\theta_i(s)$ for the rotational system below (recall Example 2-2).

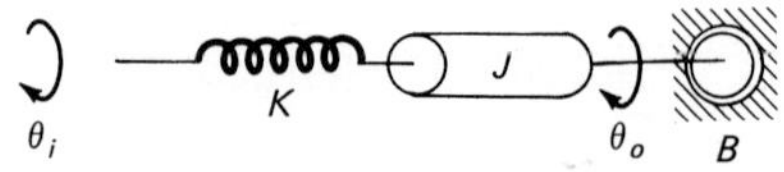

Prob. 5-12

5-13. An insulated tank of water is heated by an electric element. The water temperature is T_w, and the rate of heat flow from the element is q_h. Determine the transfer function $T_w(s)/Q_h(s)$ assuming the insulation thermal resistance is R (recall Example 2-4).

5-14. For the liquid system below, determine the transfer function $Q_o(s)/Q_i(s)$. The hydraulic resistance R is constant.

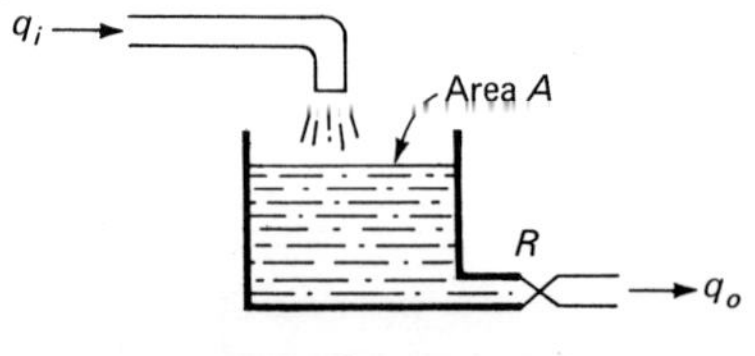

Prob. 5-14

5-15. A dc motor with field control has the following specifications:

Combined rotor and load inertia	0.03 oz-in.-s^2
Damping	0.1 oz-in./(rad/s)
Torque constant	300 oz-in./A
Field resistance	50 Ω
Field inductance	5 H

Obtain the transfer function of the motor in terms of the field voltage v_f and the velocity of the output shaft $\dot{\theta}_o$.

5-16. Determine the equation for the time response of the output speed of the motor in Prob. 5-15 if 2.5 V is suddenly applied to the field winding when the motor is at rest.

5-17. A block diagram for a velocity servomechanism is shown below. Obtain the transfer function $\dot{\theta}(s)/V_i(s)$.

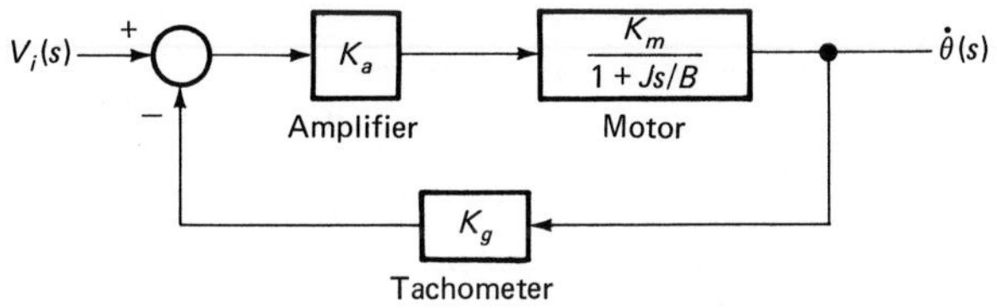

Prob. 5-17

5-18. Classify the feedback system shown in general form according to system type number if:

	$G(s)$	$H(s)$
a.	$\dfrac{10}{s}$	1
b.	$\dfrac{5}{s(0.1s + 1)}$	1
c.	$\dfrac{10}{s(s^2 + 2s + 5)}$	0.5
d.	$\dfrac{5}{(0.2s + 1)(0.01s + 1)}$	0.01s
e.	$\dfrac{5}{s(0.1s + 1)}$	$\dfrac{1}{s}$

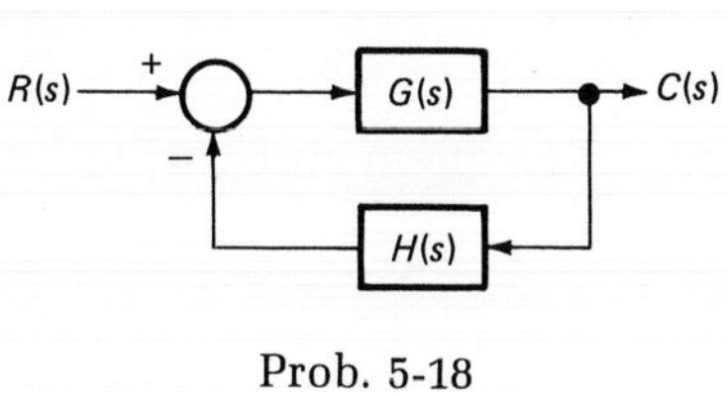

Prob. 5-18

5-19. For the systems given in Prob. 5-18a, b, and c find the position, velocity, and acceleration error constants and the steady-state error for a unit step input, unit ramp input, and unit parabolic input.

5-20. Determine the characteristic equation of each of the systems given in Prob. 5-18.

5-21. The roots of the characteristic equations of several systems are given below. For each case determine if the set of roots represents a stable, marginally stable, or unstable system.

a. $-3, -6$　　b. $-2, +1$
c. $-1, 0$　　d. $-1 + j, -1 - j$
e. $-1, 1, -j, j$　　f. $-2 + j3, -2 - j3, 1$

5-22. By means of Routh's criterion, determine if each of the following characteristic equations represents a stable system.

a. $s^3 + 4s^2 + 8s + 12 = 0$
b. $2s^3 + 4s^2 + 4s + 12 = 0$
c. $s^4 + 2s^3 + 8s^2 + 4s + 3 = 0$
d. $s^4 + 2s^3 + s^2 + 4s + 2 = 0$

5-23. The open-loop transfer function of a system is given by

$$G(s)H(s) = \frac{K}{s(s^2 + 2s + 4)}$$

where K is an adjustable loop gain. Determine the values of K for which the system is stable.

5-24. Repeat Prob. 5-23 for

$$G(s)H(s) = \frac{K}{s(s^2 + 4s + 8)}$$

5-25. Plot the root-locus diagram of the system whose open-loop transfer function is $G(s)H(s) = K/(s + 4)$.

5-26. Plot the root-locus diagram of the system whose open-loop transfer function is $G(s)H(s) = K/[(s + 1)(s + 5)]$. For what value of K will the damping ratio of the closed-loop system be $\zeta = 0.5$? What is the damped frequency of oscillation in this case?

Chapter Six

frequency-response analysis

6-1 INTRODUCTION

In the preceding chapters physical systems have been characterized by differential equations and the solution of these equations has provided a means of determining the dynamic behavior of the system. In particular the Laplace transform has been used as a method of solving differential equations and evaluating transient response.

Although solving differential equations is useful in analyzing performance in the time domain, this approach proves cumbersome for other than relatively simple systems. Therefore it is desirable to be able to predict performance by a method that does not actually require solving differential equations. Also the method should readily indicate the manner in which a system may be changed to meet performance specifications. In many cases this is not easily discerned from the system differential equations. If the equations are solved and the response does not meet specifications, it is not easy to decide the manner in which the system should be changed to obtain the desired results.

A method that overcomes the above limitations and has become well established in the design of control systems is to analyze the steady-state sinusoidal response of the system transfer function. This method is known as *frequency-response analysis,* and, by graphical means, it can indicate the manner in which a system should be modified.

The frequency-response method analyzes performance as a function of the frequency of the sinusoidal input signal rather than examining the actual time response. In addition to being advantageous in system design, this type of analysis has several other features. It allows one to evaluate the range of frequencies that affect system performance. As an illustration, consider the problem of electrical noise to which a control system may be subjected. Analyzing the frequency response permits the design of a passband of frequencies for the system that excludes the noise and thereby improves performance.

Another advantage of the frequency-response method is that the response can be determined experimentally as well as by means of analytical techniques. This is important in cases where it is difficult to describe parts of the system by differential equations.

In this chapter we will primarily concern ourselves with the graphical methods of analyzing frequency response. We will first consider frequency-response curves for the basic forms of transfer functions with emphasis on Bode (pronounced Bō′-dē) plots. These are plots that can be quickly drawn to approximate the actual frequency-response curves. Bode plots will also be used in determining closed-loop performance from open-loop response curves.

The determination of frequency-response characteristics by experimental methods will be described in Chap. 8.

6-2 AC CIRCUIT ANALYSIS

An input signal frequently used in studying electrical networks is the sinusoidal function given by Eq. (4-3). A sinusoidal voltage applied to an electric circuit may be expressed by the equation

$$e = E_m \sin \omega t \tag{6-1}$$

where E_m is the peak amplitude and ω is the angular frequency in radians per second. A voltage of this type is commonly known as an *alternating* or *ac* voltage.

A linear system has the property that the steady-state response to a sinusoidal input will also be sinusoidal with a frequency identical to that of the input but with a different amplitude and phase characteristic. Thus, the response in a linear circuit excited by the voltage of Eq. (6-1) will be an alternating current of the form

$$i = I_m \sin (\omega t + \phi) \tag{6-2}$$

where I_m is the peak value and ϕ is a phase angle or difference between the voltage and current waveforms. This relationship is illustrated by the waveforms in Fig. 6-1, where a negative or lagging phase angle is shown for the current. The phase difference is usually expressed in degrees and is taken as positive when the response leads the excitation and negative when the response lags the input.

The magnitude and phase angle of the response variable will be determined by both the parameters of the system and the frequency of the input sine wave. Since the response depends on frequency, we may regard frequency as the independent variable. In this case, it can be said that we are analyzing system performance in the *frequency*

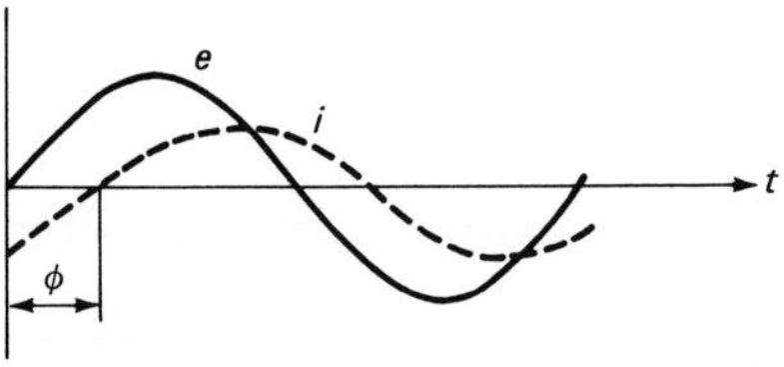

Figure 6-1 Sinusoidal voltage and current waveforms

domain. It is important to note that time does not explicitly appear as a variable in frequency-domain analysis.

The concept of impedance is widely applied in the analysis of electric circuits in the frequency domain. It provides a convenient means of determining the steady-state solution of circuit problems involving sinusoidal voltages and currents. Ac quantities are represented by complex numbers involving the operator $j = \sqrt{-1}$, and the circuit is analyzed using complex-number calculations.

As an example of this method consider the series resistance-inductance circuit shown in Fig. 6-2. If e is a sinusoidal voltage source as defined by Eq. (6-1), the current i will be of the form of Eq. (6-2).

The magnitude of an ac quantity is frequently expressed by an effective or rms[1] value. For an ac voltage, this is given as

$$E = \frac{E_m}{\sqrt{2}} \tag{6-3}$$

where E is the rms value.

The ac voltage can be written in complex polar form as a function of magnitude and phase angle. Using boldface to designate a complex

1 *Rms is an abbreviation for the root-mean-square value, which, for a sinusoidal function, is equal to $1/\sqrt{2}$ times the peak value.*

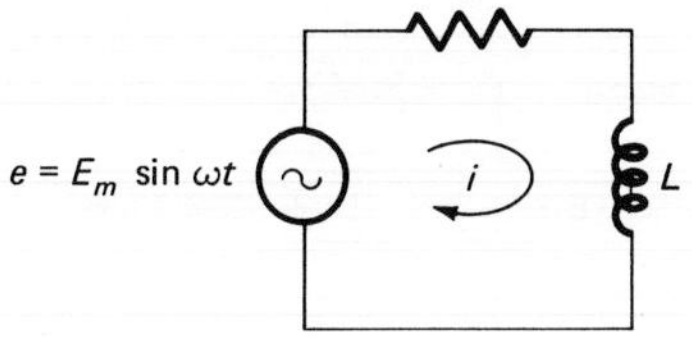

Figure 6-2 Series resistance-inductance circuit

quantity, we have

$$\mathbf{E} = E \angle \phi \tag{6-4}$$

The ac current in the circuit may be determined by applying Ohm's law. Thus

$$\mathbf{I} = \frac{\mathbf{E}}{\mathbf{Z}} = \frac{\mathbf{E}}{R + j\omega L} \tag{6-5}$$

where $j\omega L$ is the inductive reactance and $R + j\omega L$ is the complex impedance **Z**. Because of the presence of the complex operator j, the current **I** is a complex quantity. It can be written in polar form as

$$\mathbf{I} = I \angle \theta = \frac{E}{\sqrt{R^2 + (\omega L)^2}} \angle -\tan^{-1}\frac{\omega L}{R} \tag{6-6}$$

In Eq. (6-6) the current is a function of the frequency ω. As the frequency is varied, the magnitude and phase angle of the current will change. A graph of the magnitude and phase angle plotted as a function of frequency will show the frequency response of the circuit.

It should be noted that the current as given by Eq. (6-6) does not contain time as a variable. However, we should keep in mind that the current can also be expressed in the time domain according to Eq. (6-2). Thus

$$i = \frac{E_m}{\sqrt{R^2 + (\omega L)^2}} \sin\left(\omega t - \tan^{-1}\frac{\omega L}{R}\right) \tag{6-7}$$

EXAMPLE 6-1

Determine the alternating current in the series circuit shown below, where the input voltage is the sinusoidal function

$$e = 120\sqrt{2} \sin 377t \qquad \text{V}$$

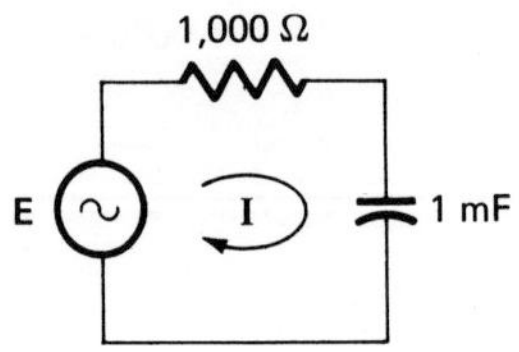

Figure 6-3 Series circuit for Example 6-1

Solution

The input voltage may be expressed in complex polar form as

$$\mathbf{E} = 120 \angle 0 \text{ V}$$

The current is then

$$\mathbf{I} = \frac{\mathbf{E}}{\mathbf{Z}} = \frac{\mathbf{E}}{R + 1/j\omega C} = \frac{120 \angle 0}{1{,}000 + 1/j377 \times 10^{-6}}$$

$$= \frac{120 \angle 0}{1{,}000 - j2{,}652} = \frac{120 \angle 0}{2{,}834 \angle -69.3^\circ}$$

$$= 42.3 \angle 69.3^\circ \text{ mA}$$

6-3 TRANSFORMING DIFFERENTIAL EQUATIONS FOR SINUSOIDAL INPUTS

Time-domain and frequency-domain analysis lead to two different though related forms of response for sinusoidal signals. If we compare the two methods of analysis, we find that the equations have similarities.

To illustrate this point let us consider the circuit that was shown in Fig. 6-2, which consists of a resistor and an inductor connected in series to an ac voltage. Referring to Eq. (2-21), the differential equation for the circuit is

$$e = L\frac{di}{dt} + Ri \tag{6-8}$$

Equation (6-8) can be treated by the Laplace transform method and the resulting equation solved for the current $I(s)$. Thus

$$I(s) = \frac{E(s)}{R + Ls} \tag{6-9}$$

Using frequency analysis, the alternating current as given by Eq. (6-5) is

$$\mathbf{I} = \frac{\mathbf{E}}{R + j\omega L}$$

or

$$I(j\omega) = \frac{E(j\omega)}{R + j\omega L} \tag{6-10}$$

where $I(j\omega)$ and $E(j\omega)$ also designate the variables as complex quantities.

It can be seen that Eqs. (6-5) and (6-9) are similar in form. However, they differ in this respect. In Eq. (6-9) the variable quantities are functions of the Laplace operator s, while in Eq. (6-5) they are functions of frequency ω. In Eq. (6-5) the Laplace operator s has been replaced by the complex operator $j\omega$. This parallelism can be formalized in the following transformation:

$$s \to j\omega$$

It is important to note the simplicity with which the steady-state solution of a differential equation may be obtained when the excitation is sinusoidal. It is only necessary to transform the equation to the Laplace domain, replace s by $j\omega$, and solve the resulting expression using complex-number calculations.

The transformation of a differential equation in the time domain to an expression in the frequency domain can be applied to any linear system. It can be shown by rigorous methods that an algebraic equation still exists when the Laplace operator s of order n is replaced by the complex operator $j\omega$ of the same order. This algebraic equation can be solved for the dependent variable of interest, and the solution yields the complex-number representation of the steady-state response to a sinusoidal input function.

This transformation procedure is outlined in the following steps:

1. Write the differential equation of the system.
2. Transform the equation to the Laplace domain.
3. Replace the Laplacian operator s^n by the complex operator $j\omega^n$ to transform the equation to the frequency domain.

EXAMPLE 6-2

Express in the frequency domain the transfer function represented by the differential equation

$$10\frac{d^2y}{dt^2} + 5\frac{dy}{dt} + 10y = x$$

Solution

Performing the Laplace transformation gives

$$(10s^2 + 5s + 10)Y(s) = X(s)$$

The transfer function is then

$$\frac{Y(s)}{X(s)} = \frac{1}{10s^2 + 5s + 10}$$

This transfer function may be transformed to the frequency domain by replacing s by $j\omega$. Thus

$$\frac{Y(j\omega)}{X(j\omega)} = \frac{1}{10(j\omega)^2 + j5\omega + 10} = \frac{1}{-10\omega^2 + j5\omega + 10}$$

6-4 POLAR FREQUENCY-RESPONSE PLOTS

We have seen that the value of a variable graphed as a function of time is a convenient way of showing behavior in the time domain. A frequency-response curve showing the magnitude and phase angle of a variable as a function of frequency is likewise useful in analyzing behavior in the frequency domain. The frequency-response curves that currently find widest use are the Bode plots mentioned in Sec. 6-1. These plots are named after H. W. Bode of Bell Telephone Laboratories, who applied them extensively in the design of feedback amplifiers. Before discussing Bode plots in detail, we will briefly describe an earlier form of frequency-response plot that has been applied in control systems analysis.

Frequency-response data may be plotted on a two-dimensional plane called the *complex* or *polar plane*, and the resulting curve is known as a *polar plot*. The x axis of this plane is identified as the *real axis*, and the y axis is called the *imaginary* or *j axis*.

If a transfer function in the Laplace domain is identified as G(s), the corresponding transfer function in the frequency domain can be expressed as $G(j\omega)$. This function can be written using polar coordinates such as $|G| \angle\phi$, where $|G|$ is the magnitude and $\angle\phi$ is the phase angle of $G(j\omega)$. For a given frequency, $G(j\omega)$ can be treated as a vector quan-

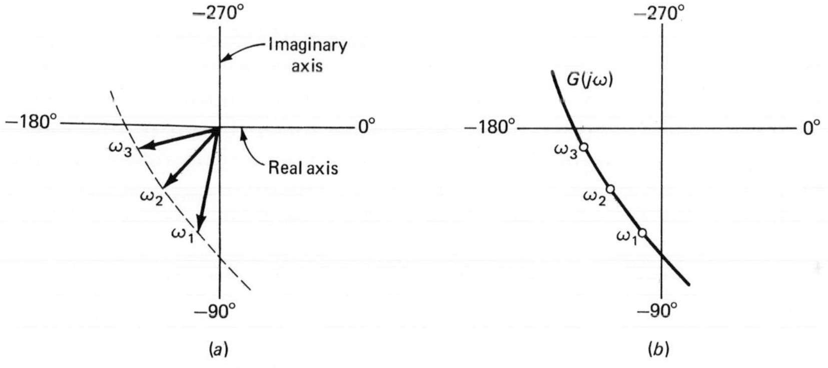

Figure 6-4 (a) Vector-locus and (b) polar plot in the complex plane

tity whose length corresponds to $|G|$ and whose angle with respect to the positive real axis of the complex plane corresponds to $\angle\phi$. This is shown in Fig. 6-4*a*, where the three vectors represent $G(j\omega)$ for three values of the frequency ω. Thus a set of values of $G(j\omega)$ for ω between 0 and ∞ yields a locus of vectors. The tips of these vectors may be connected by a curve as shown in Fig. 6-4*b*, and this curve is then the polar plot of the transfer function. This plot is sometimes identified as a *Nyquist diagram* after H. Nyquist, who did much early work relating the performance of closed-loop systems to the form of their polar plot.

EXAMPLE 6-3

Determine the polar plot for the transfer function of the circuit shown in Fig. 6-5.

Solution

Referring to Example 5-8, the transfer function of the circuit is

$$G(s) = \frac{E_o(s)}{E_i(s)} = \frac{1}{1 + RCs} = \frac{1}{1 + 0.1s}$$

Making the substitution $s = j\omega$, the sinusoidal steady-state transfer function is

$$G(j\omega) = \frac{E_o(j\omega)}{E_i(j\omega)} = \frac{1}{1 + j0.1\omega}$$

A table may be constructed giving $|G|$ and the $\angle\phi$ for various values of ω. To illustrate the calculations, consider the case $\omega = 10$. Then

$$G(j\omega) = \frac{1}{1 + j1} = \frac{1}{1.41 \angle 45^\circ} = 0.71 \angle -45^\circ$$

The data from the table are plotted on the complex plane with ω as a parameter. The resulting polar diagram is shown in Fig. 6-6.

The above example will perhaps indicate the limitations of polar

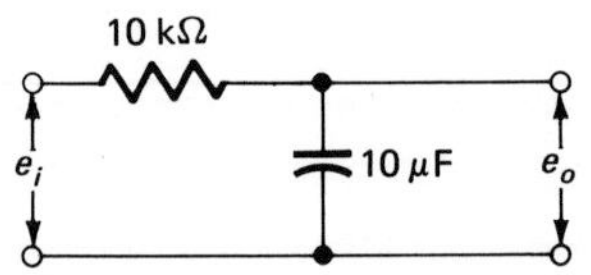

Figure 6-5 Circuit for Example 6-3

TABLE FOR EXAMPLE 6-3

ω, *rad/s*	|G|	∠φ, °
0	1.00	0.0
2	0.98	−11.3
5	0.89	−26.6
10	0.71	−45.0
20	0.45	−63.4
40	0.24	−76.0
∞	0.00	−90.0

plots. For other than simple functions, the calculations to determine the frequency response are tedious. The addition or modification of an element in a system requires the recalculation of the response data. Furthermore the polar plot does not readily indicate the effect of an individual term in the transfer function. Bode diagrams overcome many of these limitations and simplify the plotting of frequency-response data. Hence they will form the basis of the frequency-response analysis to be presented in this chapter.

6-5 BODE DIAGRAMS

Bode diagrams are two separate graphs in rectangular-coordinate form that show (1) the logarithm of the magnitude of the variable plotted against the logarithm of frequency and (2) the phase angle of the variable plotted against the logarithm of frequency. The importance of using a logarithmic frequency scale, which equally emphasizes all

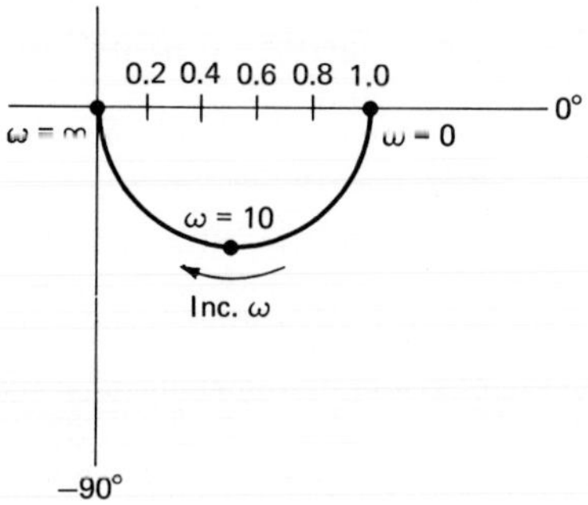

Figure 6-6 Polar plot for Example 6-3

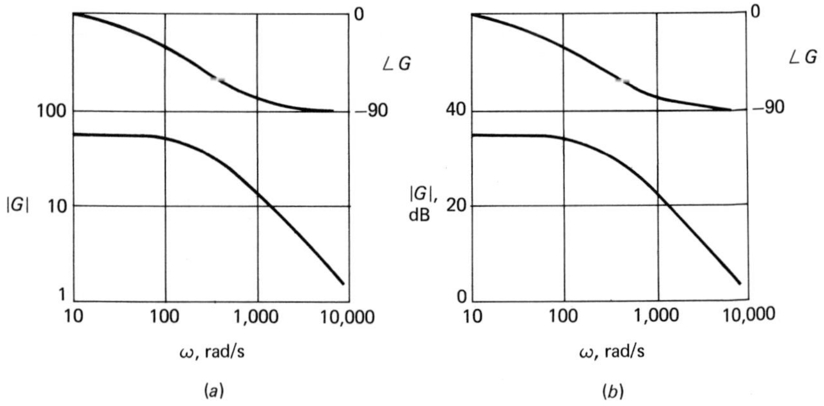

Figure 6-7 Bode diagrams with (a) logarithmic magnitude scale and (b) decibel magnitude scale

frequencies, can be indicated by a simple example. For a typical control system, the frequency band of interest may extend from 1 to 10 Hz. However, for purposes of analysis and design, it may be necessary to consider the range from 0.1 to 100 Hz. On a linear scale, the most important frequencies (1 to 10 Hz) would occupy less than one-tenth of the graph. By plotting frequency on a logarithmic scale, these frequencies can be made one-third of the range. Note that dc (frequency $= 0$) will not appear on the graph since it occurs at $\log 0 = -\infty$.

The same justification applies to using a logarithmic magnitude scale since the magnitude may also range over a number of decades. However, since the phase angle covers a smaller range, it may be plotted on a linear scale. Figure 6-7*a* illustrates a Bode diagram using logarithmic frequency and magnitude scales.

In both electronic and control systems work a common way of expressing magnitudes is in logarithmic units known as *decibels*, abbreviated dB.* The decibel is defined by the equation

$$\text{Mag} = 20 \log_{10} (G) \text{ dB} \tag{6-11}$$

where Mag is used to represent the magnitude of G expressed in decibels.

* *The term decibel as used originally in communications engineering referred to the ratio of two values of power. In recent years the term has been applied rather loosely in characterizing the ratio of other quantities. Where the two quantities have like units, the ratio is dimensionless. However, in controls work the input and output quantities defining G may not possess the same units; hence the ratio need not be dimensionless. Needless to say, an overall consistency of units within the system must be maintained.*

If the magnitude is greater than unity, the resulting decibel value is positive. If the magnitude is less than unity, the resulting decibel value is negative. A magnitude of unity gives 0 dB.

EXAMPLE 6-4

Find the magnitude in decibels of the ratio of E_o/E_i for the following values:

1. $E_o = 2$ V, $E_i = 2$ V
2. $E_o = 4$ V, $E_i = 2$ V
3. $E_o = 1$ V, $E_i = 2$ V
4. $E_o = 0.7$ V, $E_i = 2$ V

Solution

1. $\text{Mag} = 20 \log_{10} (\frac{2}{2}) = 20 \log_{10} 1 = 20 \times 0 = 0 \text{ dB}$

2. $\text{Mag} = 20 \log_{10} (\frac{4}{2}) = 20 \log_{10} 2 = 20 \times 0.3 = 6 \text{ dB}$

3. $\text{Mag} = 20 \log_{10} (\frac{1}{2})$

Here apply the logarithmic rule that $\log_{10} A/B = \log_{10} A - \log_{10} B$. Thus

$$\begin{aligned} \text{Mag} &= 20 \log_{10} 1 - 20 \log_{10} 2 \\ &= 0 - 6 \text{ dB} = -6 \text{ dB} \end{aligned}$$

4. $\text{Mag} = 20 \log_{10} (\frac{0.7}{2}) = 20 \log_{10} 0.35$

It is usually easier to express a decimal quantity as a fraction and then subtract logs.

$$\begin{aligned} 20 \log_{10} 0.35 &= 20 \log_{10} (\tfrac{35}{100}) \\ &= 20 \log_{10} 35 - 20 \log_{10} 100 \\ &= 20 \times 1.544 - 20 \times 2 = 30.9 - 40 = -9.1 \text{ dB} \end{aligned}$$

The use of the decibel eliminates the need for graph paper with both axes logarithmic since the decibel is a logarithmic unit. Thus, magnitude frequency-response curves can be plotted on semilogarithmic coordinate paper. Since the phase-shift data are plotted on a linear scale, they can also be shown on semilog paper. Quite often the amplitude and phase-shift data are plotted on the same set of coordinates.

TABLE 6-1 COMMON DECIBEL VALUES

G	Mag, dB
0.1	−20
0.5	−6
0.7	−3
1.0	0
1.4	3
2.0	6
10.0	20

This is shown in Fig. 6-7*b* in which the curves of Fig. 6-7*a* are reproduced using a decibel magnitude scale.

Bode diagrams are especially useful in the analysis and design of control systems using frequency-response methods. The response of first- and second-order systems can be quickly plotted with reasonable accuracy by means of straight-line approximations. Secondly, in a cascaded system, the overall magnitude and phase angle at a given frequency can be determined by graphically adding the ordinates associated with the individual elements in the system. This comes from the fact that the multiplication of two or more numbers may be accomplished through addition of the logarithms of the numbers.

Certain magnitudes are frequently encountered in working with Bode diagrams, and these are noted in Table 6-1.

6-6 GENERAL FREQUENCY-RESPONSE CURVES

Recall that the ratio of the output of a component or system to its input is defined as a transfer function. Although a transfer function can be exceedingly complex, it can be written in factored form as the ratio of polynomials as was indicated by Eq. (5-35). There are four basic factors that appear in a transfer function in the frequency domain. These are:

$$\text{Constant factor } K \tag{6-12}$$

$$\text{Integral or derivative factor } (j\omega)^{\pm n} \tag{6-13}$$

$$\text{First-order factor } (1 + j\omega\tau)^{\pm m} \tag{6-14}$$

$$\text{Second-order factor } \left[1 + j\frac{2\zeta\omega}{\omega_n} + \left(\frac{j\omega}{\omega_n}\right)^2\right]^{\pm p} \tag{6-15}$$

The various factors may appear in the numerator or the denominator of the transfer function in any combination. Thus a transfer function can be expressed by a general equation of the form

$$G(j\omega) = \frac{K(1 + j\omega\tau_1)(1 + j\omega\tau_2)^m \cdots}{(j\omega)^n(1 + j\omega\tau_a)[1 + j(2\zeta\omega/\omega_n) + (j\omega/\omega_n)^2]^p \cdots} \tag{6-16}$$

A positive exponent indicates that the factor appears in the numerator, while a negative exponent indicates that the factor lies in the denominator. The factors may appear raised to some integral power other than unity. If desired, the factors can be combined by simple multiplying techniques.

The advantage of identifying the individual factors in a transfer function is that the frequency response of each can be separately plotted using simple techniques. If the factors are plotted using a logarithmic magnitude scale, their product (which is the overall magnitude response) can be obtained graphically by simple addition or subtraction. The overall phase angle can also be found by graphical addition or subtraction. With this technique the response of a number of elements in a system can be readily combined to determine the response of the total system.

6-7 PLOT OF A CONSTANT FACTOR

The first type of factor to be considered is a constant. Constants are defined as terms whose magnitudes do not vary with frequency and whose phase angles do not exhibit any phase shift. Since a transfer function may be written as a product of factors, a single constant K can represent the product of all the constant terms in the transfer function. Thus, the magnitude of a constant K is given by

$$\text{Mag} = 20 \log_{10} K \text{ dB} \tag{6-17}$$

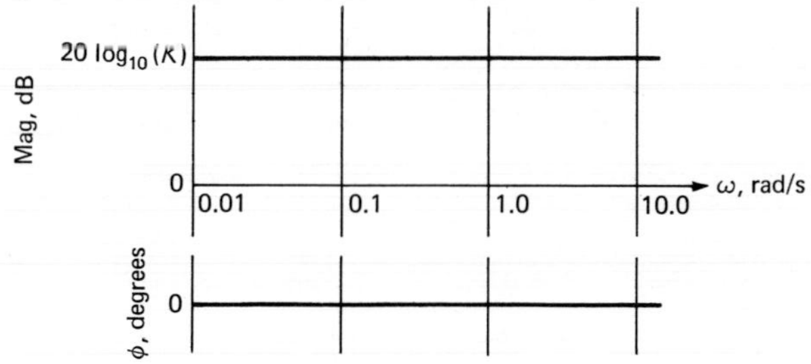

Figure 6-8 Frequency-response plot of constant factor K

and the phase angle is

$$\phi = 0° \tag{6-18}$$

Frequency-response curves represented by Eqs. (6-17) and (6-18) are plotted in Fig. 6-8.

6-8 PLOT OF AN INTEGRAL OR DERIVATIVE FACTOR

The magnitude plot of the factor $j\omega$ is given by

$$\text{Mag} = 20 \log_{10} |j\omega| = 20 \log_{10} \omega \text{ dB} \tag{6-19}$$

The magnitude plot of the factor $(j\omega)^{-1}$ is given by

$$\text{Mag} = 20 \log_{10} \left|\frac{1}{j\omega}\right| = -20 \log_{10} \omega \text{ dB} \tag{6-20}$$

Equations (6-19) and (6-20) can be extended to the general case $(j\omega)^{\pm n}$. Thus

$$\text{Mag} = \pm 20n \log_{10} \omega \text{ dB} \tag{6-21}$$

Equation (6-21) is plotted for several values of n in Fig. 6-9*a*.

Notice that the magnitude curves focus about $\omega = 1$ rad/s since the magnitude in all cases is 0 dB at this frequency. Since the magnitude plots are a function of ω and n only, they are straight lines passing through $\omega = 1$ with slopes of $\pm 20n$ dB per decade, where a decade is defined as a frequency variation by a factor of 10. Thus, plots of $(j\omega)^{\pm 1}$

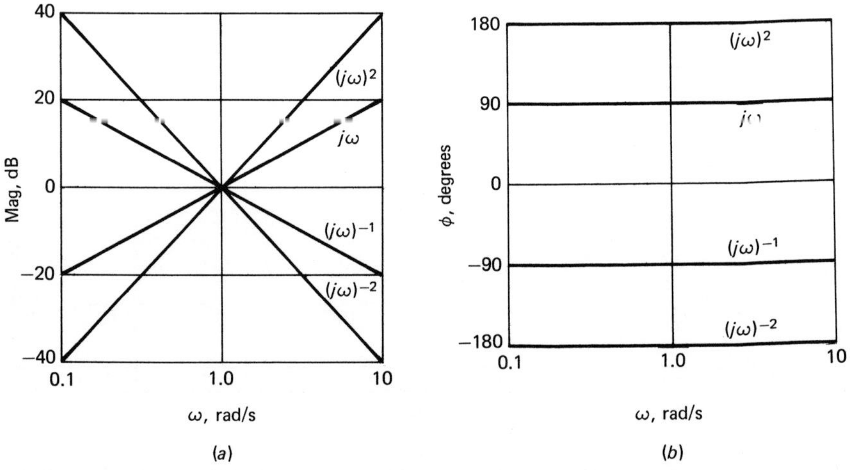

Figure 6-9 Frequency-response plot of factor $(j\omega)^{\pm n}$

have slopes of ±20 dB per decade, plots of $(j\omega)^{\pm 2}$ have slopes of ±40 dB per decade, etc.[1]

The phase angle for $(j\omega)^{\pm n}$ is given by

$$\phi = \pm n \angle \tan^{-1} j\omega = \pm n(90°) \tag{6-22}$$

The phase angle as a function of frequency for several cases is shown in Fig. 6-9*b*. Note that the phase angle is constant with frequency and is a multiple of ±90°. The power of *n* to which $j\omega$ is raised determines the multiple of 90° that represents the phase angle. The sign of the phase angle is determined by the sign of *n*.

6-9 PLOT OF A FIRST-ORDER FACTOR

The first-order factor $(1 + j\omega\tau)^{\pm m}$ has a definite magnitude and phase-shift variation with frequency. Let us first consider the basic form $(1 + j\omega\tau)$. The magnitude ratio is given by

$$\text{Mag} = 20 \log_{10} |(1 + j\omega\tau)| \text{ dB} \tag{6-23}$$

Before we can evaluate Eq. (6-23), we must change $(1 + j\omega\tau)$ to polar form. Rewriting the factor in polar form gives

$$(1 + j\omega\tau) = \sqrt{1^2 + (\omega\tau)^2} \angle \tan^{-1}(\omega\tau) \tag{6-24}$$

The square root portion of Eq. (6-24) represents the magnitude of $(1 + j\omega\tau)$, while the angle portion represents the phase angle. Thus we can write the magnitude of $(1 + j\omega\tau)$ in decibels as

$$\text{Mag} = 20 \log_{10} \sqrt{1^2 + (\omega\tau)^2} \text{ dB} \tag{6-25}$$

The phase angle is given by

$$\phi = \tan^{-1}(\omega\tau)° \tag{6-26}$$

Since the horizontal axis of the frequency-response curve is a function of frequency, it is convenient to express τ in terms of frequency. To do this we will define a cutoff frequency ω_c as

$$\omega_c = \frac{1}{\tau} \tag{6-27}$$

Thus

$$\text{Mag} = 20 \log_{10} \sqrt{1^2 + \left(\frac{\omega}{\omega_c}\right)^2} \text{ dB} \tag{6-28}$$

1 *The slopes may alternatively be expressed in terms of ±6n dB per octave, where an octave is defined as a frequency change by a factor of 2. Thus, the frequency band from 10 to 20 Hz is one octave in width.*

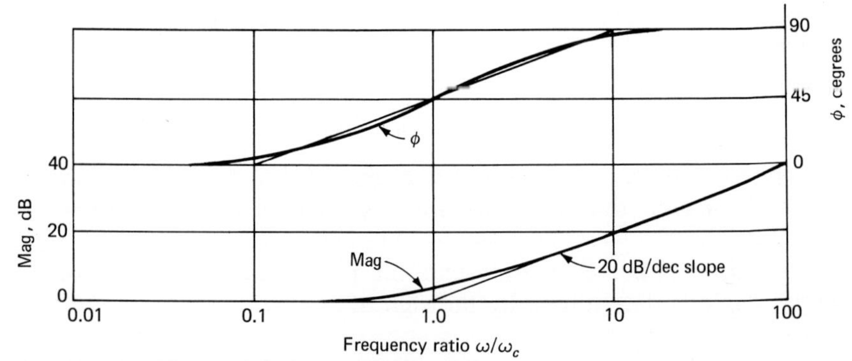

Figure 6-10 Frequency-response plot of first-order factor $(1 + j\omega\tau)$

$$\phi = \tan^{-1}\left(\frac{\omega}{\omega_c}\right)^\circ \tag{6-29}$$

Equations (6-28) and (6-29) are plotted in Fig. 6-10 using the data in Table 6-2. In this table, the magnitude and phase angle are calculated by allowing the frequency ω to take on various values as a percentage of ω_c. This gives a frequency scale normalized in terms of multiples of the cutoff frequency.

From an examination of Table 6-2 and Fig. 6-10 we can develop

TABLE 6-2 MAGNITUDE AND PHASE ANGLE OF $(1 + j\omega\tau)$ FOR SELECTED FREQUENCIES

$\frac{\omega}{\omega_c}$	$1 + \frac{j\omega}{\omega_c}$	*Magnitude* $\sqrt{1^2 + \left(\frac{\omega}{\omega_c}\right)^2}$	$20 \log_{10} \sqrt{1^2 + \left(\frac{\omega}{\omega_c}\right)^2}$, dB	*Phase Angle* $\tan^{-1} \frac{\omega}{\omega_c}$, °
0	$1 + j0$	1	0	0
0.01	$1 + j0.01$	≈1	≈0	+0.6
0.1	$1 + j0.1$	1.005	+0.04	+5.7
0.2	$1 + j0.2$	1.02	+0.17	+11.3
0.5	$1 + j0.5$	1.12	+0.97	+26.6
1	$1 + j1$	1.41	+3.0	+45.0
2	$1 + j2$	2.24	+7.0	+63.4
5	$1 + j5$	5.10	+14.1	+78.7
10	$1 + j10$	10.05	+20.0	+84.3
20	$1 + j20$	20.02	+26.1	+87.2
50	$1 + j50$	≈50	+34.0	+88.8
100	$1 + j100$	≈100	+40.0	+89.4
1,000	$1 + j1,000$	≈1,000	+60.0	≈+90

general guidelines for plotting a first-order term. Considering Fig. 6-10, the magnitude is essentially 0 dB for $\omega < 0.1\omega_c$. For $\omega > 10\omega_c$ the magnitude curve is essentially a straight line whose slope is 20 dB per decade of frequency.

Thus, for $\omega = 100$ the magnitude ratio is +40 dB, for $\omega = 1{,}000$ the magnitude ratio is +60 dB, etc. It should be noted that if the 20 dB per decade line is extended downward, it crosses the 0-dB axis at the frequency $\omega = \omega_c$. For this reason ω_c is called the *corner frequency*. Note, however, that at the corner frequency the actual magnitude ratio is +3 dB.

The magnitude of $(1 + j\omega\tau)$ can be readily plotted by noting that it is approximated by two straight lines called *asymptotes*. These asymptotes are the 0-dB axis and a line with a slope of 20 dB per decade which intersects the first asymptote at the frequency $\omega = \omega_c$. For $\omega < 0.1\omega_c$ the asymptote and actual magnitude can be considered the same. Likewise for $\omega > 10\omega_c$ the curve and the asymptote can be considered the same. Between $\omega = 0.1\omega_c$ and $\omega = 10\omega_c$ the magnitude deviates from the two asymptotes—the maximum deviation being +3 dB at $\omega = \omega_c$. The magnitude plot can be sketched with little error by correcting the straight-line approximation at the corner frequency ω_c. If more accurate results are desired, corrections of +1 dB may be made at the frequencies $\omega = 0.5\omega_c$ and $\omega = 2\omega_c$.

The phase angle as a function of frequency can also be plotted using the data of Table 6-2. However, the phase-angle curve may be approximated by three straight lines. As seen in Fig. 6-10, the phase angle is approximately 0° for $\omega < 0.1\omega_c$. For $\omega > 10\omega_c$, the phase angle is approximately 90°. For values between $\omega = 0.1\omega_c$ and $\omega = 10\omega_c$ the curve is approximated by a straight line from 0° at $\omega = 0.1\omega_c$ to 90° at $\omega = 10\omega_c$. Note that this line passes through 45° at $\omega = \omega_c$. If greater accuracy is desired, the deviations shown in Fig. 6-10 may be added.

The method developed in this section can also be used to plot frequency-response curves for $(1 + j\omega\tau)^{-1}$. The magnitude plot is obtained from

$$\text{Mag} = 20 \log_{10} |(1 + j\omega\tau)^{-1}| \text{ dB} \tag{6-30}$$

Equation (6-30) will yield the same decibel values as Eq. (6-25) except that the values will be negative. Equation (6-30) can also be written in terms of the corner frequency ω_c. Thus,

$$\text{Mag} = -20 \log_{10} \sqrt{1^2 + \left(\frac{\omega}{\omega_c}\right)^2} \text{ dB} \tag{6-31}$$

To determine the phase angle of $(1 + j\omega\tau)^{-1}$, consider this factor as the ratio of two complex numbers:

$$\frac{1}{1 + j\omega\tau} - \frac{1 \angle 0}{\sqrt{1 + (\omega\tau)^2} \angle \tan^{-1}(\omega\tau)}$$

Thus, the phase angle becomes

$$\phi = \frac{0}{\tan^{-1}(\omega\tau)} = 0 - \tan^{-1}(\omega\tau) \tag{6-32}$$

In terms of the corner frequency ω_c this becomes

$$\phi = -\tan^{-1}\left(\frac{\omega}{\omega_c}\right)^{\circ} \tag{6-33}$$

Equation (6-33) is identical to Eq. (6-29) except that the angle is negative.

EXAMPLE 6-5

Using asymptotic approximations, plot the frequency-response curves for the circuit in Example 6-3 (Fig. 6-5).

Solution

The transfer function of the circuit was found in Example 6-3 to be

$$G(j\omega) = \frac{1}{1 + j0.1\omega}$$

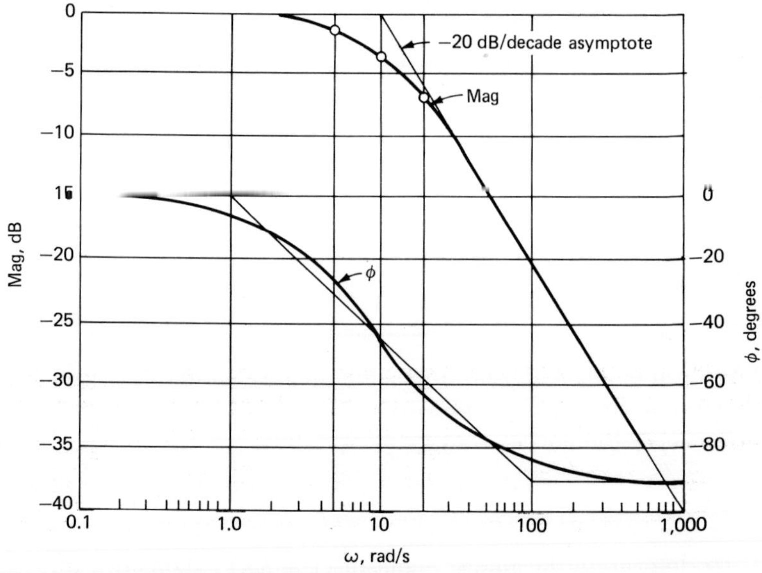

Figure 6-11 Frequency-response plot for Example 6-5

Figure 6-11 is a plot of the magnitude versus frequency. The corner frequency ω_c is 10 rad/s. For frequency values such that $\omega < 1 (\omega < 0.1\omega_c)$, the curve is the 0-dB asymptote. For values of frequency where $\omega > 100$ $(\omega > 10\omega_c)$, the curve approaches a -20 dB per decade asymptote starting from 0 dB at the corner frequency. Three correction points in the vicinity of $\omega = 10$ are located to aid in sketching the curve. When $\omega = 5$ and $\omega = 20$, the curve is 1 dB below the asymptotes, and when $\omega = 10$, the curve is 3 dB below the corner frequency.

Figure 6-11 includes the plot of phase angle versus frequency. Both the exact curve and the approximate curve are shown. Note that for $\omega < 1$ $(\omega < 0.1\omega_c)$ the phase angle approaches $0°$, while for $\omega > 100$ $(\omega > 10\omega_c)$ the phase angle approaches $-90°$.

Frequency-response curves have been considered for the factor $(1 + j\omega\tau)^{\pm 1}$. The techniques used in developing these curves can be extended to the general case $(1 + j\omega\tau)^{\pm m}$ by applying the laws of logarithms. For example, consider $(1 + j\omega\tau)^2$. The magnitude is given by

$$\begin{aligned}\text{Mag} &= 20 \log_{10} |(1 + j\omega\tau)^2| \\ &= 20 \log_{10} |(1 + j\omega\tau)(1 + j\omega\tau)| \\ &= 20 \log_{10} (\sqrt{1^2 + (\omega\tau)^2})(\sqrt{1^2 + (\omega\tau)^2})\end{aligned}$$

Applying the logarithmic identity $\log_{10} (A \times B) = \log_{10} A + \log_{10} B$, we have

$$\text{Mag} = 20 \log_{10} \sqrt{1^2 + (\omega\tau)^2} + 20 \log_{10} \sqrt{1^2 + (\omega\tau)^2}$$

or

$$\text{Mag} = 2(20 \log_{10} \sqrt{1^2 + (\omega\tau)^2}) \text{ dB} \tag{6-34}$$

In general then

$$\text{Mag} = \pm m(20 \log_{10} \sqrt{1^2 + (\omega\tau)^2}) \text{dB} \tag{6-35}$$

and the phase angle is given by

$$\phi = \pm m \tan^{-1} (\omega\tau)° \tag{6-36}$$

EXAMPLE 6-6

Draw the frequency-response curves for the transfer function

$$G(j\omega) = \left(\frac{1}{1 + j\omega/20}\right)^2$$

Solution

The frequency-response curves can be developed by first plotting the magnitude and phase angle of $1/(1 + j\omega/20)$ and then graphically doubling the curves. Figure 6-12 shows the plot of magnitude versus frequency for both $1/(1 + j\omega/20)$ and $[1/(1 + j\omega/20)]^2$. The $[1/(1 + j\omega/20)]^2$ curve is plotted by doubling the decibel values at the various frequencies. Note that the $[1/(1 + j\omega/20)]^2$ curve follows a -40 dB per decade asymptote rather than a -20 dB per decade slope at the higher frequencies. This is because the logarithm of a squared term is multiplied by 2. If a cubic term were present in the denominator, the asymptote would be -60 dB per decade. Regardless of the exponent, all the asymptotes originate at the frequency ω_c and magnitude 0 dB.

Figure 6-12 also shows the phase angle versus frequency for $[1/(1 + j\omega/20)]^2$. Included are both the actual curve and its straight-line

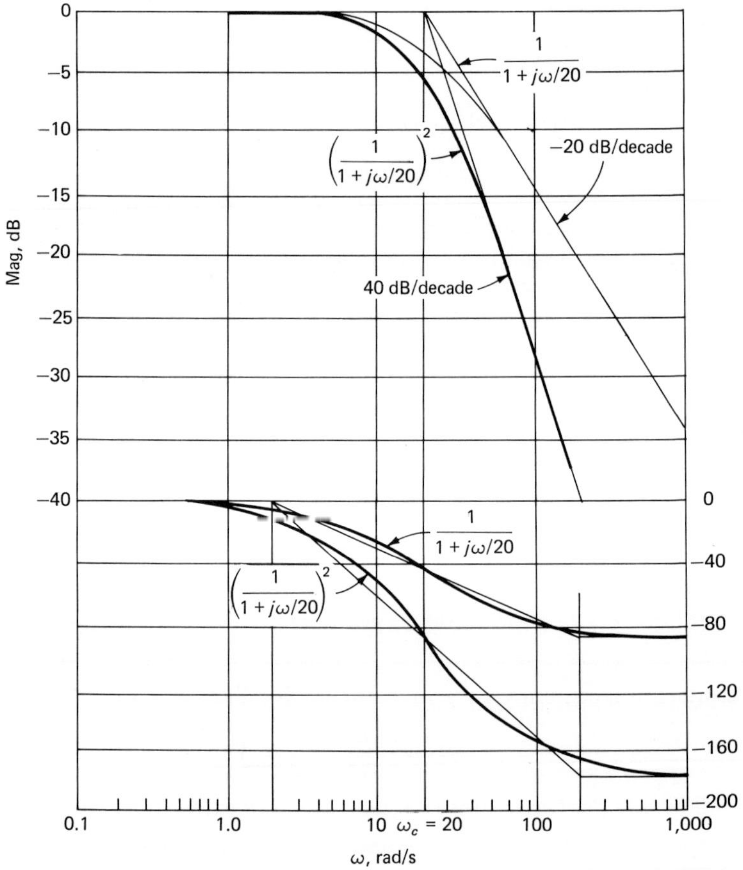

Figure 6-12 Frequency-response plot for Example 6-6

approximation. As with the magnitude plot, the phase-angle plot was developed by drawing the response for $1/(1 + j\omega/20)$ and then doubling the angle values according to Eq. (6-36). The phase angle approaches $0°$ for low values of frequency, and for large frequencies the phase angle approaches $-180°$. The upper value of the phase angle is directly related to the power of the factor and is equal to $\pm m$ times $90°$.

6-10 PLOT OF A SECOND-ORDER FACTOR

Second-order systems were introduced in Sec. 4-7, and the transfer function of a second-order system was derived in Example 5-9. If the transfer function for a second-order system is transformed to the frequency domain, the resulting transfer function will be of the form

$$G(j\omega) = \frac{1}{1 + j2\zeta\omega/\omega_n + (j\omega/\omega_n)^2} \tag{6-37}$$

The response as a function of frequency ω is determined by the two constants: ω_n (undamped natural frequency) and ζ (damping ratio).

The damping ratio determines the nature of the roots of the quadratic equation in the denominator of Eq. (6-37) and is therefore significant to the form of the response. For the overdamped case where $\zeta > 1$, the roots are real and unequal. For the critically damped case of $\zeta = 1$, the roots are real and equal. In both cases the denominator can be factored into two first-order terms. The transfer function can then be written in the form

$$G(j\omega) = \frac{1}{(1 + j\omega\tau_1)(1 + j\omega\tau_2)} \tag{6-38}$$

In this case the transfer function is expressed in terms of the product of two first-order terms. Plotting the frequency response of factors of this form has been discussed in the preceding section.

If $\zeta < 1$, the system is underdamped. The roots of Eq. (6-37) are complex conjugates, and the equation cannot be factored to give two first-order terms. In this case the factor may be plotted as a function of the frequency ratio ω/ω_n for a given value of ζ. The magnitude is given by

$$\text{Mag} = 20 \log_{10} \sqrt{\left[1 - \left(\frac{\omega}{\omega_n}\right)^2\right]^2 + \left(\frac{2\zeta\omega}{\omega_n}\right)^2} \text{ dB} \tag{6-39}$$

and the phase angle of the function is given by

$$\phi = -\tan^{-1} \frac{2\zeta\omega/\omega_n}{1 - (\omega/\omega_n)^2} \tag{6-40}$$

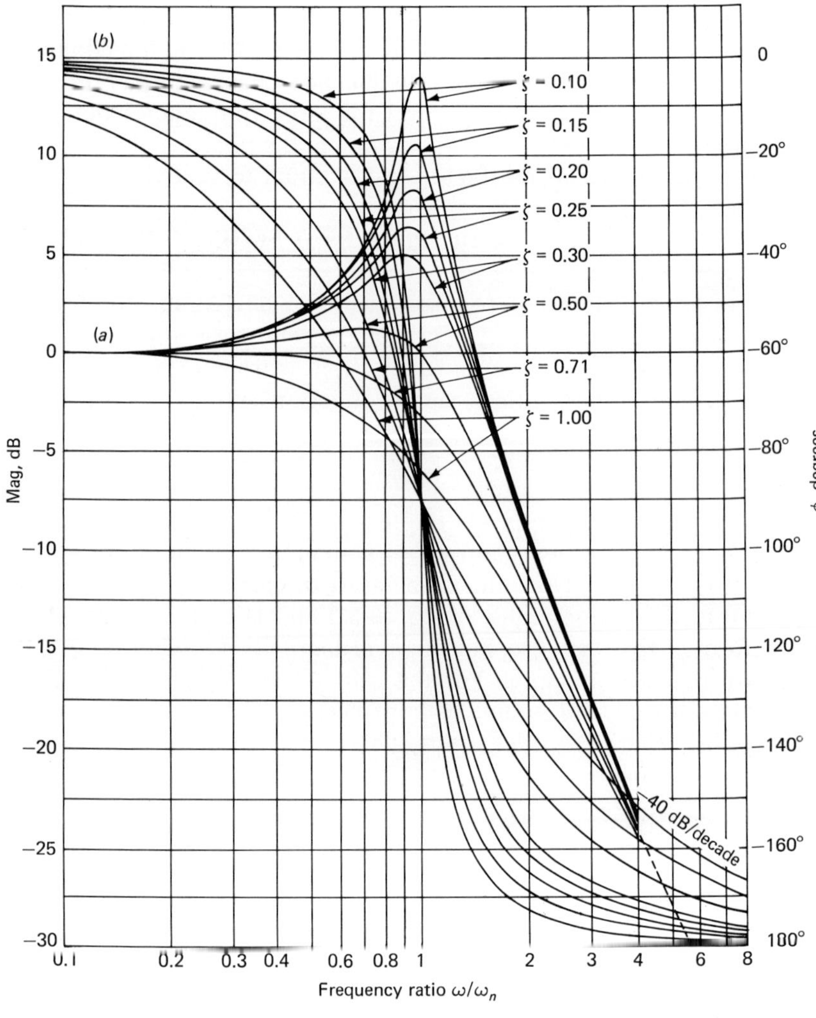

Figure 6-13 *Frequency-response plot of second-order factor* $[1 + j2\zeta\omega/\omega_n + (j\omega/\omega_n)^2]^{-1}$

The magnitude and phase plots of a second-order factor for various values of ζ are given in Fig. 6-13. Note that the curves are plotted as a function of the frequency ratio ω/ω_n. Dividing the frequency ω by the natural frequency ω_n normalizes the frequency scale, which makes the curves applicable for any value of ω_n. For a specific application in which ω_n is known, the horizontal scale can be multiplied by ω_n to obtain the actual frequency scale.

For values of ω such that $\omega < 0.1\omega_n$ the magnitude plot approaches 0 dB. For values of ω where $\omega > 10\omega_n$ the magnitude is approximated by

$$\text{Mag} = -20 \log_{10} (\omega/\omega_n)^2 = -40 \log_{10} (\omega/\omega_n) \text{ dB} \tag{6-41}$$

A slope of -40 dB per decade is to be expected since the term containing ω as a squared function will predominate at high frequencies. Magnitude values between these extremities depend upon the value of the damping ratio. Note that the magnitude in the vicinity of $\omega/\omega_n = 1$ peaks above 0 dB when the damping ratio is less than unity. This peaking is due to the resonant nature of an underdamped system.

Figure 6-13 includes a plot of the phase angle versus ω/ω_n for the various values of the damping ratio. For low values of frequency the phase angle approaches 0°, and for large values of frequency the phase angle approaches $-180°$.

Because of their complexity, Eqs. (6-39) and (6-40) are usually plotted using straight-line asymptotes corrected with the aid of Fig. 6-13. Magnitude plots may be based on an approximation of two straight-line asymptotes. For values of ω/ω_n less than 1, the approximation is the 0-dB line. For values of ω/ω_n greater than 1, the approximation is a straight line with a slope of -40 dB per decade starting from 0 dB at $\omega/\omega_n = 1$. This straight-line approximation is then graphically adjusted in the vicinity of $\omega/\omega_n = 1$ for a particular damping ratio.

The straight-line approximation for the phase angle is similar to that of a first-order system. For frequency ratios less than 0.1, the approximation is the 0° line. For frequency ratios greater than 10.0 the approximation is the $-180°$ line. For ratios between these limits the approximation is a straight line starting at 0° and $\omega/\omega_n = 0.1$ and ending at $-180°$ and $\omega/\omega_n = 10$. The actual phase curve for a particular damping ratio is obtained by correcting the approximation according to Fig. 6-13. The actual curve has a rather large deviation from the straight-line approximation. For this reason the straight-line approximation should be used only as an aid in constructing the actual phase curve and should not itself be used to represent the phase angle.

EXAMPLE 6-7

Sketch the frequency response of the following transfer function:

$$G(j\omega) = \frac{1}{1 + j0.08\omega + 0.04(j\omega)^2}$$

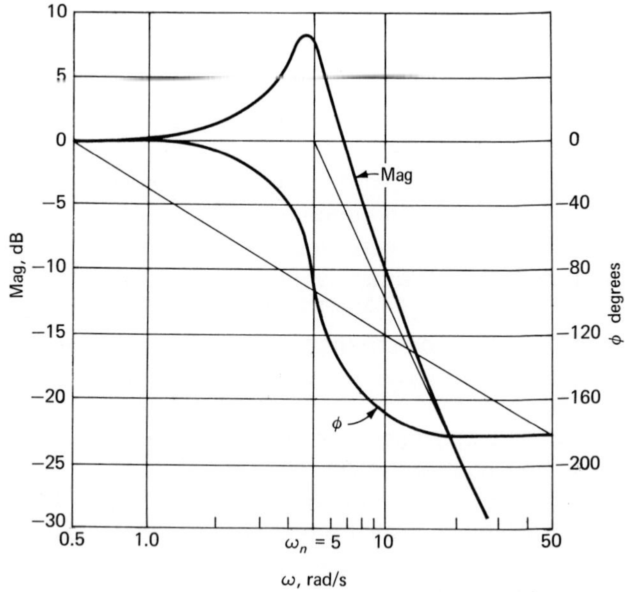

Figure 6-14 Frequency-response plot for Example 6-7

Solution

Before plotting the frequency response, ω_n and ζ must be determined. Equating like terms in the given transfer function with those defined in Eq. (6-37), we see that

$$\frac{2\zeta}{\omega_n} = 0.08 \qquad \text{and} \qquad \left(\frac{1}{\omega_n}\right)^2 = 0.04$$

Solving for the desired parameters gives

$$\omega_n = 5 \text{ rad/s}$$

$$\zeta = 0.2$$

Figure 6-14 shows the resulting frequency response. The curves are drawn first using straight-line approximations and are then corrected for $\zeta = 0.2$ with the aid of Fig. 6-13. The curves have been plotted in terms of ω since ω_n is known.

EXAMPLE 6-8

Sketch the frequency response of the transfer function

$$G(j\omega) = \frac{1}{1 + j0.55\omega + 0.025(j\omega)^2}$$

Solution

Calculating ω_n and ζ gives

$$\left(\frac{1}{\omega_n}\right)^2 = 0.025$$

and so

$$\omega_n = 6.33 \text{ rad/s}$$

In addition

$$\frac{2\zeta}{\omega_n} = 0.55$$

and so

$$\zeta = 1.74$$

Second-order factors whose damping ratios are greater than 1 can be factored into two first-order terms. These terms are found by determining the roots of the second-order factor. Applying the quadratic formula to the given function yields the roots

$$j\omega_1, j\omega_2 = \frac{-0.55 \pm \sqrt{(0.55)^2 - 4(0.025)(1)}}{2(0.025)} = -20, -2$$

which gives the factors $(j\omega + 20)$ and $(j\omega + 2)$. Thus the equation can be written as

$$G(j\omega) = \frac{40}{(j\omega + 20)(j\omega + 2)}$$

$$= \frac{1}{(1 + j\omega/20)(1 + j\omega/2)}$$

The frequency response is plotted in Fig. 6-15. The magnitude plot is constructed by sketching the plots for the first-order factors $1/(1 + j\omega/20)$ and $1/(1 + j\omega/2)$ and then graphically adding them together. The resulting curve has a high-frequency slope of -40 dB per decade, which is indicative of a second-order factor.

The phase-angle plot in Fig. 6-15 is also constructed by plotting the phase of each factor and then graphically combining them. The resultant curve is asymptotic to the 180° line.

6-11 FREQUENCY-RESPONSE SPECIFICATIONS

In Sec. 4-10 the transient response of a second-order system was specified in terms of rise time, overshoot, and settling time. As we have

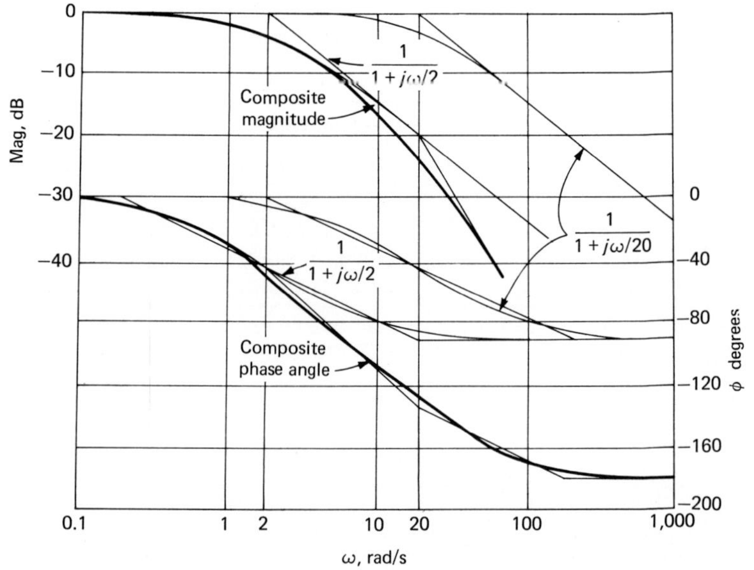

Figure 6-15 Frequency-response plot for Example 6-8

seen, frequency response is inherently related to transient response. This section will describe quantities used in specifying frequency response and their interrelationship with transient-response parameters.

Figure 6-16 shows a frequency-response curve with certain specifications identified.

The usefulness of these specifications in control systems work de-

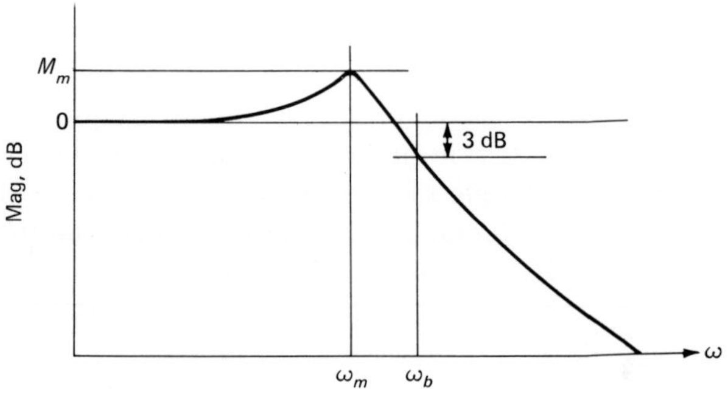

Figure 6-16 Frequency-response curve showing response specifications

pends upon the closeness of approximating the system as second order. Recall that a second-order system which has less than critical damping exhibits a pair of complex roots in its characteristic equation. Fortunately, in a large proportion of control systems with a pair of complex roots, these roots largely determine the response. In such cases a second-order approximation is valid, at least over the usable frequency range of the system. However, we should keep in mind that the approximation is only valid for specifying frequency response. Neglected terms may contribute markedly to the stability considerations to be discussed later in this chapter.

The frequency response of a second-order system is given by the equation

$$\frac{C(j\omega)}{R(j\omega)} = \frac{1}{1 + j2\zeta\omega/\omega_n + (j\omega/\omega_n)^2} \tag{6-42}$$

where $C(j\omega)$ and $R(j\omega)$ are the sinusoidal response and excitation. From the response curves in Fig. 6-13, we note that the magnitude exhibits a peak value M_m that increases as the damping ratio decreases. This parallels the increase in transient-response overshoot with decreased damping. The value of M_m can serve to indicate the relative stability of the system.

The frequency at which M_m occurs is designated as ω_m in Fig. 6-16. This frequency is given by the equation

$$\omega_m = \omega_n \sqrt{1 - 2\zeta^2} \tag{6-43}$$

The peak value M_m is

$$M_m = \frac{1}{2\zeta\sqrt{1 - \zeta^2}} \tag{6-44}$$

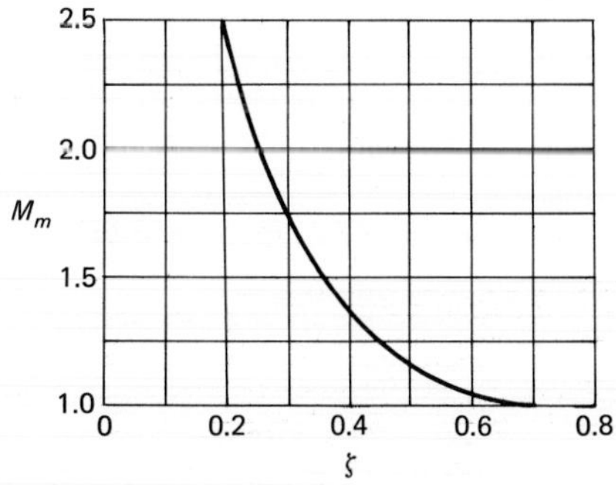

Figure 6-17 Peak magnitude versus damping ratio for second-order system

A damping ratio of $\zeta \geq 0.707$ does not produce a resonant peak; thus the above equations are valid only for an underdamped system in which $\zeta < 0.707$. Figure 6-17 shows a plot of M_m versus ζ based on Eq (6-44). For values of $\zeta < 0.3$, the value of M_m increases rapidly in magnitude. The response is therefore extremely large and might damage the system.

The bandwidth ω_b of the system is defined as the frequency at which the magnitude response is 3 dB below the low-frequency value. It can be shown that the bandwidth is given by the equation

$$\omega_b = \omega_n \sqrt{1 - 2\zeta^2 + (2 - 4\zeta^2 + 4\zeta^4)^{\frac{1}{2}}} \tag{6-45}$$

In specifying a system in the frequency domain, the following are considerations:

1. A resonant peak of about 3 dB ($M_m = 1.4$) is commonly specified.
2. As the bandwidth is increased, the time of response becomes faster. Although the bandwidth must be high enough to encompass the frequencies contained in the input, it should be no higher than necessary to minimize noise problems.

EXAMPLE 6-9

Determine the peak magnitude and bandwidth associated with the frequency response of the servomechanism in Example 4-4. Recall from the example that $\zeta = 0.182$ and $\omega_n = 52.5$ rad/s.

Solution

From Eq. (6-44)

$$M_m = \frac{1}{2 \times 0.182\sqrt{1 - 0.182^2}} = 2.79 = 8.92 \text{ dB}$$

Using Eq. (6-45),

$$\omega_b = 52.5\sqrt{1 - 2 \times 0.182^2 + (2 - 4 \times 0.182^2 + 4 \times 0.182^4)^{\frac{1}{2}}}$$
$$= 79.7 \text{ rad/s}$$

6-12 STABILITY CRITERION

The stability criterion that is applied in the frequency domain is the *Nyquist stability criterion*. Although a rigorous development of the

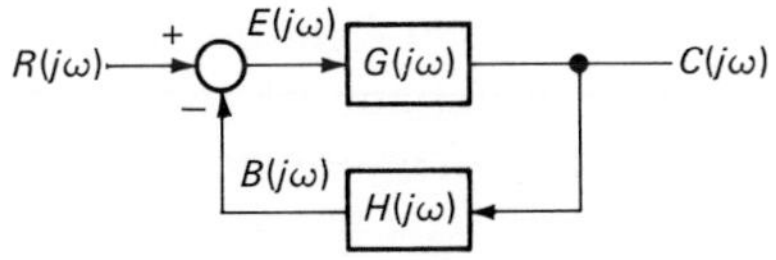

Figure 6-18 Closed-loop control system

Nyquist criterion is beyond the scope of this text, a simplified approach that is applicable to most situations will be presented.

The frequency response of the closed-loop system in Fig. 6-18 is given by the general equation

$$\frac{C(j\omega)}{R(j\omega)} = \frac{G(j\omega)}{1 + G(j\omega)H(j\omega)} \tag{6-46}$$

The error signal $E(j\omega)$ is modified in amplitude and phase angle by the loop gain $G(j\omega)H(j\omega)$ and fed back as $B(j\omega)$ to be subtracted from the input $R(j\omega)$. If $B(j\omega)$ exhibits a lagging phase angle of 180° (designated −180°), the feedback is effectively added to $R(j\omega)$, resulting in positive feedback. If the amplitude of the −180° shifted signal is less than $E(j\omega)$, the system cannot be self-sustaining and is therefore stable. If, however, the amplitude of $B(j\omega)$ is greater than $E(j\omega)$, $R(j\omega)$ is not needed to produce $E(j\omega)$ and the system will be self-sustaining or unstable. Since any practical system will generate lagging phase shifts at some frequencies that exceed 180°, there must then be at least one frequency at which the phase shift is equal to a 180° lag. If at this frequency the amplitude ratio is unity or greater, according to the Nyquist criterion, the system is unstable.

The Nyquist stability criterion may be further understood with the

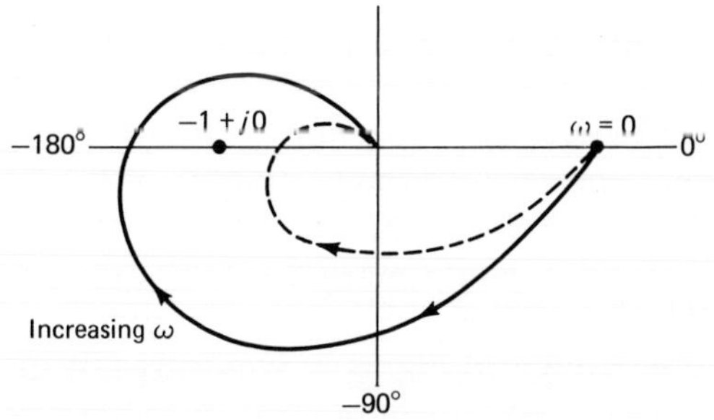

Figure 6-19 Polar diagrams illustrating Nyquist stability criterion

aid of Fig. 6-19. The polar plots in the figure represent the open-loop transfer functions of two systems. Of particular significance is the $-1 + j0$ point on the polar plane, which indicates unity gain at a $-180°$ phase shift. The transfer function given by the solid curve has a gain greater than unity at a $-180°$ phase shift. Hence we may conclude that the closed-loop system represented by this curve would be unstable. Conversely the dashed curve shows a gain less than unity at $-180°$, and this system would therefore be stable. Thus we may make the general observation that a system is unstable according to the Nyquist criterion if the complex-plane plot of its open-loop transfer function encircles the $-1 + j0$ point.

It is possible to apply the Nyquist criterion through simple graphical techniques involving Bode plots of the system's open-loop transfer function. These techniques not only will indicate the degree of stability but also can be extended to modifying or compensating the system to obtain specified closed-loop response.

Certain definitions have been developed that relate to stability in the frequency domain. These include:

1. *Gain-crossover frequency:* the frequency at which the magnitude or gain of the transfer function is unity (0 dB).

2. *Phase-crossover frequency:* the frequency at which the phase angle of the transfer function equals $-180°$.

3. *Gain margin:* the decibel value of the open-loop gain at the phase-crossover frequency. Gain margin is designated positive for negative decibels.

4. *Phase margin:* the difference between 180° and the phase angle of the transfer function at the gain-crossover frequency.

These definitions are shown in the Bode diagrams of Fig. 6-20. Note that the illustrated gain margin and phase margin are both positive.

The Nyquist criterion for stability requires that the phase angle of the open-loop transfer function be less than $-180°$ at the gain-crossover frequency. In other words, a closed-loop system is stable if a positive phase margin exists. Conversely, a negative phase margin denotes an unstable system. Applying this criterion, the transfer function shown in Fig. 6-20 would result in stable closed-loop operation.

Two theorems important to frequency-response analysis have been developed by Bode. The first states that the phase shift and the rate of change of gain of a transfer function with frequency are interdependent. For example, we have seen that a slope of -20 dB per

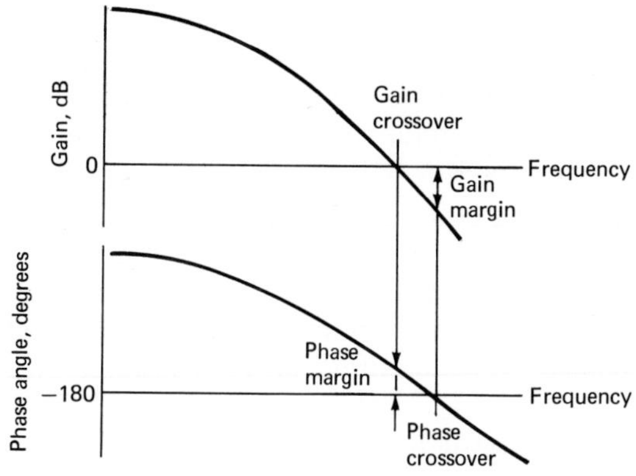

Figure 6-20 Bode diagrams showing stability-criterion definitions

decade has a 90° phase lag associated with the function, a slope of −40 dB per decade has a 180° phase lag, etc. Thus to be stable the slope of the open-loop transfer function should not exceed −40 dB per decade at the gain-crossover frequency.

The second theorem states that the gain characteristic in one portion of the frequency spectrum and the phase characteristic in another part of the spectrum may be chosen independently. However, in doing this the phase characteristic in the first range and the gain characteristic in the second range are definitely fixed in accord with the first theorem. The importance of this theorem in control system design is that it permits overall performance specifications to be met. The preceding chapters have emphasized that a high loop gain is required to obtain accurate response over a specified range of frequencies. However, stability considerations require a fairly well defined phase characteristic, including a positive phase margin, in the higher-frequency range. By selecting somewhat independently the attenuation characteristic in the low-frequency range and the phase characteristic in the high-frequency range, a compromise can be made between high loop gain and adequate stability. Methods for accomplishing this objective will be presented in Chap. 7.

6-13 CLOSED-LOOP FREQUENCY RESPONSE

Having used open-loop frequency-response plots to determine system stability, it would be desirable to use the same plots to find the resulting frequency response when the loop of a stable system is closed. In the concluding sections of this chapter we will consider ways in which closed-loop frequency response can be determined from open-loop Bode diagrams. A relatively convenient and quick method for finding the approximate response will be described in this section. The following section will discuss a more exact method that makes use of a plot of open-loop magnitude versus phase angle.

An obvious approach to determining the closed-loop response would be to substitute values of ω in Eq. (6-46). For each value of ω the transfer function reduces to a complex number. Various values would give the closed-loop magnitude and phase angle over a desired frequency range. However, the calculations would prove to be laborious and time-consuming if the transfer functions $G(j\omega)$ and $H(j\omega)$ were other than quite simple. In addition the entire computation would need to be repeated for any change made to the open-loop characteristic.

A necessary condition for good closed-loop control is that the open-loop gain be significantly greater than 1 over the working range of frequencies. In practical systems the gain will fall off at higher frequencies because of inherent lags in the system elements. With this in mind, let us consider approximations to the closed-loop response at low and high frequencies. The terms low and high frequencies are relative and for our purposes are defined as the regions below and above the gain-crossover frequency. This is illustrated in Fig. 6-21.

An open-loop transfer function is composed of a forward gain $G(j\omega)$ in cascade with a feedback function $H(j\omega)$. These functions are plotted separately and compositely in Fig. 6-21. Note that the feedback $H(j\omega)$ in this case is nonunity.

At low frequencies for which $|GH| \gg 1$, Eq. (6-46) may be approxi-

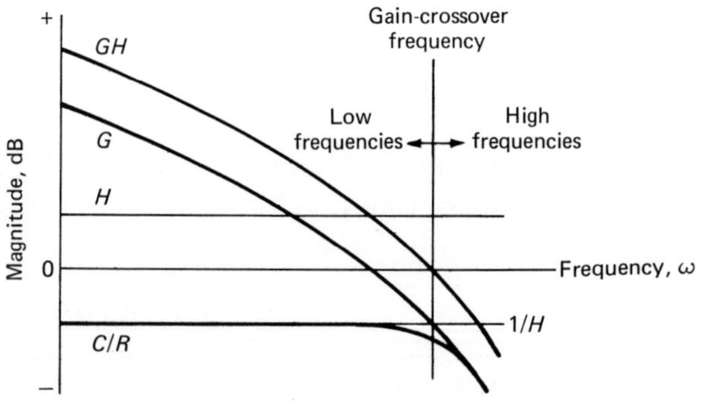

Figure 6-21 Approximations to find closed-loop response from open-loop response

mated as

$$\frac{C(j\omega)}{R(j\omega)} \cong \frac{1}{H(j\omega)} \tag{6-47}$$

Thus at low frequencies the closed-loop response is essentially determined by the feedback function.

At high frequencies where $|GH| \ll 1$, Eq. (6-46) becomes

$$\frac{C(j\omega)}{R(j\omega)} \cong G(j\omega) \tag{6-48}$$

The closed-loop approximation of C/R given by Eqs. (6-47) and (6-48) is also shown in Fig. 6-21. The quantity $1/H$ is readily plotted by taking negative decibel values of H.

It should be clear that the approximations are invalid in the region of the gain-crossover frequency, since $|GH|$ is neither much greater nor much less than 1. Nevertheless this approximation of the closed-loop response is applicable for less stringent design requirements.

If the system under consideration can be approximated by a second-order equation, as described in Sec. 6-10, the closed-loop response may be readily determined from the open-loop response. The damping ratio of an underdamped, second-order system is related to its phase margin; as the phase margin is increased, the equivalent damping in the system is increased. This relationship is shown in Fig. 6-22 in which the damping ratio is plotted as a function of the phase margin. The actual curve may be approximated by the linear relationship

$$\zeta = 0.01\phi_{PM} \tag{6-49}$$

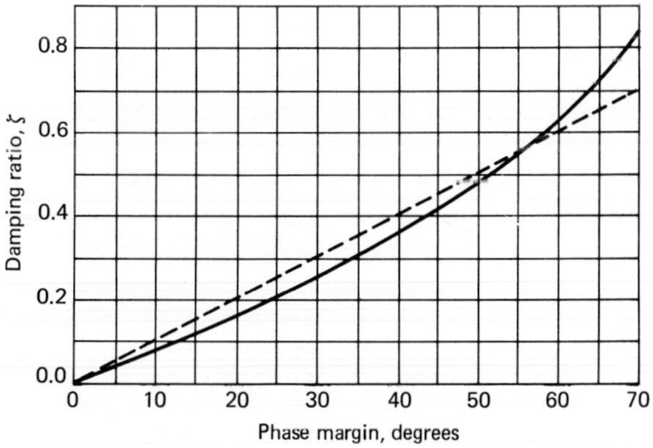

Figure 6-22 Damping ratio versus phase margin for second-order system

where ϕ_{PM} is the phase margin in degrees.

After the damping ratio is determined using Fig. 6-22, the corresponding frequency response may be found from the family of curves in Fig. 6-13. The frequency at which the maximum response M_m occurs is approximately the gain-crossover frequency.

Typical design criteria for a control system may specify a closed-loop damping ratio of 0.4 to give $M_m = 1.4$. From Fig. 6-22 this implies a phase margin of approximately 45°. A reasonable rule of thumb to meet this requirement is to provide a gain curve that cuts the 0-dB axis with a slope of −20 dB per decade over a range extending an octave above and below the gain-crossover frequency.

EXAMPLE 6-10

The open-loop transfer function of a positional servomechanism is

$$G(j\omega) = \frac{1}{j\omega(1 + j\omega)(1 + j0.2\omega)}$$

$$H(j\omega) = 1$$

Determine the gain and phase margins and the approximate closed-loop frequency response.

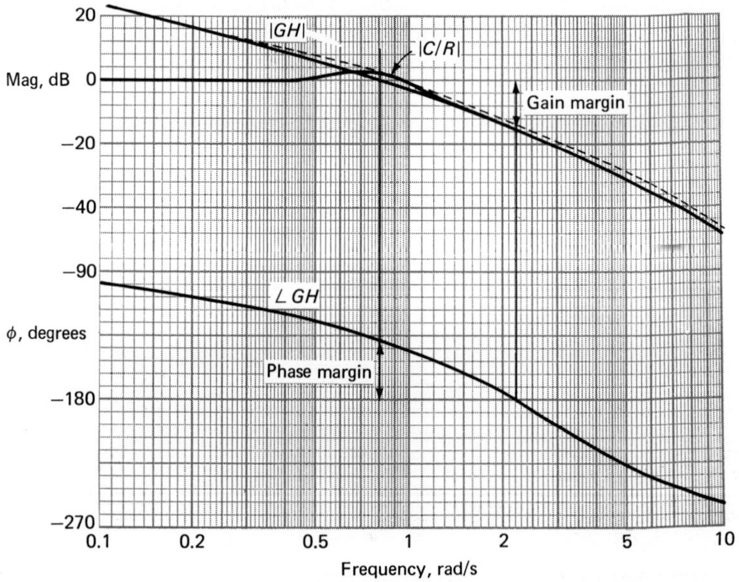

Figure 6-23 Bode diagram for Example 6-10

Solution

The Bode diagram showing the open-loop frequency response is plotted in Fig. 6-23. The phase angle at the gain-crossover frequency is 135° and so the phase margin is 180° − 135° = 45°. The magnitude of $GH(j\omega)$ at the phase-crossover frequency is −16 dB, and so the gain margin is 16 dB.

A second-order approximation of the transfer function is possible since the $1/(1 + j0.2\omega)$ factor does not significantly affect the response in the vicinity of the gain-crossover frequency. Referring to Fig. 6-22, the damping ratio corresponding to a 45° phase margin is 0.42. From Fig. 6-13, we see that the peak magnitude is 1.3 or 2.4 dB. The frequency at which M_m occurs is approximately the gain-crossover frequency. The closed-loop magnitude response C/R can then be sketched as shown in Fig. 6-23.

6-14 NICHOLS CHART

For a closed-loop control system with unity feedback, we have the relationship

$$\frac{C(j\omega)}{R(j\omega)} = \frac{G(j\omega)}{1 + G(j\omega)} = M \angle N \tag{6-50}$$

where M* is the magnitude and N is the phase angle of the closed-loop response evaluated at a given value of ω. The values of M and N taken in combination are uniquely determined by a particular transfer function $G(j\omega)$. However, an unlimited number of functions $G(j\omega)$ can produce a definite value of the magnitude M when M is considered separately. The same holds true for N when only phase angle is considered. Thus it is possible to plot curves of constant values of M and N as a function of $G(j\omega)$. The loci of points of constant values of M and N when plotted on the complex or polar plane are circles having different centers and radii.

Since it is much easier to work with Bode diagrams than with polar plots, it would be preferable if the circles of constant magnitude and phase were translated to a logarithmic gain-phase diagram in rectangular coordinates. N. B. Nichols performed this transformation, and the resulting diagram, called the *Nichols chart,* is used extensively in

* *The letter symbol M used here is not equal to and should not be confused with the manipulated variable M defined in Chap. 1. Although this dual use of M is unfortunate, it is consistent with the terminology found in most control systems texts.*

designing control systems. In particular the chart is useful when it is desired to set the system gain to meet a specific M_m criterion or when it is necessary to modify the open-loop transfer function to achieve stable operation.

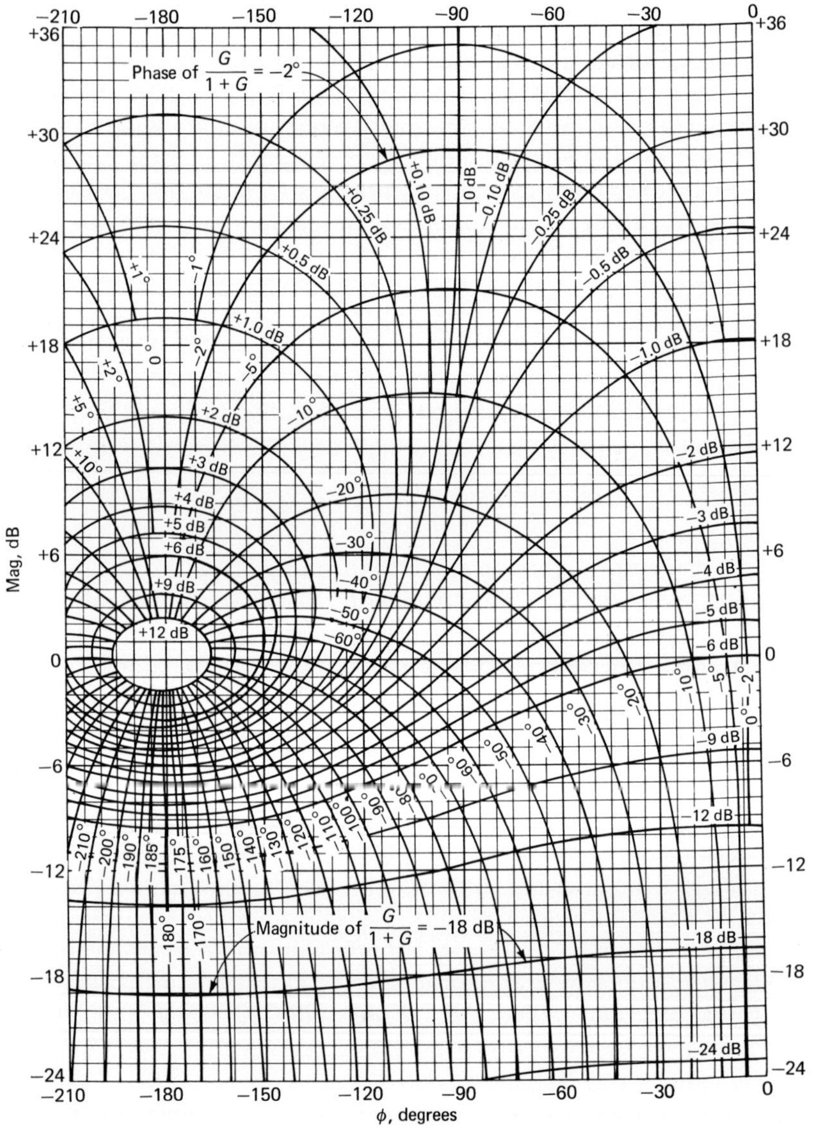

Figure 6-24 Nichols chart

Figure 6-24 shows the Nichols chart. There are two sets of coordinates: one set applies to the open-loop transfer function and the other set to the closed-loop response. Open-loop data are referred to the horizontal and vertical coordinate axes where the respective scales are phase angle in degrees and magnitude in decibels. Closed-loop data are referred to the curved lines that are superimposed on the chart. One set of lines gives constant values of the closed-loop magnitude M in decibels, and the other gives values of the closed-loop phase angle N in degrees.

To use the chart, the magnitude and phase angle of the open-loop transfer function at various frequencies are plotted on the rectangular coordinates. Connecting these points forms a line that represents the open-loop characteristic. This line also has values on the curved coordinates, and these values define the closed-loop response at the various frequencies. The values of M and N read from the Nichols chart can then be plotted in conventional Bode form to show the closed-loop frequency response.

The M contour which is just tangent to the plot of $G(j\omega)$ gives the maximum value of the closed-loop response, or M_m. The frequency at which M_m occurs is the frequency at which $G(j\omega)$ is tangent. The corresponding closed-loop phase angle is given by the N contour which passes through the tangent point.

In practice it is convenient to plot the open-loop response on semitransparent paper that overlays the Nichols chart. The closed-loop magnitude and phase data can then be read from the Nichols chart through the overlay.

We are frequently interested in determining the closed-loop response of a given system when the gain is either raised or lowered, or in determining the gain needed to provide a certain M_m value. These can be found by moving the open-loop curve up or down with respect to the Nichols chart. For example, it was previously mentioned that an M_m of 1.4 or 3 dB is frequently specified in system design. By placing the $G(j\omega)$ curve just tangent to the 3 dB ellipse of the Nichols chart, the necessary gain can be determined.

The Nichols chart is directly applicable to closed-loop systems which have unity feedback. However, it can also be used for a nonunity-feedback system in the following manner. The true closed-loop response of the system shown in Fig. 6-25*a* is given by the equation

$$\frac{C(j\omega)}{R(j\omega)} = \frac{G(j\omega)}{1 + G(j\omega)H(j\omega)} \tag{6-51}$$

By rearranging the system to that shown in Fig. 6-25*b*, the closed-loop

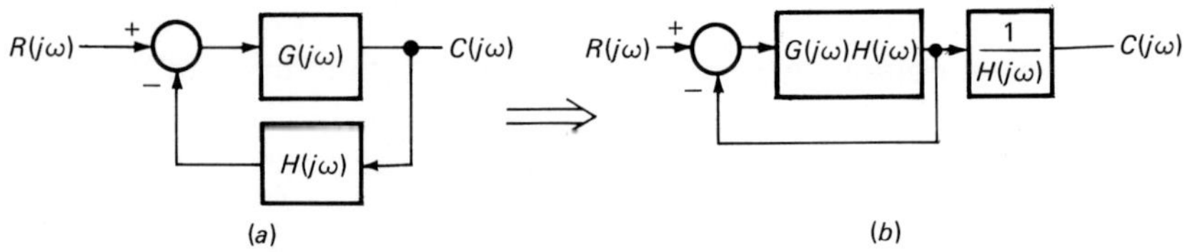

Figure 6-25 Transformation to provide unity feedback system

response may be written as

$$\frac{C(j\omega)}{R(j\omega)} = \frac{G(j\omega)H(j\omega)}{1 + G(j\omega)H(j\omega)} \frac{1}{H(j\omega)} \tag{6-52}$$

We note that the ratio $C(j\omega)/R(j\omega)$ is the same in both cases. However, Eq. (6-52) represents a unity-feedback system whose closed-loop response is multiplied by the function $1/H(j\omega)$. Thus the closed-loop response for a nonunity-feedback system may be obtained through the following procedure:

1. Plot the open-loop transfer function $G(j\omega)H(j\omega)$ on the Nichols chart.
2. Select values of M and N at particular frequencies, and plot the Bode diagram for the response function

$$\frac{G(j\omega)H(j\omega)}{1 + G(j\omega)H(j\omega)}$$

3. Plot $1/H(j\omega)$ on the Bode diagram.
4. Graphically add the magnitude and phase plots on the Bode diagram. The resultant plots are the closed-loop response $C(j\omega)/R(j\omega)$.

The application of Bode plots and the Nichols chart to the analysis and design of a control system will be illustrated by the following example.

EXAMPLE 6-11

A positional servomechanism that uses a synchro transmitter-transformer pair is shown in block-diagram form in Fig. 6-26. The parameters of the components are

K_1 = synchro constant = 0.1 V/° = 5.73 V/rad
K_2 = amplifier gain = variable V/V
K_3 = motor velocity gain constant = 0.5(rad/s)/V
K_4 = gear ratio = 10

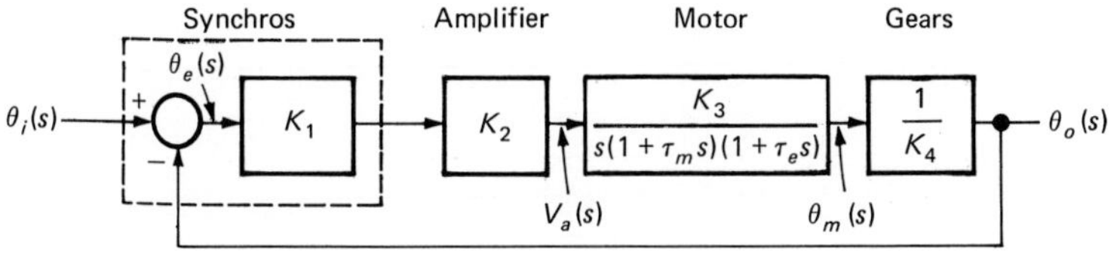

Figure 6-26 Positional servomechanism for Example 6-11

τ_m = mechanical time constant of motor and load = 0.05 s
τ_e = electrical time constant of motor = 0.02 s

Determine the amplifier gain K_2 for the system to have an M_m of 1.4 or 3 dB.

Solution

The transfer function of the motor is

$$\frac{\theta_m(s)}{V_a(s)} = \frac{K_3}{s(1+\tau_m s)(1+\tau_e s)} = \frac{0.5}{s(1+0.05s)(1+0.02s)}$$

Since this is a unity-feedback system, the open-loop transfer function is

$$G(s) = \frac{K_1K_2K_3/K_4}{s(1+\tau_m s)(1+\tau_e s)} = \frac{5.73 \times 0.5 \times 0.1K_2}{s(1+0.05s)(1+0.02s)}$$

In the frequency domain this becomes

$$G(j\omega) = \frac{K_T}{j\omega(1+j0.05\omega)(1+j0.02\omega)}$$

where K_T is the loop gain, which is equal to $0.286K_2$. Note that K_T is the velocity constant K_V of the system as defined by Eq. (5-44).

Figure 6-27 is the Bode diagram of $G(j\omega)$. For convenience in using the Nichols chart the plot is made with the loop gain K_T set at an arbitrary value of 10. The Bode curves have been plotted using the straight-line approximations described earlier in the chapter.

The information from the open-loop Bode diagram is plotted on the Nichols chart in Fig. 6-28. Specifically points have been plotted for frequencies of 1, 2, 5, 10, 20, and 40 rad/s. For the closed-loop system to have $M_m = 1.4$, the plot of $G(j\omega)$ on the Nichols chart must be tangent to the M contour marked +3 dB. The plot of $G(j\omega)$ must be translated upward by 5 dB in order to be tangent to this contour. This is indicated by the dashed curve. An increase on the vertical scale of 5 dB is equiv-

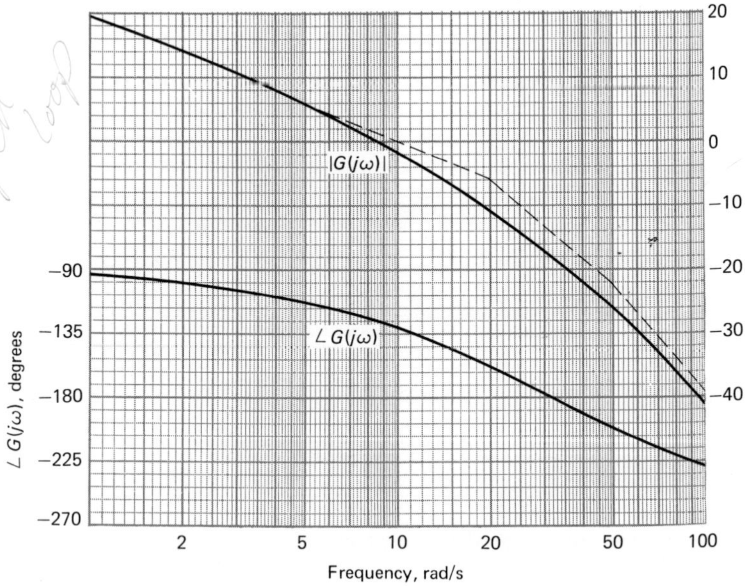

Figure 6-27 Bode diagram for Example 6-11

alent to a gain increase of 1.77. Thus $K_T = 10 \times 1.77 = 17.7 = 0.286K_2$ to provide $M_m = 1.4$. The needed amplifier gain K_2 is then 61.9 or 35.8 dB.

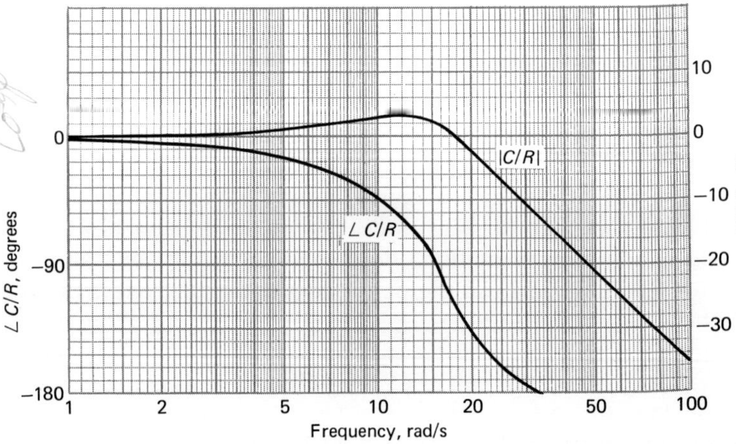

Figure 6-28 Closed-loop response for Example 6-11

The Bode diagram of the closed-loop response may be obtained by plotting the values of the M and N contours passing through the dashed plot of $G(j\omega)$ at specific frequencies. The resulting closed-loop response is shown in Fig. 6-29.

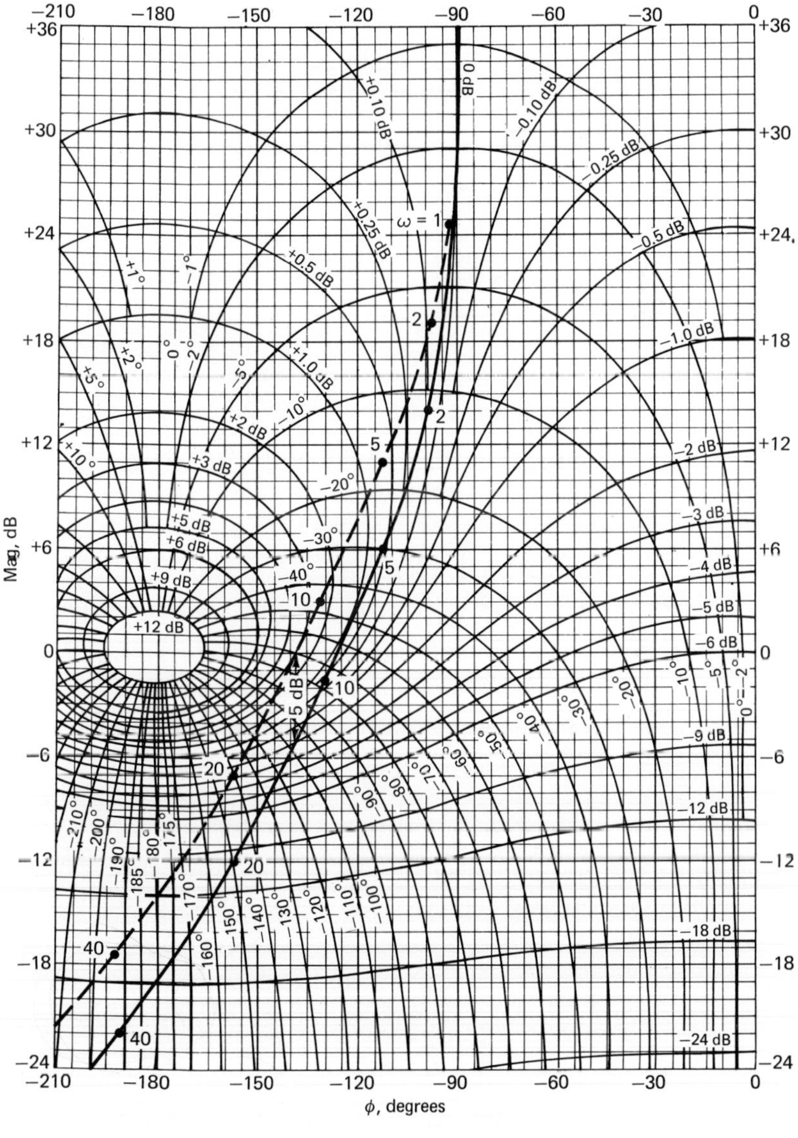

Figure 6-29 Nichols chart for Example 6-11

PROBLEMS

6-1. Express the following complex numbers in polar form.

a. $3 + j4$ b. $0.01 + j0.3$

c. $-5.4 + j6$

6-2. Determine the magnitude and phase angle of the following transfer functions at the frequencies $\omega = 2$ and $\omega = 20$.

a. $\dfrac{10}{j\omega}$ b. $\dfrac{1}{1 + j0.01\omega}$

c. $\dfrac{j\omega}{1 + j\omega}$ d. $\dfrac{1}{j\omega(1 + j0.1\omega)}$

e. $\dfrac{1}{1 + j0.1\omega + (j0.1\omega)^2}$

6-3. Sketch the approximate polar plot of the frequency response for the transfer functions given in Prob. 6-2.

6-4. Sketch the approximate polar plot of the frequency response for the following transfer functions.

a. $\dfrac{1}{(1 + 0.5s)(1 + 2s)}$ b. $\dfrac{1 + 0.5s}{s^2}$

6-5. How many octaves are between (a) 200 and 800 Hz and (b) 10 and 5 Hz? How many decades are between (a) 0.1 and 1,000 Hz and (b) 1 kHz and 1 MHz?

6-6. Express the following quantities in decibel (dB) units.

a. 2 b. 4 c. 8

d. 20 e. 25 f. 40

In Probs. 6-7 to 6-10, plot Bode diagrams for the given transfer functions. Use asymptotic approximations with corrections added to obtain a response curve.

6-7. a. 20 b. -0.5

c. $\dfrac{10}{j\omega}$ d. $(j\omega)^2$

6-8. a. $\dfrac{10}{1 + j0.1\omega}$ b. $5(1 + j2\omega)$

c. $\dfrac{1}{j\omega(1 + j0.1\omega)}$ d. $\dfrac{1 + j\omega}{1 + j5\omega}$

6-9. a. $\dfrac{1}{(1 + j0.5\omega)(1 + j2\omega)}$ b. $\dfrac{5}{(j\omega)^2(1 + j0.1\omega)}$

c. $\dfrac{1}{21s^2 + 10s + 1}$ d. $\dfrac{10(s + 2)}{s(s + 1)(s + 10)}$

6-10. a. $\dfrac{1}{1 + 0.1s + 0.01s^2}$ b. $\dfrac{800}{s(s^2 + 16s + 400)}$

6-11. Obtain the transfer function $G(s)$ for each of the magnitude-response curves shown below.

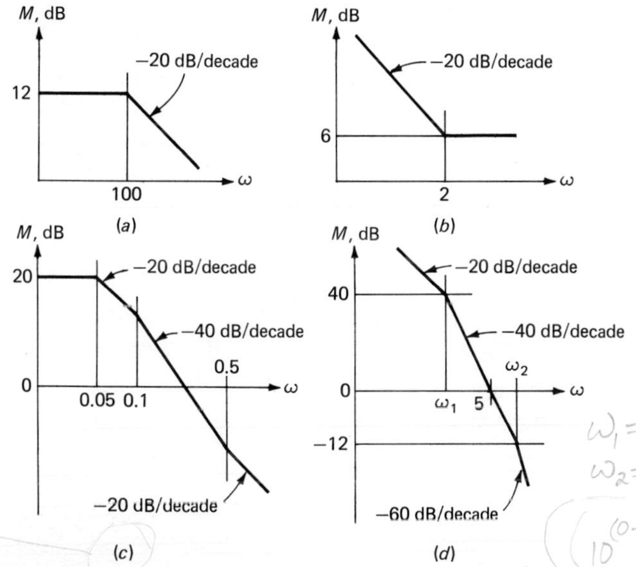

Prob. 6 11

6-12. The transfer function of a thermocouple may be represented by a first-order term of the form $K/(1 + j\omega\tau)$. If a particular thermocouple has a time constant of 5 s, at what frequency will the device attenuate by 50 percent?

6-13. A thermocouple and a thermal well (protective covering for thermocouple) have respective time constants of $\tau_1 = 10$ s and $\tau_2 = 100$ s. Plot the magnitude response of the combination using a normalized magnitude scale.

6-14. Sketch Bode diagrams for the transfer function $E_o(j\omega)/E_i(j\omega)$ for the following circuits (refer to Prob. 5-10).

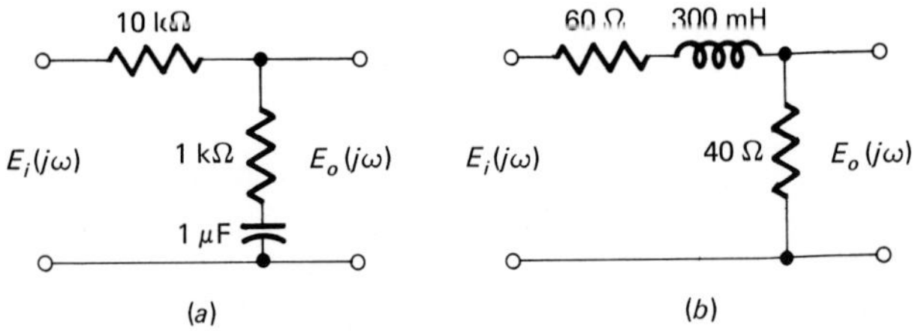

Prob. 6-14

6-15. A servomotor with field control has a mechanical time constant of 0.1 s, a field electrical time constant of 0.008 s, and a velocity gain constant of 200 rpm/V. Sketch the Bode diagram for the velocity transfer function $\dot{\theta}(j\omega)/V_f(j\omega)$ of the motor.

6-16. A servomotor with armature control has a mechanical time constant of 0.02 s and a velocity gain constant of 75 rpm/V. The electrical time constant of the armature winding can be assumed negligible. Sketch the Bode diagram for the *position* transfer function $\theta(j\omega)/V_a(j\omega)$ of the motor.

6-17. The tank shown below has a cross-sectional area of 20 ft². The inflow q_i is proportional to the valve stem position θ according to the equation $q_i = 8\theta$ ft³/min. The outflow q_o is related to the head h according to $q_o = 4h$ ft³/min. Derive the transfer function $H(s)/\theta(s)$, and plot the Bode diagram for this transfer function.

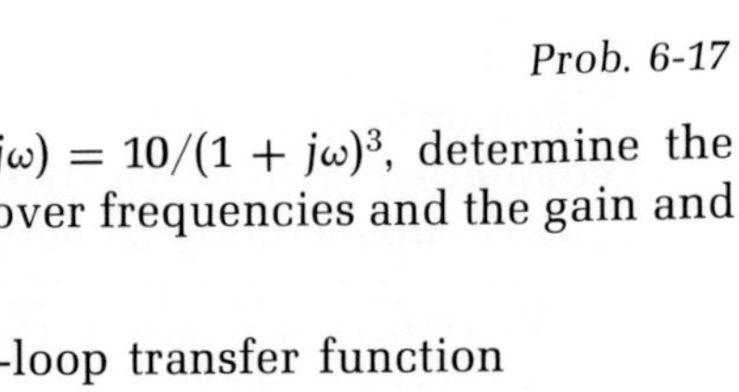

Prob. 6-17

6-18. For the system in which $GH(j\omega) = 10/(1 + j\omega)^3$, determine the gain-crossover and phase-crossover frequencies and the gain and phase margins.

6-19. A control system has the open-loop transfer function

$$GH(j\omega) = \frac{K}{(1 + j0.5\omega)(1 + j0.05\omega)(1 + j0.005\omega)}$$

where K is a variable gain in the system. Determine the value of K that will result in a phase margin of 45°.

6-20. Determine the bandwidth for the system whose transfer function is

$$\frac{C(j\omega)}{R(j\omega)} = \frac{1}{1 + j\omega}$$

6-21. Determine the resonant peak M_m, the resonant frequency ω_m, and the bandwidth for the system whose transfer function is

$$\frac{C(j\omega)}{R(j\omega)} = \frac{5}{5 + j2\omega + (j\omega)^2}$$

6-22. Using the low-frequency–high-frequency approximation method, determine the closed-loop magnitude response for the following system.

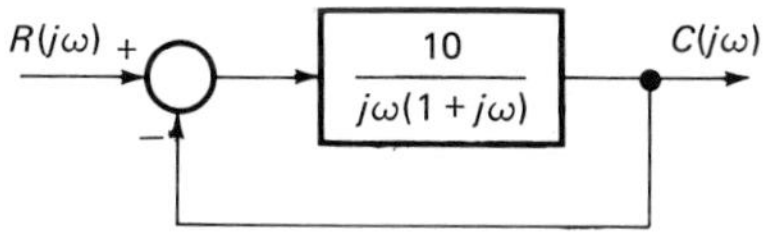

Prob. 6-22

6-23. Determine the closed-loop response for the system in Prob. 6-22 by using the phase margin to find the damping ratio. Compare the result with the response found in Prob. 6-22.

6-24. Determine the equation for the closed-loop response $C(s)/R(s)$ for the system below, and calculate the value of the gain K to provide a closed-loop damping ratio of 0.5.

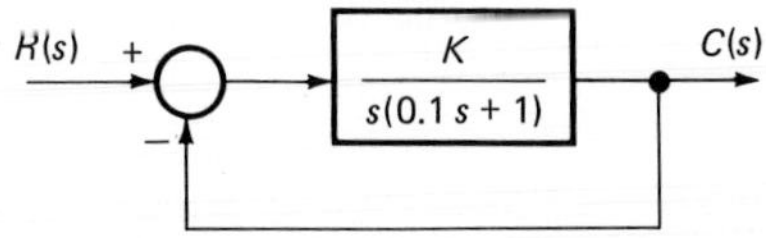

Prob. 6-24

6-25. Using the Nichols chart, determine the value of K to provide $\zeta = 0.5$ for the system in Prob. 6-24. Hint: Use Eq. (6-44) to find the appropriate M_m contour.

6-26. A unity-feedback control system has the open-loop transfer function

$$GH(s) = \frac{10K(5s + 1)}{s^2(0.25s + 1)}$$

where K represents a variable gain setting.

a. Determine the value of K that will result in an M_m of 0.2 for the closed-loop system.
b. For the gain determined in part a, make a plot of the closed-loop frequency response.

Chapter Seven

control system design

7-1 INTRODUCTION

The objective of automatic control is to perform specific tasks such as the control of position, flow, temperature, or other physical quantities. Control system design is basically concerned with the arrangement of the system and the selection of suitable elements or components to accomplish the purpose of the system. Knowing the transfer function of the various elements, a theoretical design may be established and its performance analyzed. This design is then refined through experimentation and further analysis until the desired performance is obtained.

An extensive treatment of system design is beyond the scope of an introductory controls text since a wide variety of methods exist and are available to the designer. This chapter will be limited to describing one of the more common design methods that makes use of the frequency-response techniques presented in Chap. 6.

The root-locus method described in Sec. 5-12 provides another basic approach to design. In this method the system design is accomplished by locating the roots of the characteristic equation on the complex plane to obtain the desired performance. Detailed information concerning the root-locus as well as other design methods may be obtained from references which concentrate on the design aspect of automatic control.

The first design step is to define the controlled variable and the specifications for control of that variable. These specifications pertain to the accuracy, speed of response, and stability of the control system. Transient-response specifications were considered in Chap. 4, and frequency-response specifications were described in Chap. 6.

Chapters 5 and 6 outlined methods of analyzing system performance in the time and frequency domains, including methods of investigating system stability. These methods may be extended to designing a control system to meet desired performance specifications. We have already seen that system response may be altered by adjusting a system parameter such as the gain of the controller. Frequently, however, a gain adjustment does not provide sufficient alteration of the system to meet the specifications. For example, if the gain is high enough to reduce steady-state errors to the desired value, it may cause the system to become unstable.

When a simple parameter adjustment does not achieve the desired results, it becomes necessary to modify the system in some way. Often in practice the best and simplest way to improve control system performance is to select different components if this is possible. For example, the behavior of a positional servomechanism could be improved by choosing a better motor for the system. When this is not possible, the existing system may be altered to provide satisfactory performance. This alteration of a control system is known as *compensation* since its purpose is to make up or compensate for deficiencies or inadequacies in the initial design. This chapter will consider methods of control system compensation and their effect on system behavior.

Compensation generally involves adding additional elements called *compensators* to the system controller. If the design is being performed using frequency-response methods, a compensator is selected that provides a transfer function which favorably alters the overall frequency response of the system. The compensating devices may consist of electrical networks or mechanical equipment containing springs, levers, dashpots, etc.

Two different methods of compensation are available depending on the compensating elements being located as follows:

1. In cascade with the forward transfer function—cascade or series compensation
2. In a feedback path—feedback or parallel compensation

Selection of the location of the compensator depends largely on the particular control system and the results desired. An adequately compensated system is stable, has a satisfactory transient response, and has a large enough gain to reduce the steady-state error to an acceptable level. However, as has been pointed out, it is usually necessary to compromise among conflicting performance specifications when it is not possible to meet all specifications in an optimum manner.

7-2 GENERAL CONSIDERATIONS

It is possible to make some general observations about the characteristics that control systems should possess. These characteristics may then serve as a guide for choosing specific compensation. This section will discuss general considerations for compensating a system to obtain satisfactory performance. In particular we will use compensation to reshape the frequency-response characteristic presented on a Bode diagram.

In the lower range of frequencies (those below the gain-crossover frequency) it is desirable that the output of the control system follow the input with minimum error. This requires that the system provide as high a forward gain as possible at low frequencies.

In the range of frequencies in the vicinity of the gain-crossover frequency the slope of the attenuation curve should be limited to about -30 dB per decade to ensure stability. To allow for some variation in gain, the -30 dB per decade slope should extend at least an octave above and below the crossover frequency.

At high frequencies (those above the gain-crossover frequency) it is usually desirable to attenuate the open-loop characteristic as rapidly as possible. This will reduce the effect of any high-frequency noise or disturbance which may enter the system. An example is electrical noise that may be generated internally or picked up by sensitive electronic equipment.

It was pointed out in Sec. 6-11 that performance in the frequency domain can be described in terms of the resonant peak M_m and its associated frequency ω_m. The value of M_m is essentially determined by the damping ratio and is therefore related to the amount of overshoot in the transient response. The frequency ω_m is related to the natural resonant frequency ω_n according to Eq. (6-43). The frequency ω_n in turn defines the transient-response time of the system. In addition the response at low frequencies can be related to the steady-state error specification of the system.

The situations that require compensation of the open-loop frequency-response characteristic generally fall into these categories:

1. A particular system is stable and exhibits satisfactory transient and frequency response. However, the steady-state error is excessive. Therefore the low-frequency gain must be increased to reduce the steady-state error while maintaining the high-frequency portion of the plot. This form of compensation is indicated by the dashed lines in Fig. 7-1*a*.

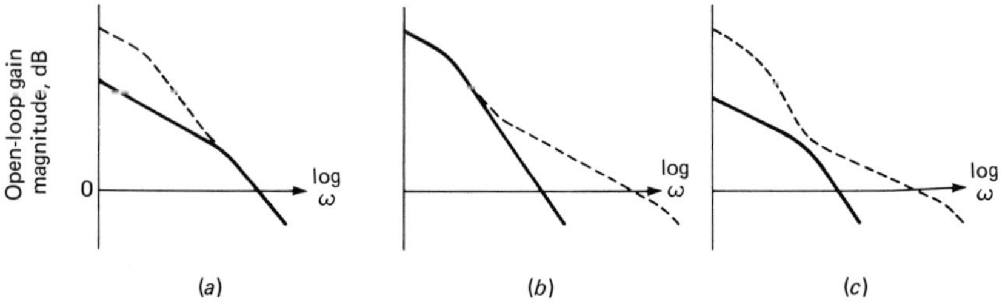

Figure 7-1 Compensation for (a) increased low-frequency gain, (b) extended high-frequency response, and (c) combination of the two forms

2. A particular system is stable with satisfactory steady-state error but the transient response is unsatisfactory. The high-frequency portion of the response plot must be altered to increase the gain-crossover frequency as shown in Fig. 7-1*b*.

3. A given system is stable but neither the steady-state nor the transient response is satisfactory. Both areas of the overall response must be improved by increasing the low-frequency gain and extending the high-frequency response. Figure 7-1*c* illustrates this type of compensation.

The compensating element may take one of several possible forms depending on the alteration of the system transfer function needed to obtain the desired performance. Three basic types of compensation corresponding to the situations described above will be described in this chapter. These are:

1. Proportional plus integral compensation, or lag compensation

2. Proportional plus derivative compensation, or lead compensation

3. Proportional plus integral plus derivative compensation, or lag-lead compensation

It is perhaps appropriate at this point to note that the design of a closed-loop system does not follow a rigid set of rules. Rather, it involves applying basic principles in reshaping the frequency-response plot to obtain a system that meets a stated set of specifications. In this chapter these principles will be described by means of illustrative examples.

7-3 PROPORTIONAL CONTROL

In the general arrangement of a feedback control system, the error signal $E(s)$ is applied to the controller $G_c(s)$. The controller output provides the correction input $M(s)$ to the system being controlled $G_s(s)$.

The basic control mode for a feedback system is *proportional control* in which the corrective action that is taken is proportional to the error. This may be expressed mathematically as

$$m = K_p e \tag{7-1}$$

where m = manipulated variable
K_p = proportional gain of G_c
e = error signal

The response of proportional control action is assumed to be instantaneous. There is no delay between a change in the error signal and the corresponding change in the manipulated variable. The transfer function for a proportional controller is

$$G_c(s) = \frac{M(s)}{E(s)} = K_p \tag{7-2}$$

which in the frequency domain is

$$G_c(j\omega) = \frac{M(j\omega)}{E(j\omega)} = K_p \tag{7-3}$$

Thus the frequency-response characteristics of proportional control are a constant magnitude ratio and a constant phase angle of 0°.

Although proportional control is the basic control mode, the fact that it cannot completely eliminate error in the controlled output is a serious disadvantage. A steady-state error is required to maintain the output at a given value. This error is often referred to as *proportional offset.*

The application of proportional control would imply that the proportional gain should be as high as possible. Unfortunately, as we are well aware, increased gain increases the tendency toward instability. It thus becomes necessary to add compensation to the basic proportional control mode.

EXAMPLE 7-1

Note: In this and the succeeding examples in this chapter, we will use the positional servomechanism described in Example 6-11 as a design example. This servomechanism is shown in block-diagram form in Fig. 7-2. The overall open-loop transfer function of this servomechanism was found to be

$$G(j\omega) = \frac{0.286K_2}{j\omega(1 + j0.05\omega)(1 + j0.02\omega)} \tag{7-4}$$

where K_2 is the amplifier transfer function.

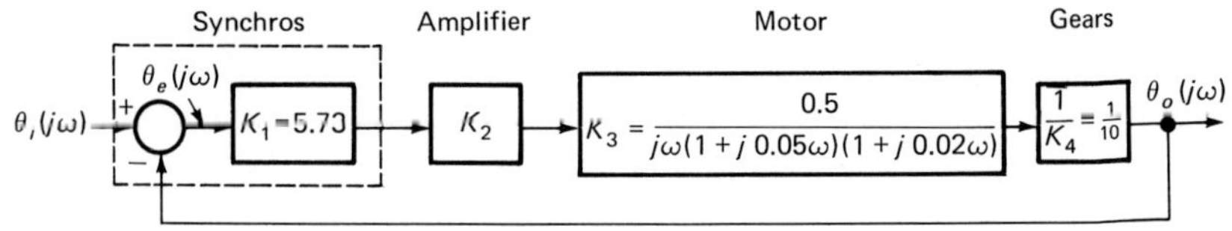

Figure 7-2 Positional servomechanism for Chap. 7 examples

The servomechanism in Example 6-11 is required to follow a ramp input of 10 rad/s. Determine the steady-state error in the output if $K_2 = 61.9$.

Solution

The positional servomechanism is a Type 1 system as indicated by the $j\omega$ term in the denominator of Eq. (7-4). Referring to Table 5-2, we note that for a Type 1 system with a ramp input

$$e_{ss} = \frac{A}{K_v}$$

where K_v is the velocity constant. Thus from Eq. (5-44),

$$K_v = \lim_{s\to 0} sG(s) = 0.286 \times 61.9 = 17.7\ \text{s}^{-1}$$

The steady-state error is then

$$e_{ss} = \frac{10}{17.7} = 0.565\ \text{rad}$$

7-4 PROPORTIONAL PLUS INTEGRAL CONTROL

When the transient response of a feedback control system is considered satisfactory but the steady-state error is excessive, the error can be reduced by providing increased system gain at low frequencies. However, this form of compensation must be accomplished without appreciably changing the open-loop response in the vicinity of the gain-crossover frequency.

If the compensating element introduces an integral control action into the forward gain path, the correction m will be proportional to the time integral of the error e. This is given by the equation

$$m = K_i \int e\,dt \tag{7-5}$$

or

$$M(s) = \frac{K_i E(s)}{s} \tag{7-6}$$

where K_i = integral gain constant, s^{-1}

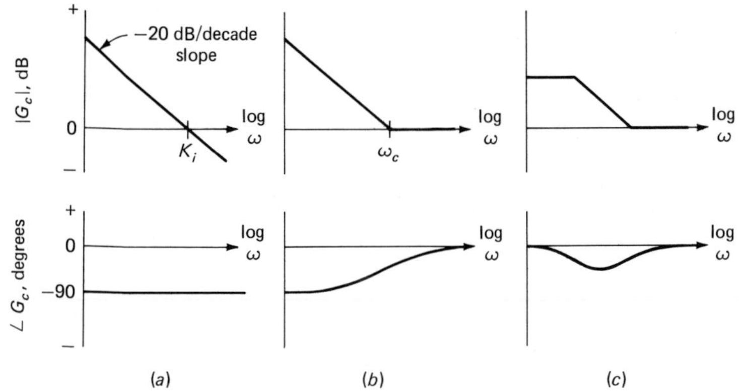

Figure 7-3 Bode diagram of (a) integral control, (b) proportional plus integral control, and (c) lag compensation

Equation (7-5) indicates that control action continues to increase as long as any error is present. When sufficient actuating signal has accumulated, the error signal will have decayed to 0.

If the control function represented by Eq. (7-6) is expressed in the frequency domain, we have

$$G_c(j\omega) = \frac{M(j\omega)}{E(j\omega)} = \frac{K_i}{j\omega} \tag{7-7}$$

This transfer function is plotted in Fig. 7-3*a*. The magnitude ratio falls off at a slope of −20 dB per decade, and at a frequency $\omega = K_i$, the ratio is unity. At 0 frequency or direct current, the integrating action produces infinite gain. Although the gain does increase at lower frequencies, the function is also effective at high frequencies. Hence, although pure integral control would eliminate steady-state errors, it will also adversely affect the transient response and, most likely, lead to instability.

A close approximation to integral control can be obtained from active networks. For example, the operational-amplifier circuit shown in Fig. 3-47 will provide an output that is proportional to the time integral of the input.

The steady-state error can be reduced without appreciably altering the high-frequency characteristic by combining the integral control function with a proportional function. This is shown by the block diagram in Fig. 7-4. The transfer function of the controller is

$$G_c(s) = \frac{M(s)}{E(s)} = K_p\left(1 + \frac{K_i}{s}\right) \tag{7-8}$$

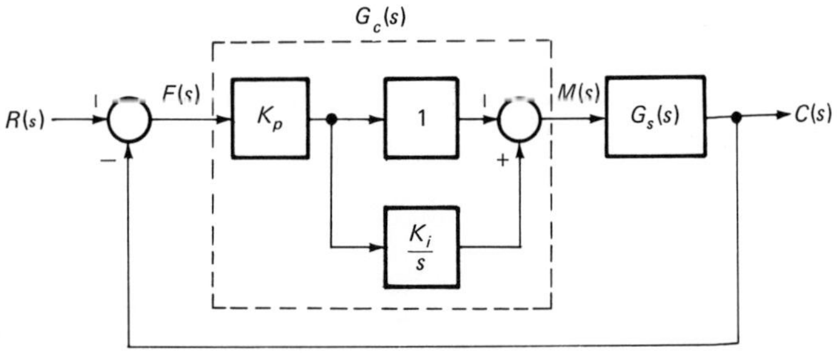

Figure 7-4 Proportional plus integral control action

which in the frequency domain becomes

$$G_c(j\omega) = K_p\left(1 + \frac{K_i}{j\omega}\right) = K_p\left(1 + \frac{1}{j\omega/\omega_c}\right) \tag{7-9}$$

where ω_c is a corner frequency equal to the value of K_i. The Bode diagram of this function is shown in Fig. 7-3*b*. The constant K_i must be selected to maintain satisfactory transient response.

The operational-amplifier circuit shown in Fig. 7-5 can be used to obtain the proportional plus integral control function shown in Fig. 7-3*b*.

Following the method of analysis used in Sec. 3-18, it can be shown that the generalized transfer function of an operational amplifier with a sinusoidal input signal is given by the equation

$$\frac{E_o(j\omega)}{E_i(j\omega)} = -\frac{Z_f}{Z_i} \tag{7-10}$$

where Z_f and Z_i are the impedances of the feedback and input circuits, respectively. Applying this relationship to the circuit in Fig. 7-5 gives

$$\frac{E_o(j\omega)}{E_i(j\omega)} = -\frac{R_f + 1/j\omega C}{R_i} = -\frac{R_f}{R_i}\left(1 + \frac{1}{j\omega CR_f}\right) \tag{7-11}$$

Since this equation is of the form of Eq. (7-9), the circuit does provide

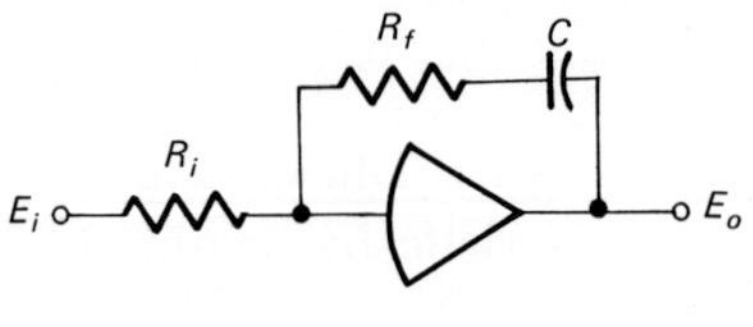

Figure 7-5 Operational-amplifier circuit for proportional plus integral control

a proportional plus integral function. Equating like terms, we have

$$\omega_c = \frac{1}{CR_f} \qquad \text{and} \qquad K_p = \frac{R_f}{R_i} \tag{7-12}$$

The product CR_f is a time constant having units of seconds where R is in ohms and C in farads.

7-5 LAG COMPENSATION

Proportional plus integral control is frequently approximated by a transfer function that can be conveniently obtained with passive electrical networks. This compensating function is shown in Fig. 7-3c. Because it introduces a lagging phase angle over a range of frequencies, the function is known as *lag compensation*. The design of lag networks and their application will be covered in this section.

Lag compensation can be provided by the circuit shown in Fig. 7-6. Assuming that the network is driven by a source with 0 impedance and the load is infinite, the transfer function $G_c(j\omega)$ is

$$G_c(j\omega) = \frac{E_o(j\omega)}{E_i(j\omega)} = \frac{R_2 + 1/j\omega C}{R_1 + R_2 + 1/j\omega C} = \frac{1 + j\omega CR_2}{1 + j\omega C(R_1 + R_2)} \tag{7-13}$$

Equation (7-13) can be written in the form

$$G_c(j\omega) = \frac{1 + j\omega\tau}{1 + j\omega\alpha\tau} \tag{7-14}$$

where $\tau = CR_2$

$$\alpha = \frac{R_1 + R_2}{R_2}$$

Figure 7-7 shows the Bode diagram of this function for various values of α. The frequency scale is normalized in units of $\omega\tau$. The magnitude plot exhibits two corner frequencies. The asymptote changes

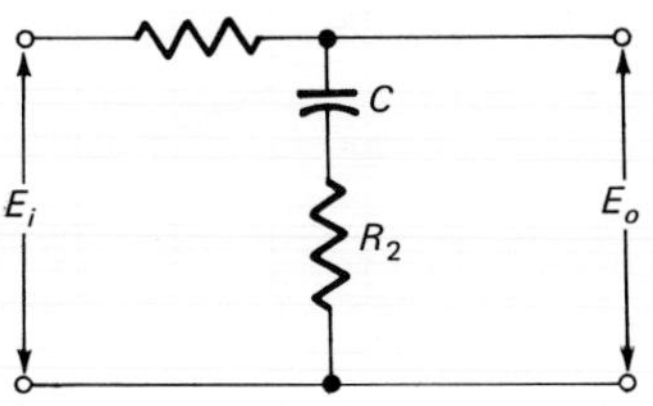

Figure 7-6 Network for lag compensation

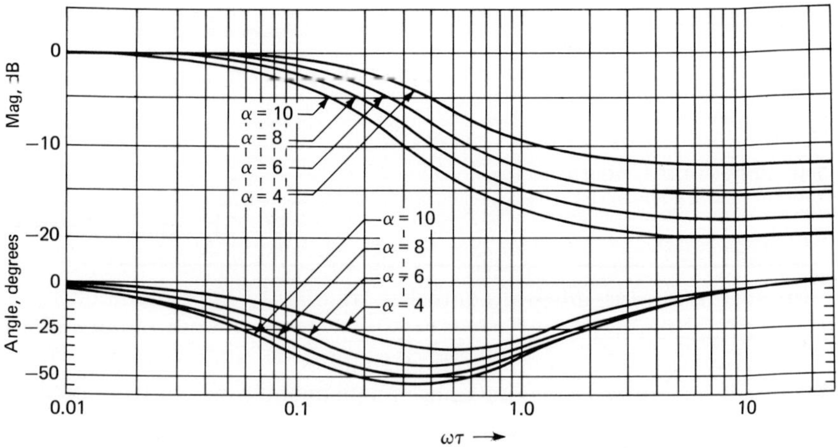

Figure 7-7 Bode diagram for lag compensation

from 0 dB per decade to −20 dB per decade at the lower corner frequency $\omega\tau = 1/\alpha$ and back to 0 dB per decade at the upper corner $\omega\tau = 1$. At frequencies above $\omega\tau = 1$ the network introduces an attenuation equal to $20 \log_{10} \alpha$ dB. The maximum phase-angle lag ϕ_m occurs at a frequency ω_{ϕ_m}, which is the geometric mean between the two corner frequencies. Thus

$$\omega_{\phi_m} = \frac{1}{\tau\sqrt{\alpha}} \tag{7-15}$$

The maximum lag is a function of α, and this relationship is shown in Fig. 7-8.

The design of a phase-lag network for compensation can be readily accomplished on the Bode diagram since the frequency response of

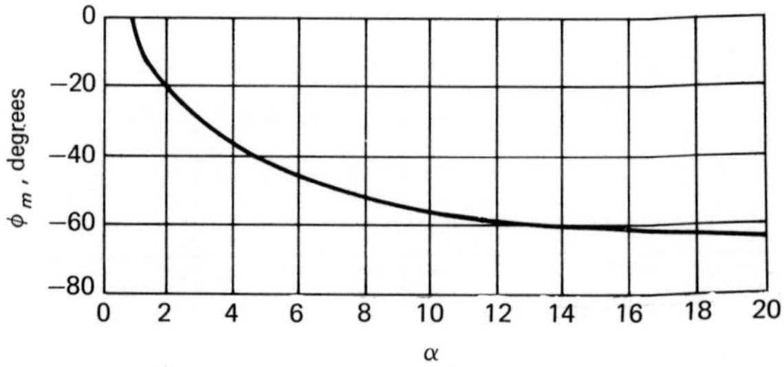

Figure 7-8 Maximum phase angle versus α for lag network

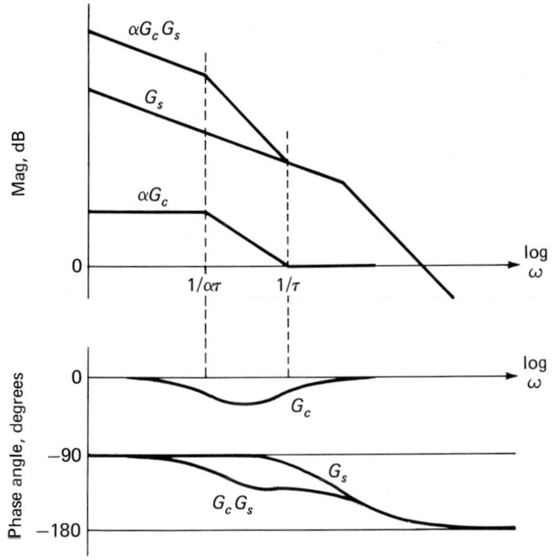

Figure 7-9 Original (G_s) and lag-compensated (G_cG_s) response curves

the compensator can be simply added to the response of the uncompensated system. The procedure for designing lag compensation is given in the following steps as illustrated by Fig. 7-9.

1. Draw the Bode diagram of the uncompensated system $G_s(j\omega)$ with the gain adjusted for the desired transient response.

2. Determine the additional gain that is necessary to meet the steady-state error specifications, and calculate the value of α associated with this gain increase.

3. Locate the appropriate compensator phase curve on the Bode diagram so that about 5° of additional phase lag appears at the gain-crossover frequency.

4. Add the compensator magnitude curve to that of the uncompensated system so that the low-frequency gain is *increased* by the value of α.

5. Determine the closed-loop response of the compensated system from the Nichols chart.

In summary, α is chosen to provide the required low-frequency gain and τ is selected so that the phase lag is essentially unaffected around the gain-crossover frequency. There is generally no practical limit on the maximum value of α. However, for large values such as $\alpha > 20$, it may be preferable to use an active network such as an operational amplifier.

If the required compensation introduces excessive phase lag over a portion of the frequency range, two lag networks may be cascaded to provide the same overall attenuation with less phase lag but with extension over a wider frequency band.

To illustrate the above design procedure as well as provide a basis for comparing the various methods of compensation, the servomechanism of Example 6-11 will be used in the following example.

EXAMPLE 7-2

The servomechanism in Example 6-11 is required to follow a ramp input of 10 rad/s with a maximum steady-state error of 5° (0.0873 rad). Determine the amplifier gain required to meet this error specification. Design a suitable series compensating network, assuming that the existing transient response of the servomechanism is satisfactory.

Solution

It is necessary to first determine the loop gain to meet the error specification. Referring to Table 5-2, we note that for a Type 1 system with a ramp input

$$e_{ss} = \frac{A}{K_v}$$

where K_v = velocity constant = required loop gain

Thus

$$K_v = \frac{A}{e_{ss}} = \frac{10}{0.0873} = 114.5 \text{ s}^{-1}$$

The required amplifier gain is then

$$K_2 = \frac{114.5}{0.286} = 400 = 52.0 \text{ dB}$$

The frequency response specified in Example 6-11 was obtained by setting the amplifier gain at 61.9 or 35.8 dB. To meet the 5° tracking-error specification the low-frequency amplifier gain must be increased by 52.0 − 35.8 = 16.2 dB or a factor of 6.45. The addition of lag compensation with $\alpha = 6.45$ will introduce 16.2 dB of attenuation at high frequencies to maintain the existing transient response.

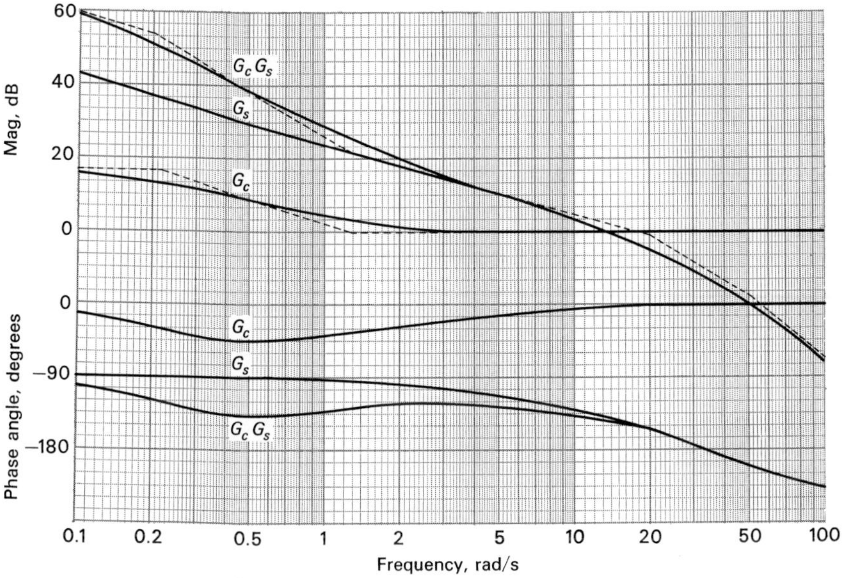

Figure 7-10 Bode diagram for Example 7-2

Figure 7-10 is the Bode diagram of the open-loop system. The response of the uncompensated servo G_s is shown with the system gain $K_T = 17.7 = 25$ dB as determined in Example 6-11. The lag compensator G_c has been located on the diagram so that it increases the overall phase angle by about 5° at the gain-crossover frequency of 14 rad/s. The compensated magnitude curve G_cG_s shows the gain increase of 16.2 dB at low frequencies. Note that this increase is removed at higher frequencies by the compensating network. The resulting velocity constant is $17.7 \times 6.45 = 114\text{s}^{-1}$.

The open-loop response is plotted on a Nichols chart in Fig. 7-11. The closed-loop response as read from this chart is shown in Fig. 7-12. Comparing this response with that of the uncompensated system in Fig. 6-28 indicates that the frequency response (and in turn the transient response) has been essentially unchanged by the addition of the lag compensation.

The compensating network can be designed with the aid of Eqs. (7-14) and (7-15). From Fig. 7-10, we note that $\omega_{\phi_m} = 0.55$ rad/s. Then, for $\alpha = 6.45$,

$$\tau = \frac{1}{\omega_{\phi_m}\sqrt{\alpha}} = \frac{1}{0.55\sqrt{6.45}} = 0.72 \text{ s}$$

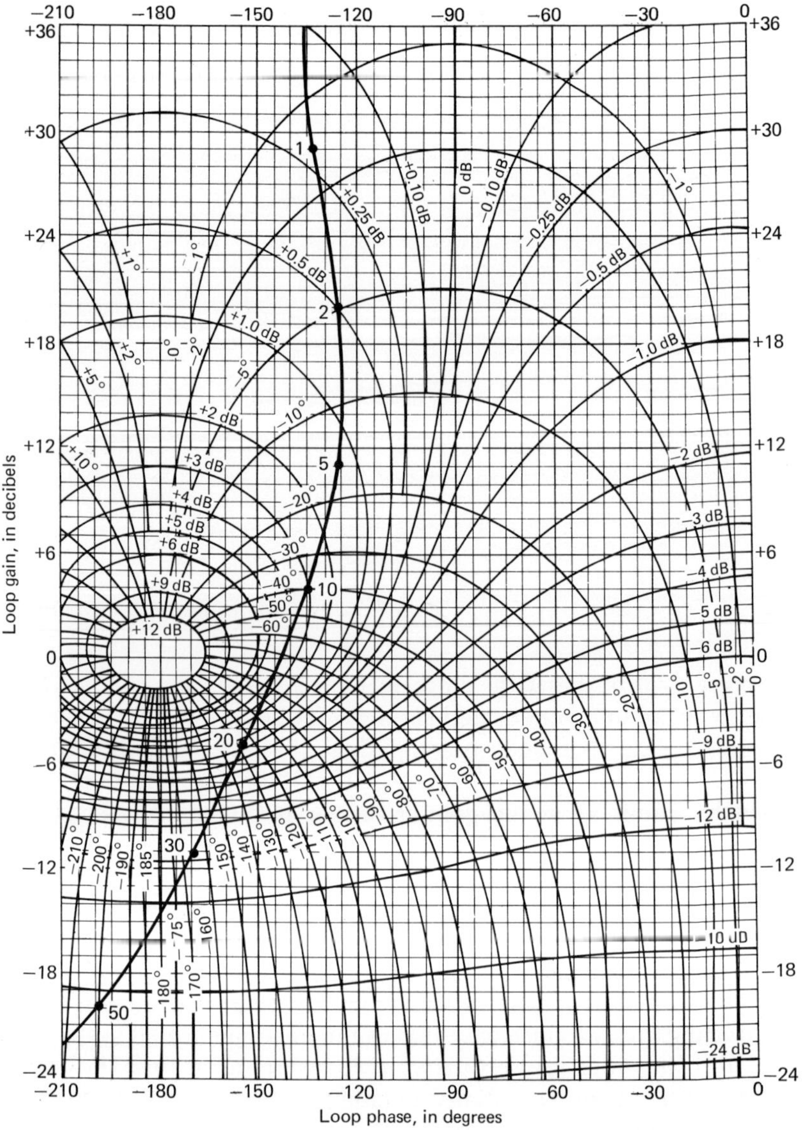

Figure 7-11 Nichols chart for Example 7-2

and the transfer function of the compensating network is

$$G_c(j\omega) = \frac{1 + j0.72\omega}{1 + j4.64\omega}$$

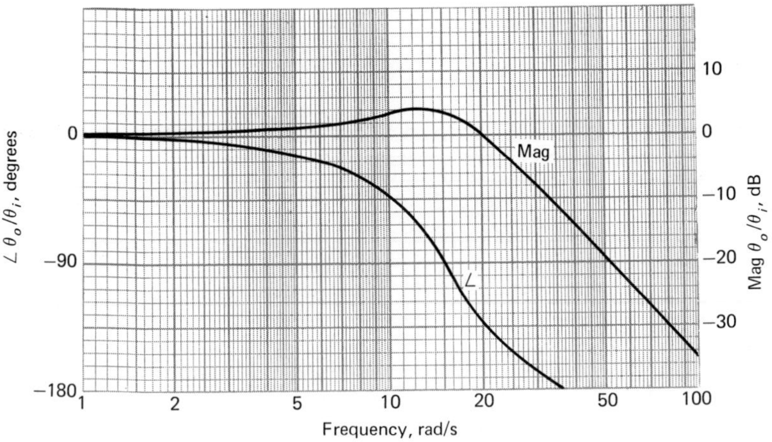

Figure 7-12 Closed-loop response for lag-compensated system in Example 7-2

The component values in a specific network are necessarily related to the source and load impedances. If, for example, R_2 should be chosen as 10 kΩ,

$$C = \frac{\tau}{R_2} = \frac{0.72}{10^4} = 72\mu \text{ farads}$$

and
$$R_1 = R_2(\alpha - 1) = 10^4(6.45 - 1) = 54.5 \text{ k}\Omega$$

Finally the overall open-loop transfer function of the servomechanism is

$$G(j\omega) = \left[\frac{5.73 \times 400 \times 0.5/10}{j\omega(1 + j0.05\omega)(1 + j0.02\omega)}\right]\left[\frac{1 + j0.72\omega}{1 + j4.64\omega}\right]$$

$$= \frac{114.5(1 + j0.72\omega)}{j\omega(1 + j0.05\omega)(1 + j0.02\omega)(1 + j4.64\omega)}$$

7-6 PROPORTIONAL PLUS DERIVATIVE CONTROL

We have seen how proportional plus integral control or lag compensation can be employed to improve the steady-state response of a control system. When we want to improve the transient performance, proportional plus derivative control may be used.

If a series compensating element that introduces derivative control action is added to the controller, a correction is made that is proportional to the rate of change or derivative of the error signal. This control action is given by the equation

$$m = K_d \frac{de}{dt} \tag{7-16}$$

or

$$M(s) = sK_d E(s) \tag{7-17}$$

where K_d = derivative gain constant, s

The frequency-response transfer function of derivative control may be obtained from Eq. (7-17) as

$$G_c(j\omega) = \frac{M(j\omega)}{E(j\omega)} = j\omega K_d \tag{7-18}$$

This function is shown in Fig. 7-13*a*.

Derivative control is useful because in responding to the rate of change of the error, it can produce a significant correction before the magnitude of the error becomes large. Thus derivative control may be regarded as anticipatory in nature since it has the effect of introducing immediate control action.

Derivative action is useful in controlling systems with sudden input or load changes because it produces a control signal while the error is changing. Since the derivative signal opposes changes in the controlled output, it provides a stabilizing effect in a closed-loop system which can damp a tendency toward oscillation.

It is important to note that derivative control cannot be used alone since it will not respond to a steady-state error. It must therefore be used in combination with the proportional or proportional plus integral control modes. Figure 7-14 shows a series compensator that provides proportional plus derivative control action. The transfer function of the controller is now

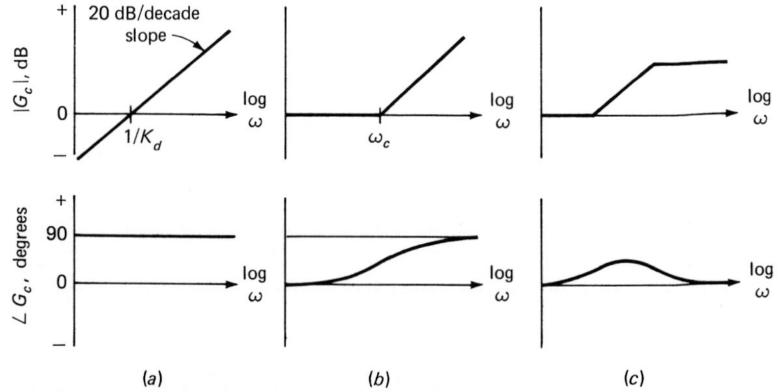

Figure 7-13 Bode diagram of (a) derivative control, (b) proportional plus derivative control, and (c) lead compensation

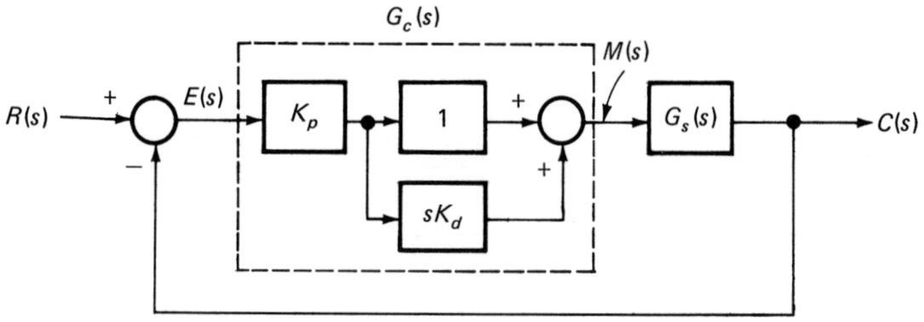

Figure 7-14 Proportional plus derivative control action

$$G_c(s) = \frac{M(s)}{E(s)} = K_p(1 + K_d s) \tag{7-19}$$

which in the frequency domain is

$$G_c(j\omega) = K_p(1 + j\omega K_d) = K_p(1 + j\omega/\omega_c) \tag{7-20}$$

where ω_c = corner frequency = $1/K_d$

The proportional plus derivative function is shown in the Bode diagram in Fig. 7-13*b*. However, since a gain approaching ∞ is required at high frequencies, it should be apparent that this function cannot be realized in practice. In addition, derivative action amplifies unwanted noise or spurious signals that are frequently present in the error signal. This can result in an overload or saturation of the controller, which effectively blocks control action. Therefore an approximate derivative function is preferred in practice. The frequency-response characteristic of this function is shown in Fig. 7-13*c*. Because a leading phase angle is introduced over a range of frequencies, it is often referred to as *lead compensation*.

7-7 LEAD COMPENSATION

As mentioned above, phase-lead compensation is a practical form of proportional plus derivative control. Figure 7-15 shows a lead compensating network that is made up of passive circuit elements. If the

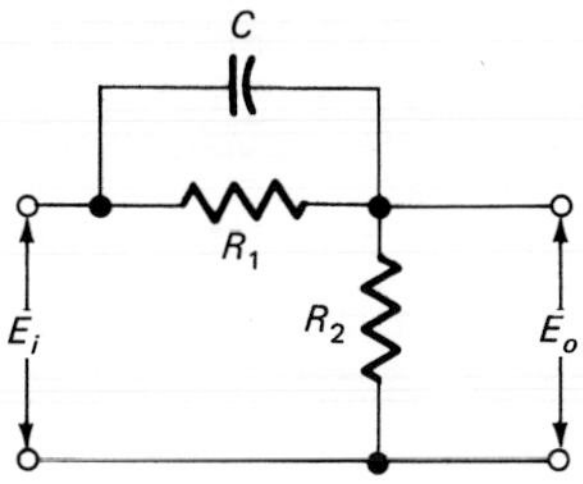

Figure 7-15 Network for lead compensation

source impedance is assumed to be 0 and the load impedance infinite (or part of R_2),

$$G_c(j\omega) = \frac{E_o(j\omega)}{E_i(j\omega)} = \frac{R_2}{R_2 + \dfrac{R_1/j\omega C}{R_1 + 1/j\omega C}}$$

$$= \left(\frac{R_2}{R_1 + R_2}\right) \frac{1 + j\omega C R_1}{1 + \dfrac{j\omega C R_1 R_2}{R_1 + R_2}}$$

$$= \frac{1}{\alpha}\left(\frac{1 + j\omega\tau}{1 + j\omega\tau/\alpha}\right) \tag{7-21}$$

where $\tau = CR_1$

$$\alpha = \frac{R_1 + R_2}{R_2}$$

Figure 7-16 shows the Bode diagram of this transfer function for various values of α.

Comparing Fig. 7-16 with the lag characteristic shown in Fig. 7-7, we note that the curves have an inverse relationship. The magnitude asymptote of the plot of the lead function changes from 0 dB per decade to 20 dB per decade at the lower corner frequency $\omega\tau = 1$ and back to 0 dB per decade at the upper corner frequency $\omega\tau = \alpha$. An attenuation equal to $20 \log_{10} \alpha$ dB is introduced at frequencies below $\omega\tau = 1$.

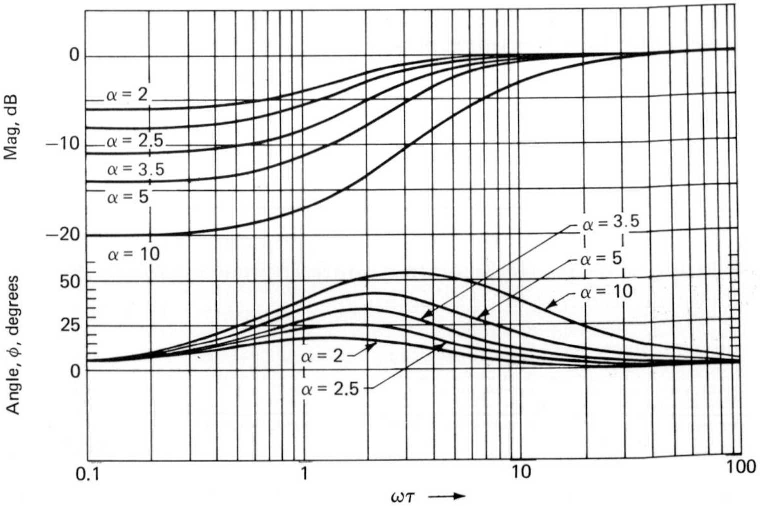

Figure 7-16 Bode diagram for lead compensation

The lead compensator introduces a leading or positive phase angle over a relatively narrow band of frequencies. The maximum phase angle occurs at a frequency

$$\omega_{\phi_m} = \frac{\sqrt{\alpha}}{\tau} \tag{7-22}$$

The maximum phase-angle curve shown in Fig. 7-8 is also applicable to lead compensation with the difference that ϕ_m will be positive rather than negative.

The purpose of lead compensation is to increase the frequency ω_m at which M_m occurs, which in turn results in improved transient response. The gain-crossover frequency can be increased by properly selecting the value of τ. In practice, the value of α should be limited to a maximum of 20, and is usually 10 to 15 because of the problem of the high-frequency noise increasing to unacceptable proportions. If a phase lead of more than 60° is desired, it is best achieved by cascading two or more lead networks.

Design of the compensating network can be accomplished using the guidelines outlined below and illustrated in Fig. 7-17.

1. Draw the Bode diagram for the uncompensated system $G_s(j\omega)$, and adjust the system gain to satisfy the steady-state-error requirements.

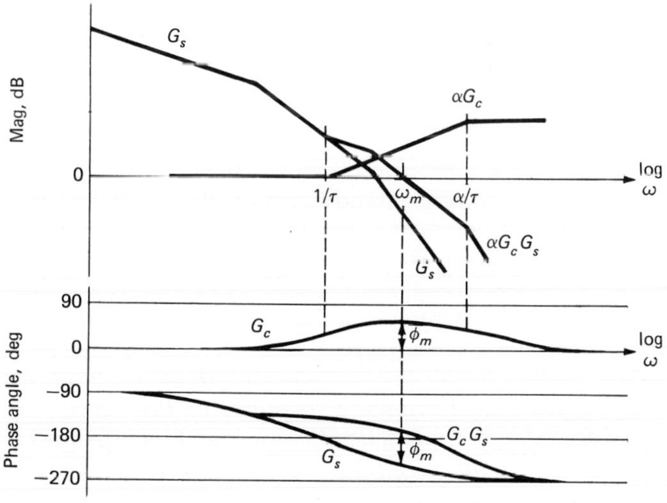

Figure 7-17 Original (G_s) and lead-compensated (G_cG_s) response curves

2. Find the phase angle for the uncompensated system at the desired ω_m of the closed-loop system.

3. From the Nichols chart, find the minimum value of the open-loop phase angle corresponding to the specified value of M_m (this is the phase angle of a vertical line tangent to the M_m ellipse). For example, with $M_m = 3$ dB, the angle is 135°.

4. The compensator phase lead ϕ_m is the difference between the angles found in steps 2 and 3.

5. From Fig. 7-8 determine the value of α corresponding to ϕ_m. If ϕ_m exceeds about 60°, several stages of compensation should be used.

6. Plot $G_c(j\omega)$ on the Bode diagram with ϕ_m located at the frequency ω_m. Graphically combine $G_s(j\omega)$ and $G_c(j\omega)$ to obtain the compensated open-loop curve. [In Fig. 7-17 the magnitude of $G_c(j\omega)$ has been multiplied by α to better illustrate its effect on the response.]

7. Plot the open-loop curve on the Nichols chart, and adjust the loop gain so that the curve is tangent to the desired M_m contour. The frequency at the point of tangency will not be exactly ω_m, since the point of tangency may not be at the extreme of the contour as assumed in step 3 above. However, the values obtained will be close.

8. Determine the component values for the compensating network, and increase the gain of the amplifier to provide the loop gain found in step 7.

The use of lead compensation will be illustrated by applying the above procedure to the servomechanism of Example 6-11.

EXAMPLE 7-3

It is desired that the bandwidth of the servomechanism in Example 6-11 be increased so that an M_m of 3 dB occurs at or slightly above 30 rad/s frequency. Provide series compensation to meet this specification.

Solution

The open-loop frequency response of the servo was developed in Fig. 6-27. The uncompensated response G_s is repeated in Fig. 7-18 with $K_T = 17.7 = 25$ dB. From the phase curve for G_s we note that the angle is 175° at $\omega = 30$ rad/s. From the Nichols chart we see that the minimum open-loop phase angle for $M_m = 3$ dB is 135°. Hence the compensation should provide $175° - 135° = 40°$ of phase lead at $\omega_{\phi_m} = 30$ rad/s.

Referring to Fig. 7-8, a compensator with $\alpha = 5$ provides approximately

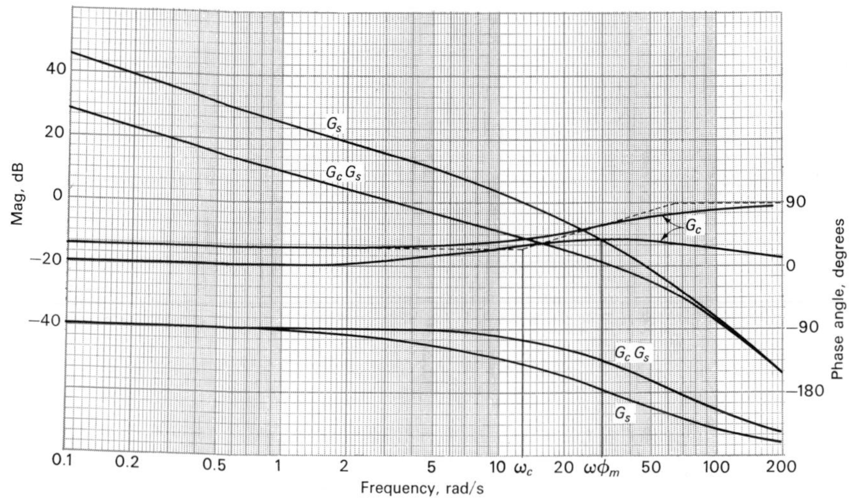

Figure 7-18 Bode diagram for Example 7-3

40° of maximum phase lead. The transfer function G_c of this compensator is added to the G_s curves giving the compensated response G_cG_s. The G_c function is located so that $\omega_{\phi_m} = 30$ rad/s. Note that the compensated magnitude response is reduced by $1/\alpha = 0.2 = -14$ dB at low frequencies.

The compensated function is plotted as the solid line on the Nichols chart shown in Fig. 7-19. The gain is then increased by 17 dB to make the curve tangent to the 3-dB contour. This is shown by the dashed line in Fig. 7-19.

The closed-loop response as read from the Nichols chart is shown in Fig. 7-20. We see that the response peaks to a 3 dB value at a frequency of 35 rad/s, which slightly exceeds the minimum specified frequency.

Since the loop gain was increased by 17 dB or a factor of 7.1, the velocity constant K_v is $17.7 \times 0.2 \times 7.1 = 25.1\ \text{s}^{-1}$. The required amplifier gain is

$$K_2 = \frac{25.1}{0.286} = 87.8 = 38.9\ \text{dB}$$

A suitable compensating network may now be designed. From Eq. (7-22)

$$\tau = \frac{\sqrt{\alpha}}{\omega_{\phi_m}} = \frac{\sqrt{5}}{30} = 0.0745\ \text{s}$$

or

$$\omega_c = \frac{1}{0.0745} = 13.4\ \text{rad/s}$$

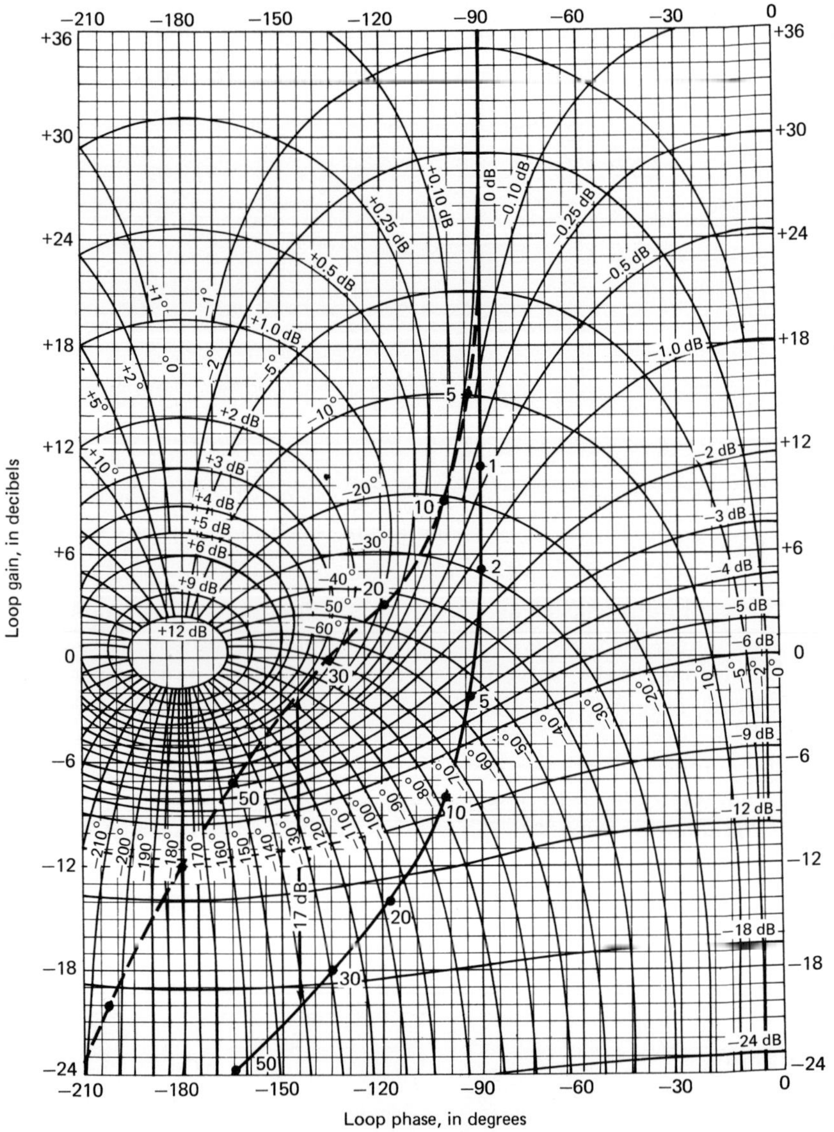

Figure 7-19 Nichols chart for Example 7-3

which agrees with the asymptotes of the G_c magnitude curve in Fig. 7-18. The transfer function of the compensating network is

$$G_c(j\omega) = \frac{1}{5}\left(\frac{1 + j0.0745\omega}{1 + j0.0149\omega}\right)$$

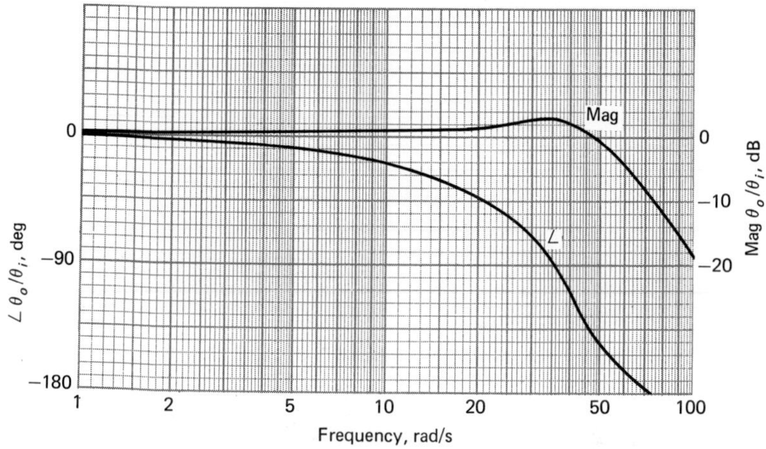

Figure 7-20 Closed-loop response for lead-compensated system in Example 7-3

The component values may be computed with the aid of Eq. (7-21). If R_1 is selected as 100 kΩ,

$$C = \frac{\tau}{R_1} = \frac{0.0745}{10^5} = 0.745\mu \text{ farad}$$

$$R_2 = \frac{R_1}{\alpha - 1} = \frac{10^5}{5 - 1} = 25 \text{ k}\Omega$$

The functions of system gain and lead compensation may be combined in the operational amplifier circuit shown in Fig. 7-21. The sinusoidal transfer function of this circuit is

$$\frac{E_o(j\omega)}{E_i(j\omega)} = -\frac{Z_f}{Z_i} - \frac{R_3}{R_2 + (R_1/j\omega C)/(R_1 + 1/j\omega C)}$$

$$= \frac{R_3}{R_1 + R_2}\left(\frac{1 + j\omega C R_1}{1 + (j\omega C R_1 R_2)/(R_1 + R_2)}\right)$$

$$= \frac{R_3}{R_1 + R_2}\left(\frac{1 + j\omega\tau}{1 + j\omega\tau/\alpha}\right) \tag{7-23}$$

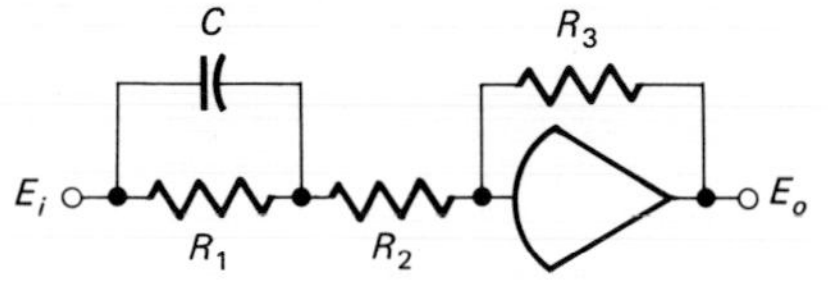

Figure 7-21 Operational-amplifier circuit for lead compensation

where, as previously defined in Eq. (7-21),

$$\tau = CR_1$$

$$\alpha = \frac{R_1 + R_2}{R_2}$$

We see that the amplifier transfer function is of the same form as Eq. (7-21) for the passive lead network. Thus the frequency-response curves shown in Fig. 7-16 may be applied with the difference that the magnitude response has a low-frequency gain equal to $R_3/(R_1 + R_2)$ rather than an attenuation of $1/\alpha$.

7-8 LAG-LEAD COMPENSATION

The preceding sections have shown that the introduction of phase-lag compensation improves the steady-state response, while lead compensation increases bandwidth and improves the transient response. Thus each form of compensation provides certain advantages. If both changes are required simultaneously, individual lag and lead networks may be inserted in series. However, it is also possible to combine both functions into a single network. Such a network, shown in Fig. 7-22, is known as a *lag-lead compensator* because the phase of the response compared to the input varies from a lagging angle to a leading angle as the frequency is increased from 0 to ∞.

The transfer function for the lag-lead network is

$$G_c(j\omega) = \frac{E_o(j\omega)}{E_i(j\omega)} = \frac{R_2 + 1/j\omega C_2}{R_2 + 1/j\omega C_2 + (R_1/j\omega C_1)/(R_1 + 1/j\omega C_1)} \tag{7-24}$$

which simplifies to

$$G_c(j\omega) = \frac{(1 + j\omega R_2 C_2)(1 + j\omega R_1 C_1)}{(1 + j\omega R_2 C_2)(1 + j\omega R_1 C_1) + j\omega R_1 C_2} \tag{7-25}$$

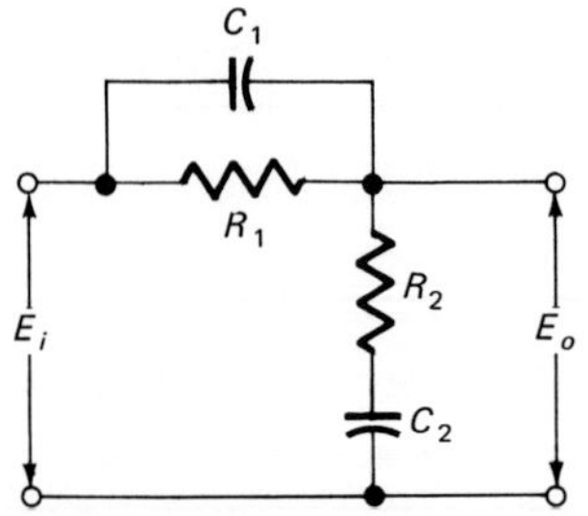

Figure 7-22 Network for lag-lead compensation

Let us define the time constants $\tau_1 = R_1C_1$, $\tau_2 = R_2C_2$, and $\tau_{12} = R_1C_2 = \tau_2R_1/R_2$. In terms of these time constants the transfer function is

$$G_c(j\omega) = \frac{(1 + j\omega\tau_2)(1 + j\omega\tau_1)}{(1 + j\omega\tau_2)(1 + j\omega\tau_1) + j\omega\tau_{12}}$$

$$= \frac{1 + j\omega(\tau_1 + \tau_2) + (j\omega)^2\tau_1\tau_2}{1 + j\omega[\tau_1 + (1 + R_1/R_2)\tau_2] + (j\omega)^2\tau_1\tau_2} \tag{7-26}$$

The denominator of Eq. (7-26) can be factored. Assume factors of the form

$$(1 + j\omega\alpha\tau_2)(1 + j\omega\tau_1/\alpha)$$

where, as previously defined, $\alpha = (R_1 + R_2)/R_2$. The denominator is then

$$1 + j\omega(\alpha\tau_2 + \tau_1/\alpha) + (j\omega)^2\tau_1\tau_2 \tag{7-27}$$

Comparing Eq. (7-27) with Eq. (7-26), we see that

$$\tau_1 + \alpha\tau_2 \cong \tau_1/\alpha + \alpha\tau_2$$

if $\tau_2 > \tau_1$ and $\alpha > 1$ by factors of several times or more. Thus the network transfer function can be approximated as

$$G_c(j\omega) = \left(\frac{1 + j\omega\tau_2}{1 + j\omega\alpha\tau_2}\right)\left(\frac{1 + j\omega\tau_1}{1 + j\omega\tau_1/\alpha}\right) \tag{7-28}$$

We see that Eq. (7-28) combines the characteristics previously described for the individual lag and lead networks. The first factor is the transfer function of a lag network as given by Eq. (7-14) and the second is that of a lead network as given by Eq. (7-21).

For $\tau_2 > \tau_1$, the frequency response of the lag-lead network is shown in the Bode diagram in Fig. 7-23. From Eq. (7-26) we see that the real part of the equation is 0 for

$$(j\omega)^2\tau_1\tau_2 = -1$$

Hence at

$$\omega = \frac{1}{\sqrt{\tau_1\tau_2}} \tag{7-29}$$

the magnitude of the transfer function is

$$|G_c(j\omega)| = \frac{\tau_1 + \tau_2}{\tau_1 + \alpha\tau_2} \tag{7-30}$$

which is its minimum value, and the phase angle is 0, as indicated in Fig. 7-23.

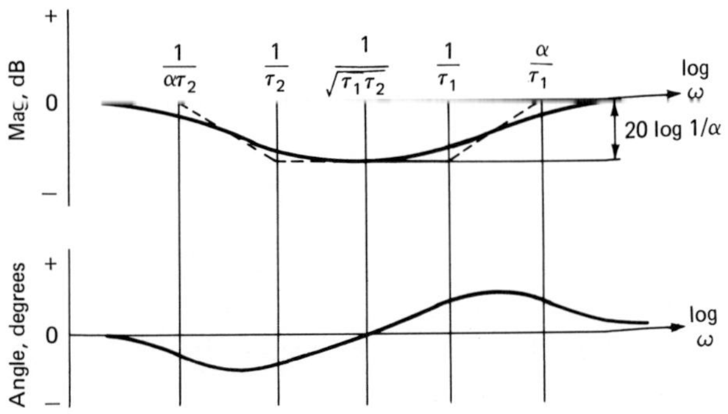

Figure 7-23 Bode diagram for lag-lead compensation

The design procedure for lag-lead compensation combines the procedures given for applying lag and lead networks individually. The magnitude and phase-angle curves of the lag compensation can be located as described in Sec. 7-5. This permits the determination of τ_2 and α. The curves for the lead term can be located as outlined in Sec. 7-7. The location chosen for the angle curve permits the determination of τ_1. To utilize the lag-lead network shown in Fig. 7-22, the value of α used for the lead term must be the same as that chosen for the lag term (or vice versa).

After the compensator has been added, the new overall transfer function $G_s(j\omega)G_c(j\omega)$ may be plotted on the Nichols chart and the gain adjusted for the desired M_m. The additional amplifier gain is then determined from this gain value.

EXAMPLE 7-4

Design a lag-lead compensating network so that the servomechanism in Example 6-11 will meet the steady-state and transient-response specifications of Examples 7-2 and 7-3, respectively.

Solution

In Example 7-2 it was determined that lag compensation with $\alpha = 6.45$ and a time constant $\tau = 0.72$ s would satisfy the steady-state specification. Thus the lag-compensation network that was used has the transfer function

$$G_c(j\omega) = \frac{1 + j0.72\omega}{1 + j4.64\omega}$$

where the time constants in the denominator and numerator correspond to lower and upper corner frequencies at 0.22 and 1.38 rad/s, respectively.

From Fig. 7-18 in Example 7-3, we found that lead compensation with $\alpha = 5$ and a lower corner frequency $\omega = 13.4$ rad/s would provide the specified transient response. The corresponding lead transfer function is

$$G_c(j\omega) = \frac{1 + j0.075\omega}{1 + j0.015\omega}$$

Since the value of α must be the same for both lag and lead terms in a lag-lead network, we will slightly modify the lag factor to incorporate $\alpha = 5$. The overall lag-lead compensator transfer function then becomes

$$G_c(j\omega) = \left(\frac{1 + j0.89\omega}{1 + j4.45\omega}\right)\left(\frac{1 + j0.075\omega}{1 + j0.015\omega}\right)$$

The frequency response of the uncompensated system with a loop gain of 17.7 is shown by the magnitude and phase curves marked G_s in Fig. 7-24. The response of the lag-lead compensator defined by the

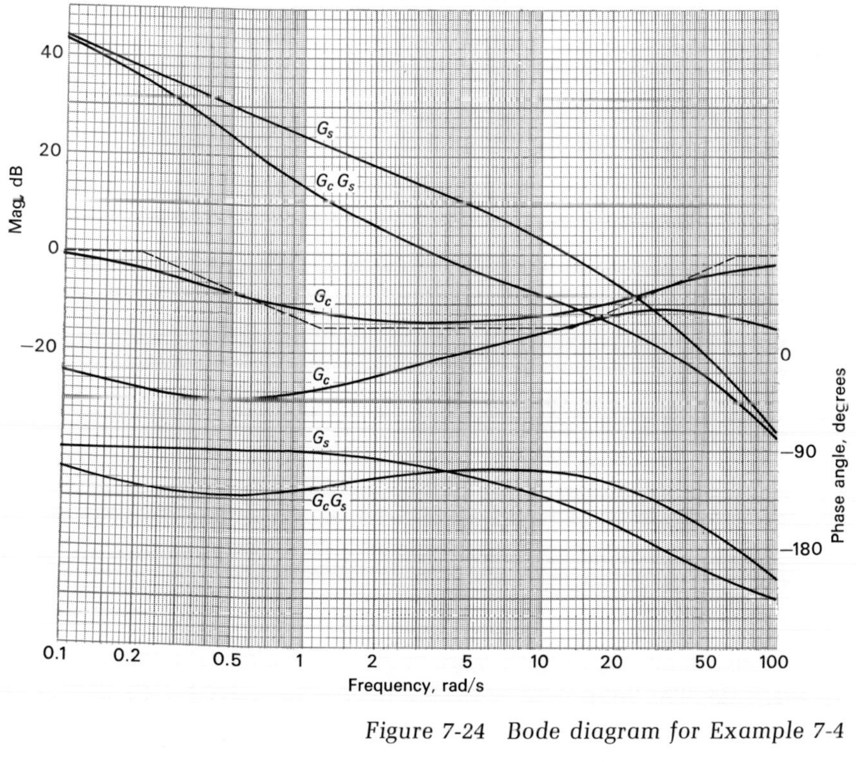

Figure 7-24 Bode diagram for Example 7-4

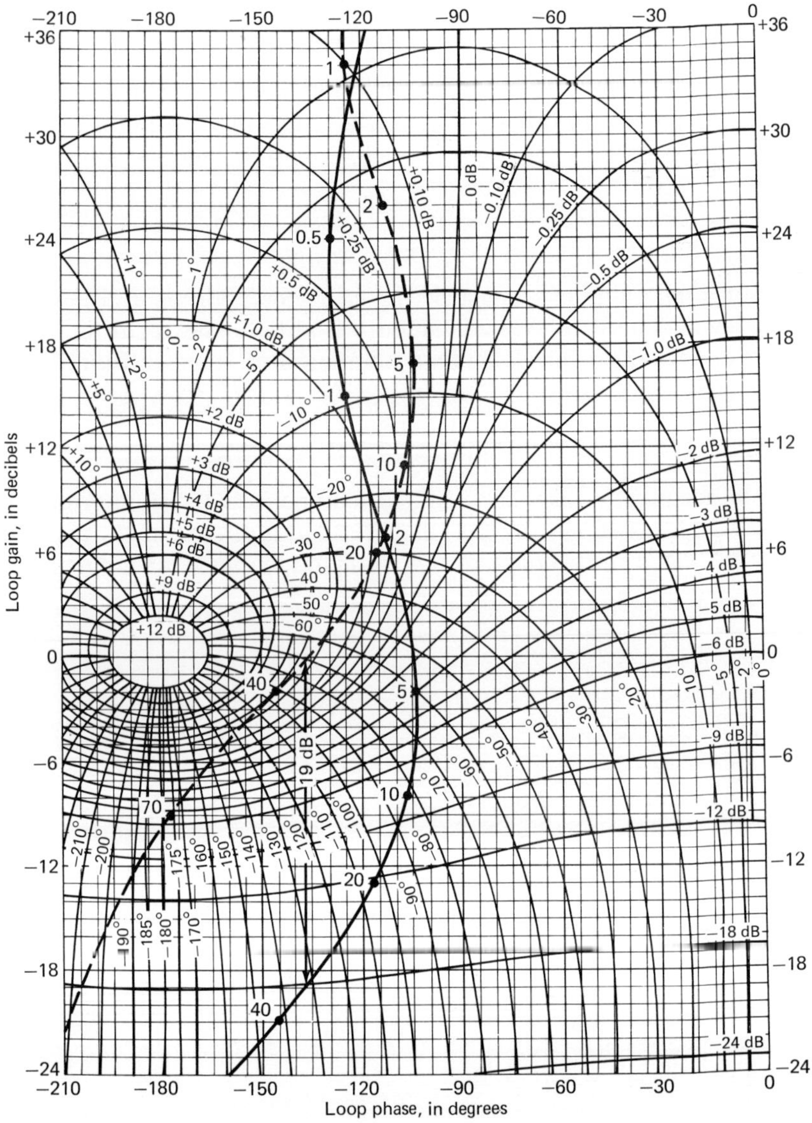

Figure 7-25 Nichols chart for Example 7-4

above equation is shown as G_c. The compensated system response is shown by the curves marked G_cG_s.

The open-loop response from Fig. 7-24 is plotted on the Nichols chart in Fig. 7-25. The loop gain is then adjusted upward by 19 dB (factor of 8.9) so that the response curve is tangent to the $M_m = 3$-dB contour, as shown by the dashed line. Since the velocity constant (loop gain) of the compensated system is $17.7 \times 8.9 = 157\ s^{-1}$, the steady-state specification is met. We also note that the frequency at the point of tangency with the 3-dB contour is approximately 40 rad/s. Hence we may conclude that the frequency response specification has been met.

Referring to Fig. 7-22, component values for the compensating network can be determined as follows. If resistor R_2 is taken as 10 kΩ, $R_1 = R_2(\alpha - 1) = 40$ kΩ. The capacitor values are then

$$C_1 = \frac{\tau_1}{R_1} = \frac{0.075}{40 \times 10^3} = 1.88\mu \text{ farads}$$

$$C_2 = \frac{\tau_2}{R_2} = \frac{0.89}{10 \times 10^3} = 89\mu \text{ farads}$$

7-9 THREE-MODE CONTROLLER

Although proportional plus integral or proportional plus derivative control is adequate for many applications, most manufacturers of process control equipment market a standard controller which combines proportional, integral, and derivative control actions. This device is known as a *three-mode controller* and finds widest application in controlling processes which have large transfer lags.

The principles involved in each of the control actions have been described in preceding sections. The equation describing the performance of the three-mode controller is a combination of Eqs. (7-8) and (7-19). Thus

$$G_c(s) = \frac{M(s)}{E(s)} = K_p\left(1 + \frac{K_i}{s} + K_d s\right) \tag{7-31}$$

The effect of integral control action is to eliminate proportional offsets, especially those due to large disturbances, by continuously adjusting or resetting the controlled output. For this reason it is sometimes known as *reset action*. The rate at which the reset action duplicates or repeats the proportional action is defined as the *reset rate* T_i. This reset rate is the reciprocal of the integral gain K_i; that is, $T_i = 1/K_i$. On some controllers the reset adjustment is calibrated in repeats per minute.

As described in Sec. 7-6, derivative control action tends to anticipate changes in the error signal and thereby reduces the tendency toward oscillation. Since the derivative action responds to the rate of change of the controlled output, it is known as *rate action*. Some controller

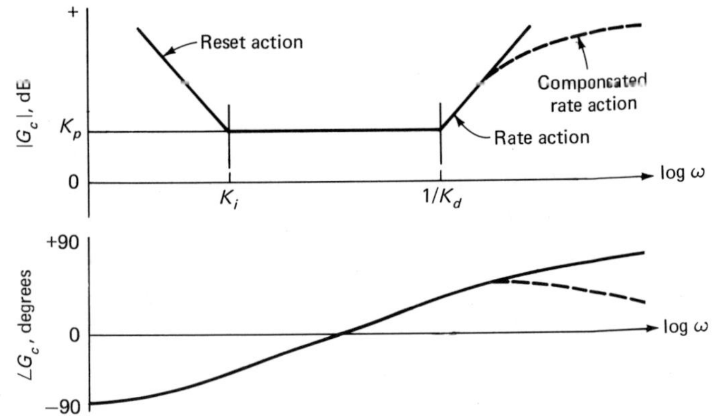

Figure 7-26 Bode diagram of three-mode control

manufacturers define the derivative gain K_d as the interval T_d by which the derivative action leads the action resulting from the proportional control. This interval is called the *rate time* or lead time and is generally expressed in minutes.

The frequency-response characteristic of proportional plus integral plus derivative control is shown by the Bode diagram in Fig. 7-26. This diagram combines the response characteristics of Figures 7-3*b* and 7-13*b*. The gain at high frequencies is usually limited in commercial three-mode controllers to minimize the effects of noise in the system. This limited derivative action is called *compensated rate action*, and the response characteristic is indicated by the dashed lines in Fig. 7-26.

If we compare Fig. 7-26 with the frequency response of the lag-lead compensator shown in Fig. 7-23, certain similarities are noted. Lag-lead compensation represents an approximate form of proportional plus integral plus derivative control. The frequency response methods presented in the preceding section are also applicable in determining the settings for the three-mode controller.

7-10 FEEDBACK COMPENSATION

The preceding sections of this chapter have described the use of series or cascade compensation to improve the performance of a feedback control system. As we shall see, the feedback principle not only may be applied to the complete control loop but also can be used to advantage on an individual control element or group of elements. This can produce improved time response from a given element. In addition the performance of an element can be less dependent on the characteristics of the element itself.

This form of compensation is known as *feedback* or *parallel compensation*. The feedback path may provide a simple proportional relationship or may include a compensator of the lag, lead, or lag-lead form as previously mentioned with series compensation.

Figure 7-27*a* illustrates the general case of feedback compensation. A feedback path is constructed around the system element $G_2(s)$ with a compensating element $H_c(s)$ inserted. In this case $G_2(s)$ and $H_c(s)$ are considered as an inner loop that modifies the transfer function $C(s)/M(s)$ and hence reshapes the overall open-loop function. The transfer function of this inner loop may be determined in the usual manner for a closed-loop system, and the block diagram of Fig. 7-27*a* simplified to that shown in Fig. 7-27*b*. The compensated open-loop transfer function then becomes

$$\frac{B(s)}{E(s)} = \frac{G_1(s)G_2(s)H(s)}{1 + G_2(s)H_c(s)} \tag{7-32}$$

Unfortunately the simple series compensation approach of summing the log magnitude and phase angle of the compensator with that of the original system is not directly applicable to feedback compensation. However, the nature of the inner-loop transfer function $C(j\omega)/M(j\omega)$ can be determined using the low-frequency–high-frequency approximation previously considered in Sec. 6-13. Consider the case where $G_2(j\omega)H_c(j\omega) \ll 1$. The inner-loop transfer function can then be approximated by

$$\frac{C(j\omega)}{M(j\omega)} \cong G_2(j\omega) \tag{7-33}$$

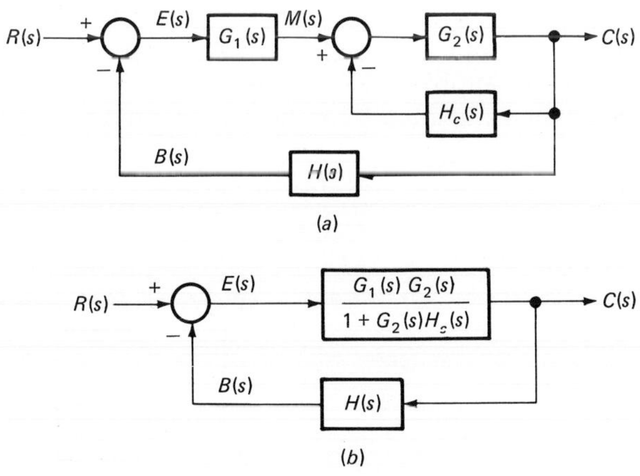

Figure 7-27 Feedback compensation: (a) general arrangement and (b) simplified diagram

For the condition $G_2(j\omega)H_c(j\omega) \gg 1$,

$$\frac{C(j\omega)}{M(j\omega)} \doteq \frac{1}{H_c(j\omega)} \qquad (7\text{-}34)$$

Still undefined is the condition $G_2(j\omega)H_c(j\omega) \cong 1$, for which neither Eq. (7-33) nor Eq. (7-34) is applicable. If this condition is neglected, we can approximate the transfer function to obtain at least qualitative results. Where a more exact transfer function is desired, the open-loop function $G_2(j\omega)H_c(j\omega)$ may be plotted on the Nichols chart and the closed-loop response $C(j\omega)/M(j\omega)$ read from the chart.

The following example illustrates the use of these approximations in applying feedback compensation.

EXAMPLE 7-5

A magnetic amplifier has a transfer function that can be represented by

$$G(j\omega) = \frac{100}{1 + j0.04\omega}$$

where the 0.04-s time constant is due to the inductance of the control winding. Determine the compensated transfer function $G_c(j\omega)$ when feedback is used around the amplifier as shown in Fig. 7-28.

Solution

The frequency response of the uncompensated amplifier is shown by the curve $G(j\omega)$ in Fig. 7-29. The amplifier cutoff frequency ω_c is $1/0.04 = 25$ rad/s. The loop gain $G(j\omega)H(j\omega)$ is also plotted for $H = 0.1$. It is seen that $G(j\omega)H(j\omega) > 1$ for frequencies below $\omega_1 = 250$ rad/s. Using the approximation given by Eq. (7-34), the value of $G_c(j\omega)$ for frequencies below ω_1 can be represented by

$$G_c(j\omega) = \frac{E_o(j\omega)}{E_i(j\omega)} \cong \frac{1}{H(j\omega)} = 10$$

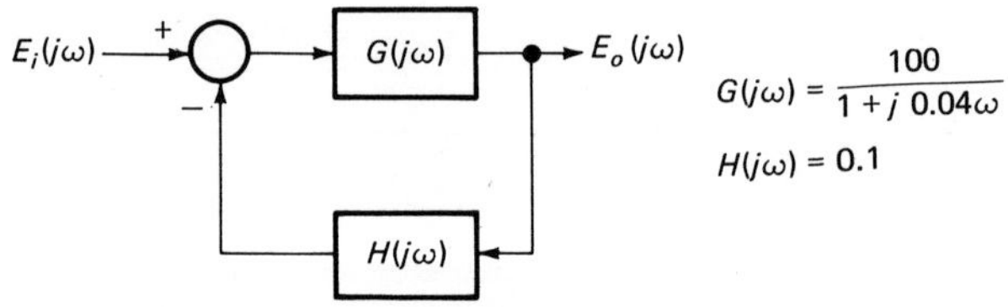

Figure 7-28 Feedback-compensated amplifier for Example 7-5

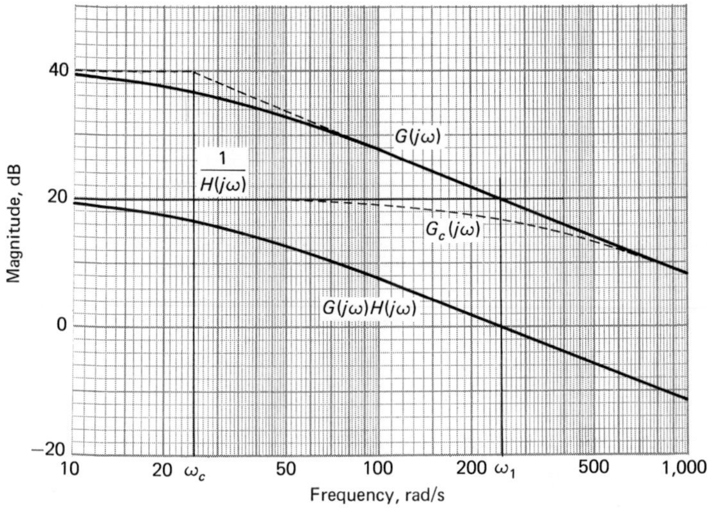

Figure 7-29 Bode diagram for Example 7-5

For frequencies above ω_1, $G(j\omega)H(j\omega) < 1$. From the approximation given by Eq. (7-33), $G_c(j\omega)$ can be represented by $G(j\omega)$ for frequencies above ω_1. $G_c(j\omega)$ can therefore be represented by a line of zero slope and 20 dB for frequencies up to ω_1 and a line of -20 dB per decade slope for frequencies above ω_1. The compensated transfer function of the amplifier can then be written as

$$G_c(j\omega) = \frac{10}{1 + j0.004\omega}$$

Although the feedback compensation has improved the frequency response of the amplifier, we should note that this has been at the expense of reduced low-frequency gain.

7-11 COMPENSATION BY DERIVATIVE FEEDBACK

The principle of derivative or velocity feedback was described in Sec. 4-11, where we saw that such feedback could improve the transient performance of a positional servomechanism. At that time our discussion was limited to analyzing the effect on the damping ratio of a second-order system. Because derivative feedback is commonly used in control systems, it is worth considering from a frequency-response standpoint as a form of feedback compensation.

A control system incorporating derivative feedback compensation is shown in block-diagram form in Fig. 7-30. The compensation is applied to the element represented by $G_2(j\omega)$. The feedback element $H_c(j\omega)$

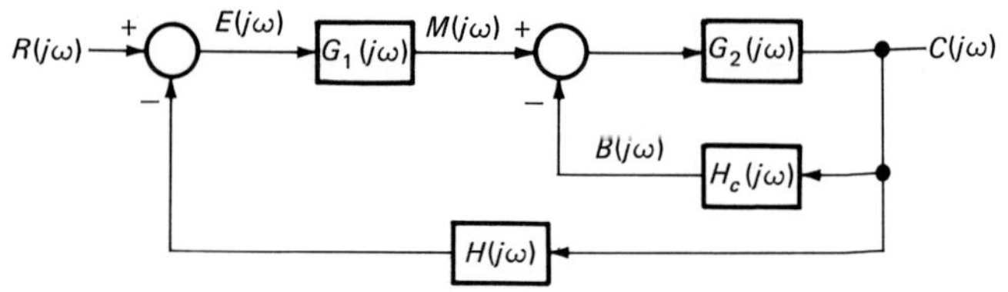

Figure 7-30 Control system including derivative feedback compensation

provides a compensating signal proportional to the derivative or rate of change of the output. Thus its transfer function is given by

$$\frac{B(j\omega)}{C(j\omega)} = H_c(j\omega) = j\omega K_d \tag{7-35}$$

where K_d is the gain constant of the derivative feedback element.

The effect of derivative feedback is shown in the frequency-response plot of Fig. 7-31. The forward gain $G_2(j\omega)$ and the feedback response $H_c(j\omega)$ are combined to give the gain $G_2(j\omega)H_c(j\omega)$ of the inner feedback loop shown by the dashed line. The compensated response $C(j\omega)/M(j\omega)$ shown by the bold line is obtained using the low-frequency–high-frequency approximation technique. Comparing the uncompensated response $G_2(j\omega)$ with the compensated response, we note that the cutoff frequency has been extended from ω_{c1} to ω_{c2}, which improves the frequency response of the element. We also see that the low-frequency gain has been reduced. What is not shown on the diagram is that the dc gain is not reduced (as would be the case with proportional feedback compensation) since there is no derivative feedback with an unvarying

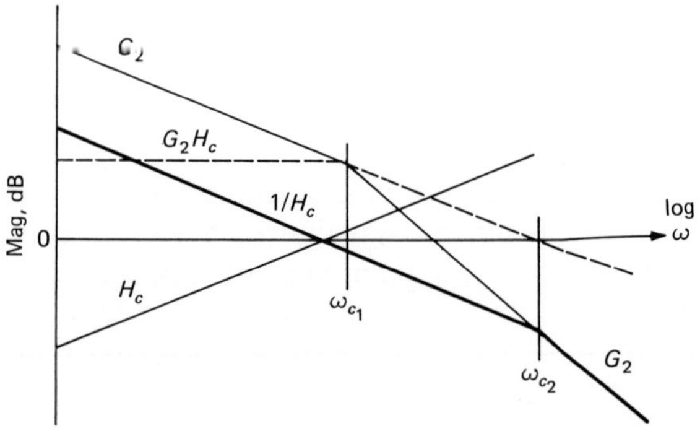

Figure 7-31 Bode diagram for compensation by derivative feedback

output. Thus the static accuracy of the system is unaffected by derivative compensation.

We should note that systems higher than second order can just as readily be analyzed using this technique. However, with higher-order systems care must be taken to maintain the stability of the inner loop. Also it is apparent that the compensated response can be obtained more accurately using a Nichols chart. Finally, after the compensated response has been established, the overall system response $C(j\omega)/R(j\omega)$ may be determined following the usual procedure.

EXAMPLE 7-6

The uncompensated positional servomechanism in Example 6-11 is modified to include tachometer feedback compensation as shown in the block diagram of Fig. 7-32.

Determine the velocity constant and closed-loop response when the amplifier gain K_2 is adjusted so that $M_m = 3$ dB.

Solution

The first step in evaluating the closed-loop performance of the compensated servomechanism is to determine the response of the motor with tachometer feedback, $\theta_m(j\omega)/E_1(j\omega) = G_c(j\omega)$. This may be obtained on an approximate basis as shown in the Bode diagram in Fig. 7-33.

The magnitude of the frequency response of the motor with a power amplifier is the asymptotic plot marked G_m. The tachometer response H_c is then combined with G_m to give the loop response G_mH_c. Note that the loop gain decreases through unity at $\omega = 70$ rad/s. The low-

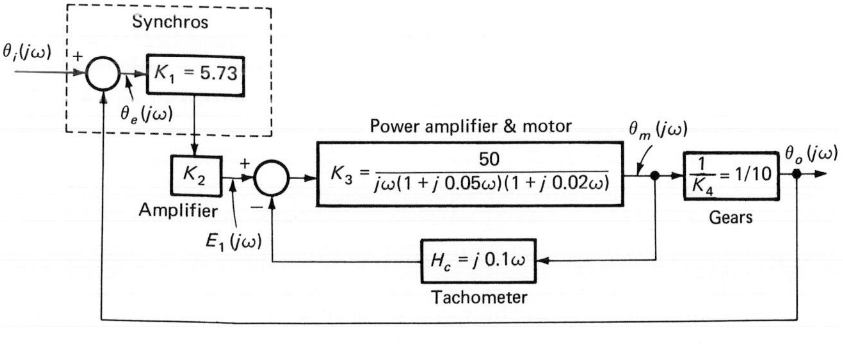

Figure 7-32 Positional servomechanism including tachometer feedback for Example 7-6

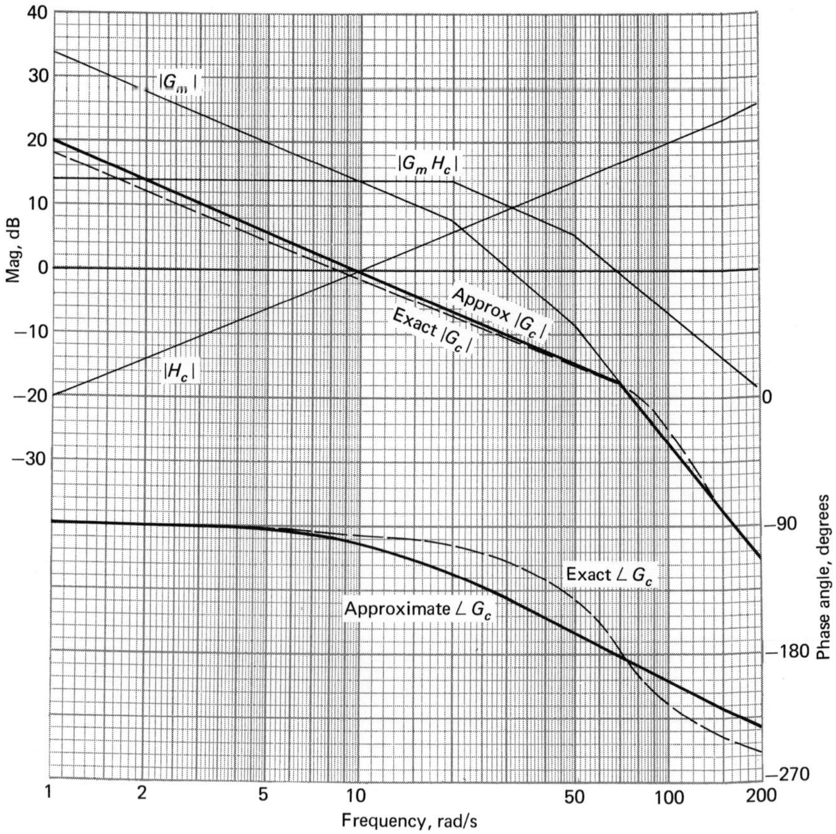

Figure 7-33 Bode diagram for Example 7-6

frequency–high-frequency approximation is applied to determine the magnitude of the compensated motor response G_c. For frequencies below 70 rad/s $G_c \cong 1/H_c$, and above 70 rad/s $G_c \cong G_m$. This response, shown in Fig. 7-33 by the bold-faced line, is represented by the approximate transfer function

$$G_c(j\omega) = \frac{10}{j\omega(1 + j0.014\omega)^2}$$

The phase-angle curve corresponding to this function is also plotted in Fig. 7-33.

The transfer function of the overall loop is

$$G(j\omega) = K_1K_2[G_c(j\omega)]/K_4 = \frac{5.73 \times K_2 \times 10 \times \frac{1}{10}}{j\omega(1 + j0.014\omega)^2}$$

This function is plotted on the Nichols chart in Fig. 7-34 with the amplifier gain K_2 initially taken as unity. The plot of $G(j\omega)$ is then translated upward by 15 dB as shown by the dashed curve to establish $M_m = 3$ dB. The gain increase of 15 dB becomes the necessary amplifier

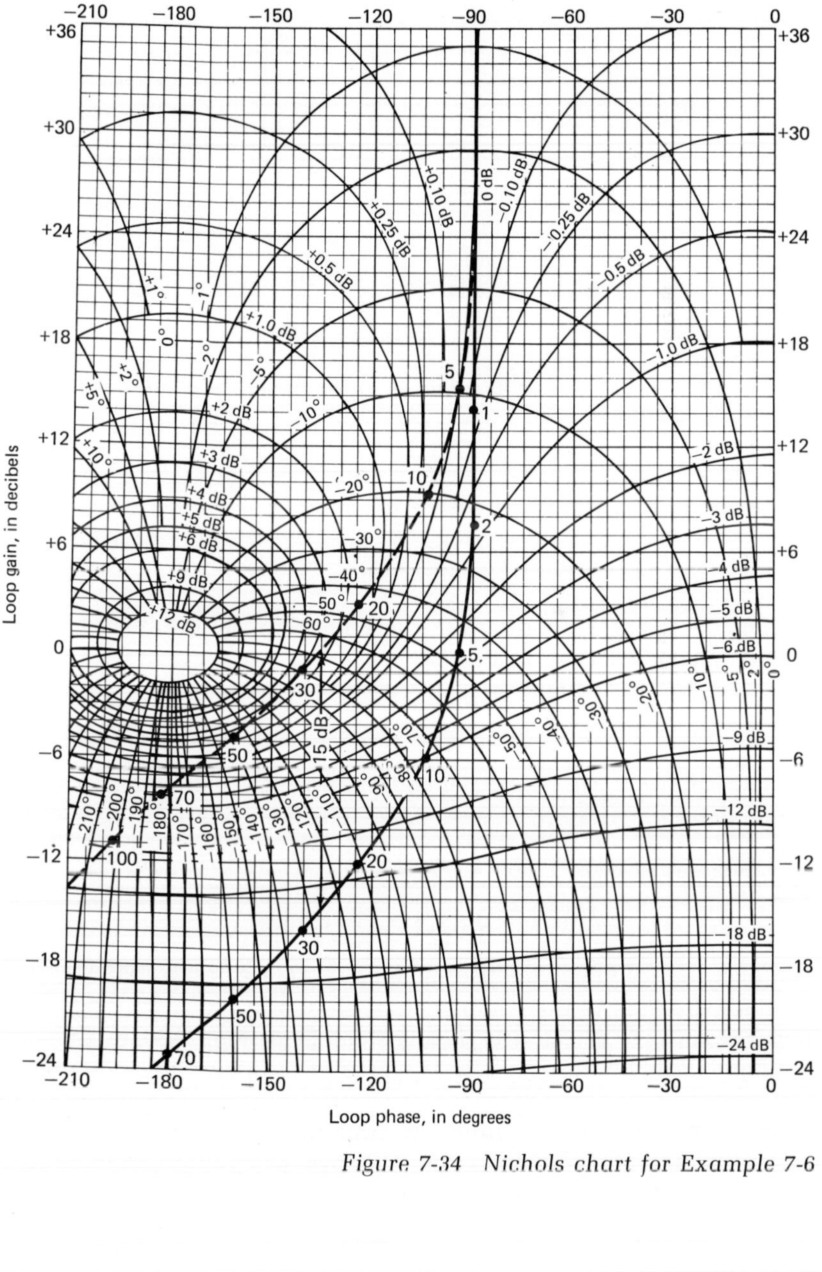

Figure 7-34 Nichols chart for Example 7-6

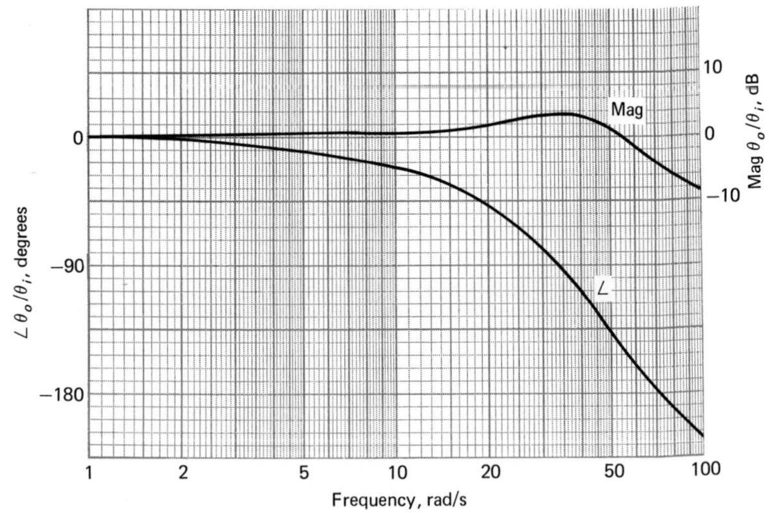

Figure 7-35 Closed-loop response for tachometer-compensated system in Example 7-6

gain (since K_2 was initially taken as unity) so that $K_2 = 5.6$.

The velocity constant is then

$$K_v = \frac{5.73 \times 5.6 \times 10}{10} = 32.1\ \mathrm{s}^{-1}$$

The closed-loop response of the servomechanism obtained from the Nichols chart is shown in Fig. 7-35.

Alternate Solution

The response $\theta_m(j\omega)/E_1(j\omega)$ of the motor with tachometer feedback can be determined more exactly using a Nichols chart. In this case we take

$$G_c(j\omega) = \frac{1}{H_c(j\omega)}\left[\frac{G_m(j\omega)H_c(j\omega)}{1 + G_m(j\omega)H_c(j\omega)}\right]$$

Response data of the compensation loop G_mH_c are first transferred from the Bode diagram to the Nichols chart shown in Fig. 7-36. The closed-loop response $G_mH_c/(1 + G_mH_c)$ is read from Fig. 7-36 in the usual manner and then multiplied by $H_c(j\omega) = j0.1\omega$ to give the exact magnitude and phase curves for $G_c(j\omega)$. The exact response curves are shown in Fig. 7-33 by the dashed lines. We note that the magnitude response is close to that obtained using the low-frequency–high-frequency ap-

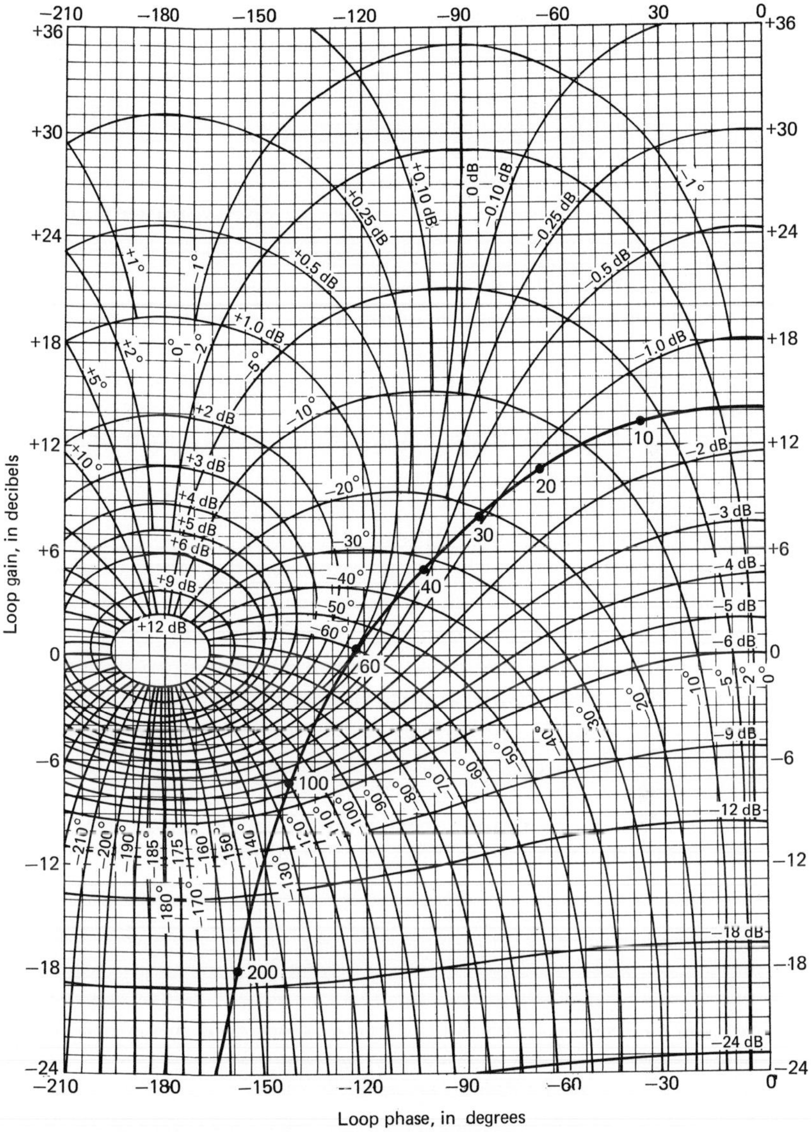

Figure 7-36 Nichols chart for alternate solution for Example 7-6

proximation method. However, the phase curves differ noticeably using the exact method with less phase shift present in the lower-frequency range.

The closed-loop response $C(j\omega)/R(j\omega)$ may be determined by plotting the overall transfer function on the Nichols chart. Although this step is not included here, the response does not differ significantly from that shown in Fig. 7-35.

Although exact results are obtained, it is apparent that more work is required as compared to the first method of solution.

7-12 FEEDFORWARD CONTROL

Throughout this text we have been concerned with using negative-feedback signals as a means of automatic control. There are circumstances under which a feedforward of certain signals provides a useful addition to the feedback loop. This arrangement is known as *feedforward control,* and it is primarily applied to reduce or eliminate the response of a system to disturbance input. Feedforward control has found widest application in the control of industrial processes where elements downstream are subject to disturbances and long time constants or delays are present in the system.

The use of feedforward control to minimize the effect of a disturbance input is illustrated by the block diagram in Fig. 7-37. If there are significant lags in the response of the control elements $G_1(s)$, corrective action in response to a disturbance input $U(s)$ would not be felt for some time. With feedforward control the disturbance input is measured as it enters the system and adjustment is made to compensate for its effect before it appears in the output. In the system shown in Fig. 7-37 the disturbance signal is fed forward through a control element $G_c(s)$ and then subtracted from the error signal $E(s)$. The manipulated variable as a function of the disturbance input is now

$$\begin{aligned} M(s) &= U(s) - G_c(s)G_1(s)U(s) \\ &= U(s)[1 - G_c(s)G_1(s)] \end{aligned} \tag{7-30}$$

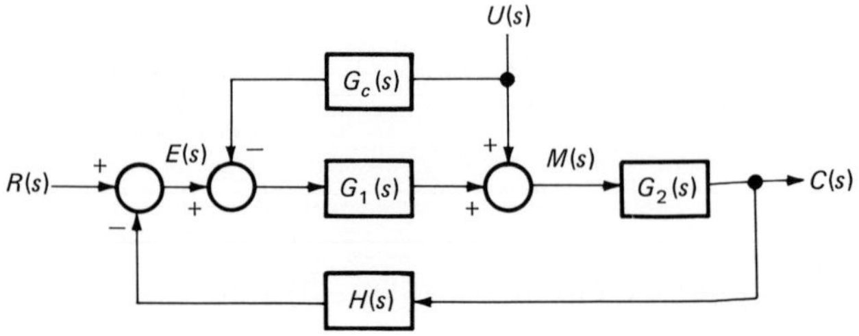

Figure 7-37 Feedforward control to minimize effect of disturbance input

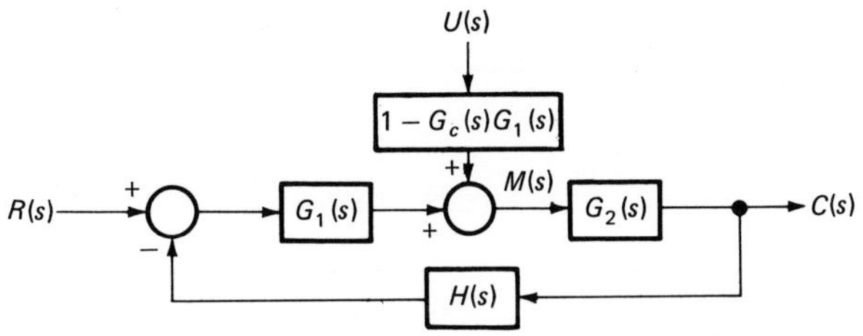

Figure 7-38 Rearrangement of Fig. 7-37

If $G_1(s)$ is known and the feedforward control is designed such that $G_c(s) = 1/G_1(s)$, we see that the effect of the disturbance is totally eliminated. Because of the delays that are likely to be present in the system, perfect feedforward compensation is virtually impossible to obtain in practice. However, a simple feedforward path combined with feedback control can substantially improve the performance of a system with regard to disturbances.

The design of the feedback loop for a system incorporating feedforward control is the same as for the basic feedback system. Using the results of Eq. (7-36), the disturbance input may be applied to the system through an element whose transfer function is $1 - G_c(s)G_1(s)$. This is shown in Fig. 7-38, which is a rearrangement of the block diagram in Fig. 7-37. The transfer function for the disturbance input is

$$\frac{C(s)}{U(s)} = \frac{[1 - G_c(s)G_1(s)]G_2(s)}{1 + G_1(s)G_2(s)H(s)} \tag{7-37}$$

The system characteristic equation [denominator of Eq. (7-37)] is unchanged by the addition of the feedforward element. Hence the stability of the closed loop is independent of the feedforward control.

An example of feedforward control is a home-heating system in which an outdoor weather thermometer measures the ambient temperature and changes the furnace control in anticipation of the change in heat load as yet unfelt by the thermostat inside the house. As indicated by the curves in Fig. 7-39, the steady-state deviation from the desired temperature is reduced and the transient response is improved by the addition of feedforward compensation.

7-13 SUMMARY

The need for compensation of an automatic control system arises because a given set of control elements is not always capable of meeting

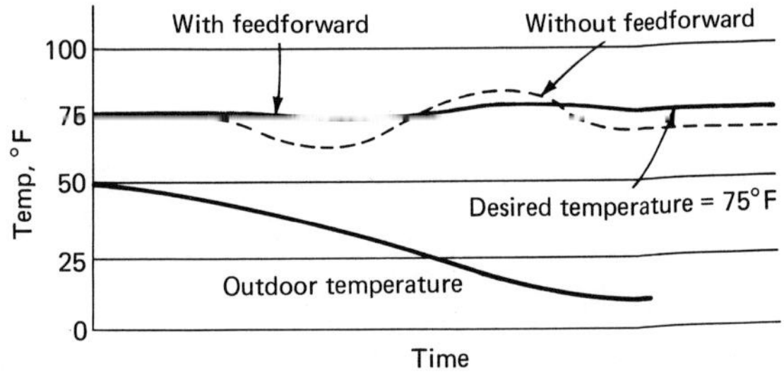

Figure 7-39 Effect of feedforward compensation on home heating system

the specifications for control. The performance of a system may be inadequate from the standpoint of stability, speed of response, or accuracy. Because it is not always desirable or even possible to change the basic system, compensation techniques are employed.

Several methods of compensation to improve control system performance have been described in this chapter. Various forms of compensating elements may be placed in series with system elements or in an inner feedback path around one or more system elements.

Table 7-1 presents a comparison of the results of the methods of applying compensation to a basic control system—the servomechanism first described in Example 6-11. From this summary we see that:

1. Lag compensation results in a large increase in gain, which means a smaller steady-state error with little change in ω_m or the transient response.

2. Lead compensation results in a moderate increase in gain and a larger increase in ω_m, thereby improving the transient response.

3. Lag-lead compensation combines the benefits of both lag and lead compensation, producing a significant increase in both gain and ω_m.

4. Derivative feedback compensation also improves both steady-state and transient performance.

This discussion of control system compensation is by no means exhaustive; numerous other methods and devices are available. It is important to emphasize that compensation cannot be applied indiscriminately. The necessary improvements in system performance may not be possible by the basic means listed in Table 7-1. In such cases, two or more compensators may be located in series and/or in parallel in the system. Choice of compensation depends upon the nature of the system and practical considerations such as performance limitations, space, weight, and cost.

TABLE 7-1 SUMMARY OF RESULTS OF COMPENSATION OF A BASIC CONTROL SYSTEM

Compensation	Ex. No.	K_v, s^{-1}	ω_m, *rad/s*	M_m, *dB*
Uncompensated	6-11	17.7	13	3
Lag: $\dfrac{1 + j0.72\omega}{1 + j4.64\omega}$	7-2	114.5	13	4
Lead: $\dfrac{1}{5}\left(\dfrac{1 + j0.0745\omega}{1 + j0.0149\omega}\right)$	7-3	25.1	35	3
Lag-lead: $\left(\dfrac{1 + j0.89\omega}{1 + j4.45\omega}\right)\left(\dfrac{1 + j0.075\omega}{1 + j0.015\omega}\right)$	7-4	157	40	3
Derivative feedback: $j0.1\omega$	7-6	32.1	33	3

The reader has perhaps sensed that control system compensation involves a certain amount of trial-and-error effort and the use of judgment gained from experience. This is indeed true, and, as experience is acquired, the methods described in this chapter may be extended and refined. In particular the use of templates representing the frequency-response characteristics of various compensators can be a substantial aid in the design process.

PROBLEMS

7-1. A control system has the open-loop transfer function

$$G(j\omega) = \frac{10}{j\omega(1 + j\omega/4)(1 + j\omega/16)}$$

Determine the velocity constant, gain margin, and phase margin of this system.

7-2. Repeat Prob. 7-1 if series lag compensation having the transfer function

$$G_c(j\omega) = \frac{10(1 + j4\omega)}{1 + j40\omega}$$

is added to the system. Comment on the effect of this compensation.

7-3. Repeat Prob. 7-1 if series lead compensation having the transfer function

$$G_c(j\omega) = \frac{1 + j\omega/4}{1 + j\omega/16}$$

is added to the system. Comment on the effect of this compensation.

7-4. Plot the closed-loop frequency response of the system in Prob. 7-1.

7-5. Plot the closed-loop frequency response of the system in Prob. 7-2.

7-6. Plot the closed-loop frequency response of the system in Prob. 7-3.

7-7. The open-loop transfer function of a servomechanism is given by

$$G(s) = \frac{7}{s(1 + 0.5s)(1 + 0.15s)}$$

a. Plot the Bode diagram, and determine the gain and phase margins and the velocity constant of the system.
b. Determine the values of M_m and ω_m associated with the closed-loop frequency response.

7-8. Add series lag compensation to the servomechanism in Prob. 7-7 to give a gain margin of at least 15 dB and a phase margin of at least 45°. Find suitable values of τ and α for the compensator.

7-9. It is required to reduce the steady-state error of a control system whose open-loop transfer function is given by

$$G(j\omega) = \frac{0.8}{j\omega(1 + j\omega/2)(1 + j\omega/3)}$$

Assuming the transient response is satisfactory, introduce series lag compensation to provide a velocity constant of 4. Find suitable values of τ and α.

7-10. It is desired to improve the transient response of the system in Prob. 7-9 by adding lead compensation having the transfer function

$$G_c(j\omega) = \frac{0.1(1 + j\omega/2)}{1 + j\omega/20}$$

The gain of the system is to be adjusted to maintain the same phase margin as for the uncompensated system. Evaluate and compare the phase-crossover frequency and velocity error constant of the uncompensated and compensated system.

7-11. A control system has the open-loop transfer function

$$G(j\omega) = \frac{0.25}{(j\omega)^2(1 + j0.25\omega)}$$

Add lead compensation to provide approximately a 45° phase margin at the frequency $\omega = 1$ rad/s.

7-12. The open-loop transfer function of a positional servomechanism is

$$G(j\omega) = \frac{K}{j\omega(1 + j\omega/100)(1 + j\omega/5)}$$

It is desired that the closed-loop M_m of the servo be 1.3. Determine the value of K to meet this specification.

7-13. The servomechanism in Prob. 7-12 is additionally required to have a velocity constant $K_v = 50$ while maintaining $M_m = 1.3$. Provide suitable compensation to meet these specifications.

7-14. Determine the transfer function $E_o(j\omega)/E_i(j\omega)$ of the operational amplifier circuits below. Sketch the approximate frequency response in each case.

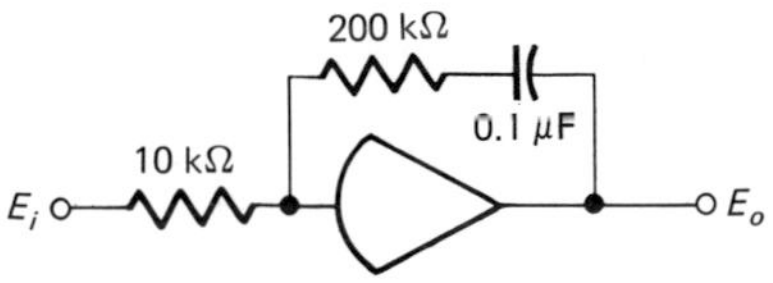

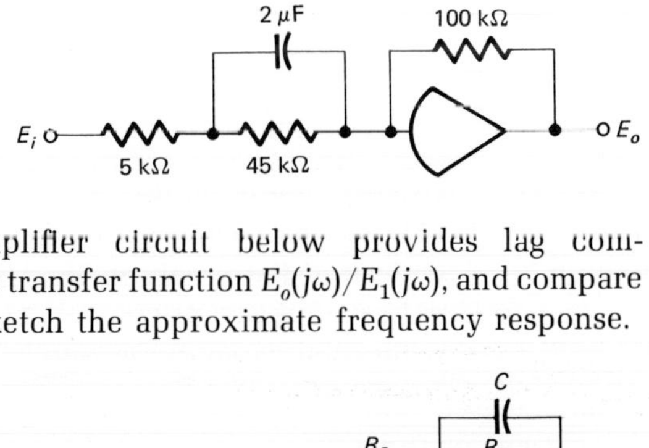

Prob. 7-14

7-15. The operational-amplifier circuit below provides lag compensation. Derive the transfer function $E_o(j\omega)/E_1(j\omega)$, and compare it with Eq. (7-14). Sketch the approximate frequency response.

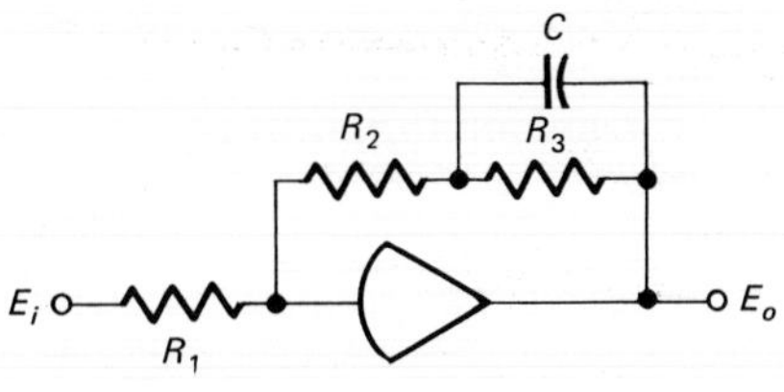

Prob. 7-15

7-16. The operational-amplifier circuit below provides lag-lead compensation. Derive the transfer functions $E_o(j\omega)/E_i(j\omega)$, and compare it with Eq. (7-28).

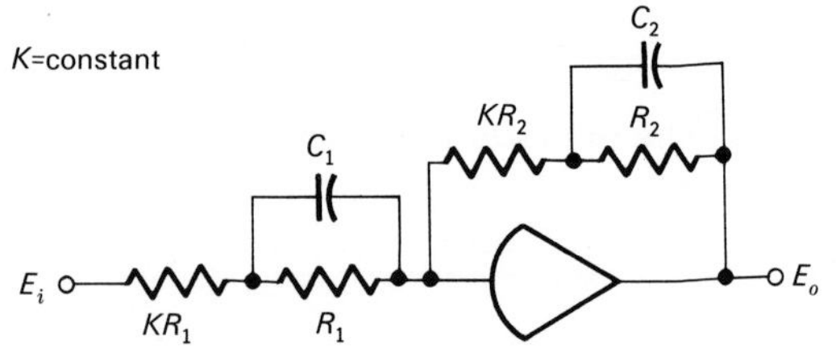

Prob. 7-16

7-17. A closed-loop system for heating a constant flow of water from an inlet temperature T_i to a desired temperature T_d is shown in the diagram below. Assume a negligible time lag in the heating process so that the temperature change due to heating is simply proportional to the controller output signal. If a controller which has a proportional gain $K = 10$ is used, determine the proportional offset in the output for $T_d = 120°F$ and $T_i = 70°F$.

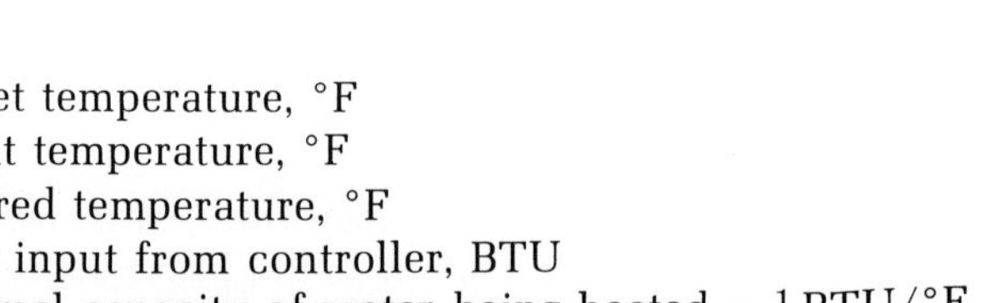

Prob. 7-17

T_o = outlet temperature, °F
T_i = input temperature, °F
T_d = desired temperature, °F
Q = heat input from controller, BTU
C = thermal capacity of water being heated = $\frac{1}{3}$ BTU/°F

7-18. Integral control is added in Prob. 7-17 so that the controller transfer function is

$$\frac{Q(s)}{T_e(s)} = 10\left(1 + \frac{1}{30s}\right)$$

Determine the transient response of T_o if the inlet temperature T_i suddenly changes from 70 to 60°F.

7-19. The block diagram of a positional servomechanism incorporating derivative feedback compensation is shown below. Graph the closed-loop frequency response of the system.

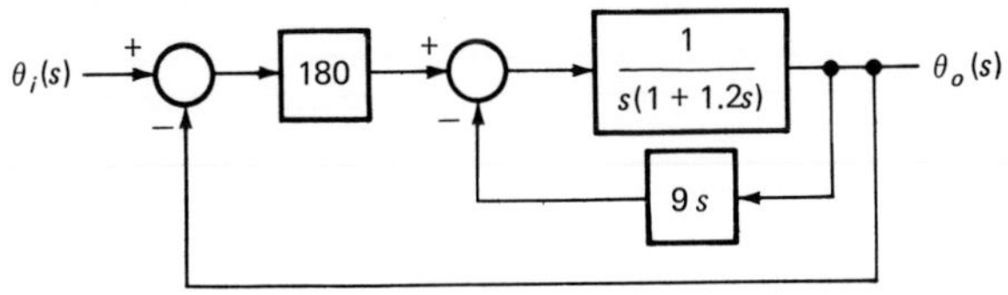

Prob. 7-19

7-20. A closed-loop system for controlling the temperature T of a furnace is shown in the block diagram below.

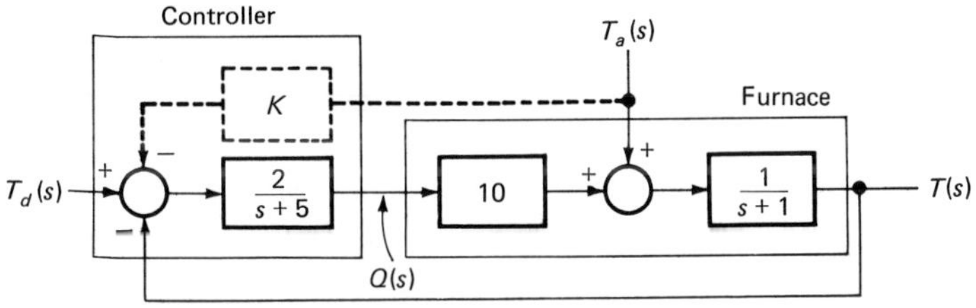

Prob. 7-20

T = furnace temperature
T_d = desired temperature (set point)
T_a = ambient temperature surrounding furnace
Q = heat input from controller

With respect to the basic feedback system determine:
a. The transfer function $T(s)/T_a(s)$
b. The steady-state change in T if T_a changes by 10°F

7-21. Feedforward control with a proportional gain K is added to the system in Prob. 7-20 as indicated by the dotted path.
a. Repeat part a of Prob. 7-20.
b. What is the optimum value of K for the minimum effect of T_a on the furnace temperature?

7-22. In the unity-feedback system shown below, it is desired to provide feedforward control to obtain the ideal transfer function $C/R = 1$ (zero error in the output). Determine the transfer function for G_c in order to accomplish this.

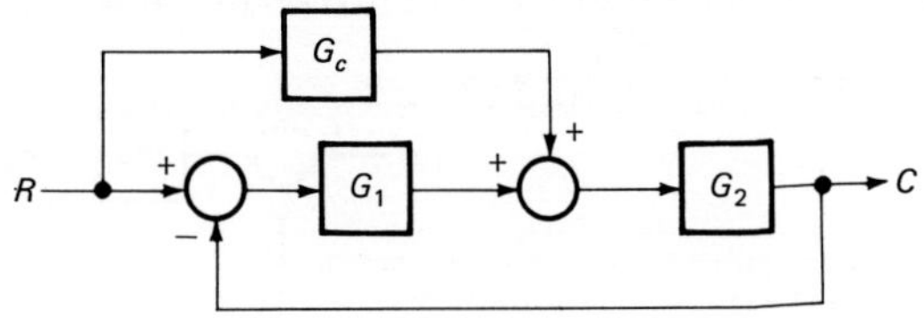

Prob. 7-22

Chapter Eight

control system testing

8-1 INTRODUCTION

An indispensable part of the design and application of automatic control systems is testing. This chapter considers the general problem of testing control components and systems. As such it is directed toward individuals who are involved in the design, installation, and maintenance of control equipment.

Control system testing may be divided into two basic classes of tests depending upon the test's objectives. These are:

1. Tests to determine the characteristics of the system and its components
2. Tests to establish system operation by examination of the system response

The first class of tests is generally used during the design process to determine system transfer functions. Unfortunately there are no general methods for direct measurement of a transfer function. Hence some form of testing that permits a transfer function to be derived from the test data is required. Frequency-response measurements are widely used for this purpose.

The second group of tests is generally used in evaluating system performance in the final design phase or in the production and operational check-out of the system. For this purpose the system response is often compared with a standard or reference.

The testing of control systems involves measuring the response of the system to various types of input signal. The most commonly used inputs are a step function for transient-response testing and a sine wave of varying frequency for frequency-response testing.

In this chapter we will consider various objectives in control-system testing, methods of performing tests and deriving transfer functions, and specifications to which test results are applied.

The calibration and adjustment of control systems is interrelated with testing since these tasks are frequently performed by the same people and since parameter settings are dependent on test results. Hence some discussion of these topics will be included.

For testing purposes, an extensive knowledge of control system design theory is not necessary. A general knowledge of the fundamental principles (as presented in this text) is generally adequate in interpreting specifications, planning and executing tests, and analyzing test data. As is so frequently the case, an understanding of the underlying theory greatly aids in achieving success in practice.

8-2 TEST OBJECTIVES

During the evolution of a control system from initial design to end use, different types of tests are carried out for different purposes. The test method, variables to be measured, test equipment, and data-reduction techniques should be selected with the overall test objectives in mind.

This section discusses the testing requirements of different phases of control system development and application. These phases include design testing, system performance testing, production-acceptance testing, operational-check-out testing, and maintenance and troubleshooting testing.

The general objectives and criteria for each phase of testing are presented. In some cases these are the same, while in other cases they are widely different.

Design Testing

Design tests are carried out in conjunction with theoretical analysis and are usually performed under laboratory or pilot-plant conditions. The primary purpose of these tests is to obtain data for use in developing the initial analytical design and in refining the design. Gathering data to support the design involves establishing the transfer functions of the system to be controlled and/or the control elements to be used in the system. These data are necessary to perform the theoretical design.

A second purpose is to provide a rough evaluation of the design to check validity of data and assumptions and to obtain more detailed information to indicate the direction to follow in further refinement of the design.

For both purposes the tests must provide enough information to define the tested components so that they may be described in mathematical form for use in the analytical design.

System Performance Testing

System performance tests are usually conducted to completely check that system operation is within the specifications under required conditions of loads, input signals, and environment. During this phase of testing, the system usually operates in its normal condition, i.e., closed-loop. The test methods should be selected to make the important changes in operation readily apparent.

Production-Acceptance Testing

In production-acceptance testing, the principal purpose is to check the operation of a manufactured system as a whole. The design is assumed to be finalized, and the performance is established. For production testing, a valid set of performance data is ordinarily available for use as a reference to compare with test data for a production unit. Production tests for control systems and components range from tests which provide a simple visual check of the response to completely instrumented tests. The choice of tests is determined by sensitivity of system performance to manufacturing tolerances, complexity of the system, economy, available facilities, and, last but not least, customer requirements.

Operational–Check-out Testing

Operational check-out is done to determine the readiness of a control system to perform its assigned task. It is almost always desirable to perform operational–check-out tests in a short time period immediately preceding actual use. These tests are generally as simple as is necessary to ensure proper functioning and may take the form of a single response measurement in the operational configuration. In some cases rather elaborate check-out systems are provided to minimize the possibility of unsatisfactory performance. The thorough prelaunch testing of rocket controls is an example of this extreme.

Maintenance and Troubleshooting Testing

Maintenance or troubleshooting tests are performed on operational equipment to find the cause of a malfunction or to detect deterioration of performance that may lead to a future failure. For preventative maintenance or periodic maintenance testing, tests are used that will detect changes in system functioning. For example, an open-loop test will clearly show changes in loop gain due to deterioration of components before closed-loop performance is significantly affected. In maintenance testing, the tests are conducted on a routine, planned basis and the test data are compared with previously established performance criteria.

Troubleshooting tests, on the other hand, are generally not routine. If the equipment is "off-line" (not in actual use), the conditions are similar to those in other phases of testing. However, if the troubleshooting is conducted "on-line," the trouble must be found and repaired quickly to minimize loss of production time or delay of an operation. Few rules can be laid down on the type of tests that are performed during troubleshooting. Much depends on the type of system, the maintainability built into the system, the available test equipment, and, perhaps most importantly, the ability and experience of the test personnel.

8-3 TEST METHODS

In previous chapters control system response was discussed in terms of certain standard inputs. Since the response is specified in terms of these inputs, it is natural to expect that they will be used for test purposes, and such is the case. The most commonly used test inputs are:

1. The step input for transient-response testing
2. The sine-wave input for frequency-response testing
3. The ramp (triangular wave) input for testing of velocity response

A generalized arrangement for testing is shown in Fig. 8-1. Since measurements are generally performed on electrical signals, transducers such as those described in Chap. 3 are required to convert the input and output signals to the electrical form required by the display equipment. Obviously the characteristics of the transducers must be known to determine the performance of the system under testing.

The choice of whether to use sine-wave or step-type input functions is a matter of convenience and is left to the discretion of the person performing the test. Either type input, when properly used, will yield the same information about the system. The step input does offer a

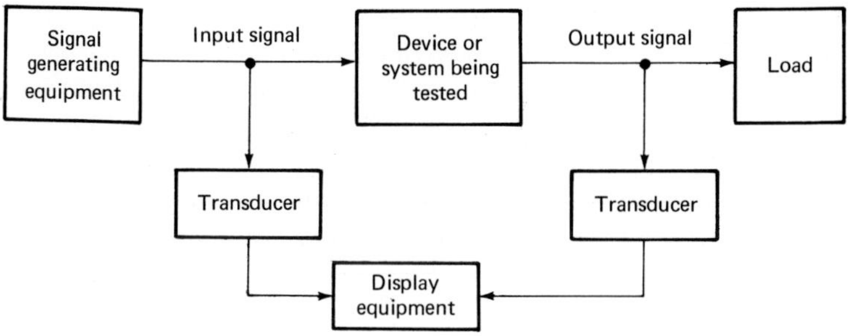

Figure 8-1 Generalized testing arrangement

distinct advantage when the generation of a sinusoidal input signal is difficult. However, transient-response data do require fairly extensive data reduction to provide much detail. It is felt that the greatest potential use of transient-response testing is not on control system components but on system processes and systems themselves.

By comparison the frequency-response test yields data such that the contributing effects of each part of the system are more readily seen with simpler means of data reduction. From graphs of frequency-response data it is often possible to directly determine the effects of changing system parameters.

Another testing consideration is open-loop versus closed-loop testing. The primary purpose for using feedback is to make the closed-loop operation largely independent of the characteristics of the system elements. Thus the use of feedback tends to mask the details of the various components. If the purpose of testing is to determine what these characteristics are, making closed-loop measurements is self-defeating. Open-loop measurements make the response of each part of the system apparent. However, it is sometimes not possible to test a system in an open-loop configuration because the nominal operating point may shift significantly. In this case the system may be operated with the minimum feedback which will stabilize the output without substantially changing the response in the frequency range of interest.

The testing of control systems can be further subdivided into two categories of tests: steady-state tests and dynamic tests. Steady-state tests are defined by the final-value response after all transients have decayed. Since steady-state performance characteristics are based on the final ratio of output to input of the system, the primary testing requirement is to measure the input function and the resulting output.

A steady-state test that is frequently performed is the determination of the static positioning accuracy of a positional servomechanism. For small-instrument-type servomechanisms, the output shaft may be manually moved in both directions from the null position until the torque developed by the motor resists further motion. The overall angle of rotation represents a band within which the controlled output will ultimately settle. The static positioning accuracy can be taken as one-half of this dead band. In a second form of this test the input signal is increased slightly in each direction until motion of the output is detected. The positioning accuracy is taken as one-half of the overall range of the input signal.

8-4 TRANSIENT-RESPONSE TESTING

Transient-response measurements using a step input are important since control system dynamic specifications are often written in this form. The transient response will frequently resemble that of an underdamped second-order system. In this case the measured response can be analyzed to determine such parameters as rise time, overshoot, settling time, and frequency of damped oscillation.

The equipment for measuring transient response can be relatively simple. An electrical step-input function can be readily introduced by means of a switch. An electronic square-wave generator may be used to obtain a periodic step input. It is common practice to use a step-input signal that encompasses either 10 or 90 percent of the operating span of the input.

The transient response may be displayed on a pen recorder or a cathode-ray oscilloscope. The multichannel high-speed recorder has proven to be a fundamental item of equipment for dynamic test work. Such recorders are available for use at frequencies up to 100 Hz, which is adequate for most control system testing. The multichannel recorder permits several variables to be recorded on the same time axis for direct comparison.

System response times are frequently too slow for a conventional oscilloscope to be used with a step input. To overcome this limitation a long-persistence or storage type oscilloscope can be used. If a periodic input function is applied at a rate slow enough to allow system transients to decay between steps, the oscilloscope can be synchronized to the repeating input to provide a more consistent display.

8-5 FREQUENCY-RESPONSE TESTING

Sine-wave test data are useful for mathematical analysis, for graphical solution of control problems, and for characterization of dynamic performance. However, more elaborate equipment is required than for transient tests. The system is excited by a sinusoidal signal, and the resulting sinusoidal output is compared in amplitude and phase with the input. Methods of plotting and analyzing frequency-response data have already been described in Chap. 6. It is generally recommended that the sine-wave signal have a peak-to-peak magnitude of 10 percent of the input range of the device being tested. The sine-wave signal should have the arithmetical midpoint of the input range as its center.

The test should start at a frequency low enough so that the magnitude of the output is essentially unaffected by frequency and there is essentially no phase shift between input and output (except of course if the device has pure integral or derivative functions). This may be as low as 0.001 Hz for some devices. The frequency should be increased until the magnitude of the output sine wave is so small as to be lost in the system noise. Generally speaking, a frequency-response test should cover a frequency range of at least 2 or 3 decades. Obviously care should be taken to vary the frequency in small enough increments to detect sudden changes in the response curve.

Conventional low-frequency electronic oscillators that give a choice of different output waveforms, e.g., sine, square, or triangular, are readily available. Although the input will be electrical in nature in nearly all cases, special equipment can be obtained to generate mechanical or pneumatic sine-wave signals. One approach is to use an electrical sine-wave generator with a signal-converting device (such as an electromechanical servo operating a pilot valve) whose output signal is pneumatic or hydraulic.

The output sine wave can be displayed on a long-persistence or storage-type cathode-ray oscilloscope or a pen recorder. Due to the low frequencies involved, the oscilloscope should provide dc inputs. In addition the use of a conventional short-persistence tube may prove impractical.

To obtain maximum information about the frequency response, the output response should be measured and compared to the input in both amplitude and phase. It was stated in Sec. 6-12 that for linear systems there is a fixed relationship between phase shift and rate of change of amplitude. Thus for many systems, particularly simple ones, enough information can be obtained by measuring the output amplitude in rms, peak, or peak-to-peak values and comparing it with the input.

More accurate frequency-response testing requires the measurement of both amplitude and phase shift. A dual-beam oscilloscope or two-channel pen recorder provides a convenient means of displaying input and output waveforms on a common time base. Figure 8-2 shows the

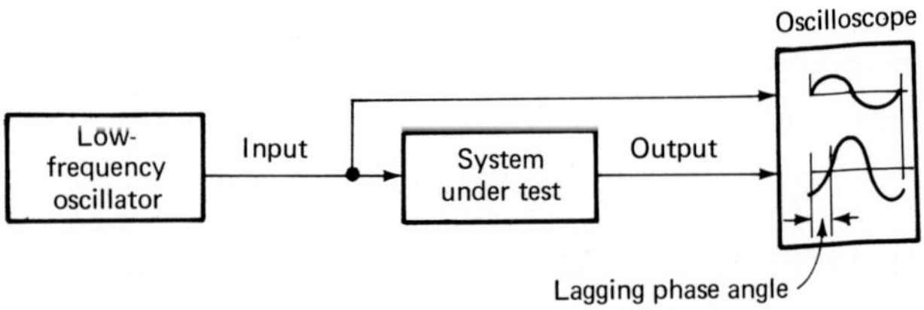

Figure 8-2 Frequency-response measurement using dual-beam oscilloscope

use of a dual-beam oscilloscope. Phase shift is determined by measuring relative displacement on the time axis and converting to angular units.

Two sine waves of the same frequency can be readily compared in phase by using the horizontal (X) and vertical (Y) inputs of a conventional oscilloscope. For two waves not in phase, an elliptical pattern called a *Lissajous figure* will result as shown in Fig. 8-3. The phase difference ϕ can be calculated from the equation

$$\sin \phi = \frac{\text{Y deflection when X is 0}}{\text{maximum Y deflection}} = \frac{A}{B} \tag{8-1}$$

Note that as ϕ increases through 90°, the major axis of the ellipse shifts from the first and third to the second and fourth quadrants.

More precise phase measurements can be made using a specially designed oscillator that provides two signal outputs. These outputs are ganged in frequency but one can be varied in phase with respect to the other. Applying the test setup shown in Fig. 8-4, the variable-phase output is adjusted until there is zero phase difference between the two inputs to the oscilloscope. The system phase shift can then be read as equal to the phase shift introduced in the variable-phase output from the oscillator. Phase measurement accuracies of 1 to 3° can be obtained from this method.

Frequency-response curves may be plotted based on a mathematical

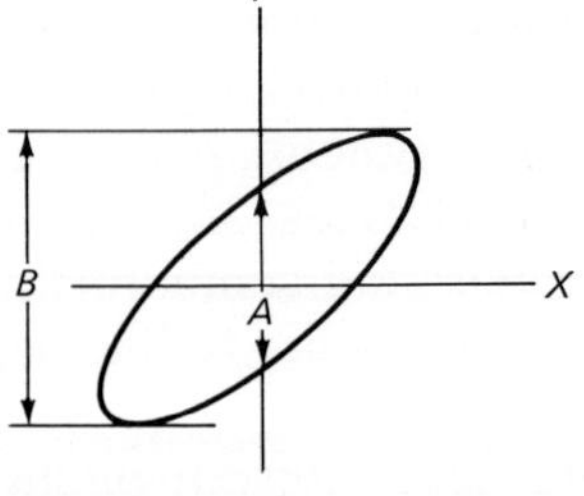

Figure 8-3 Phase measurement by Lissajous figure

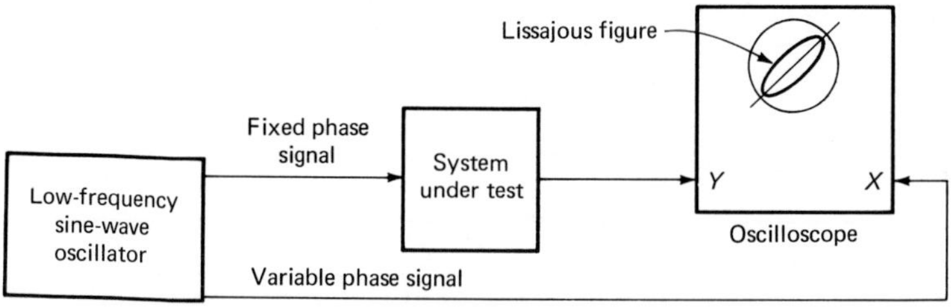

Figure 8-4 Frequency-response measurement using variable-phase-shift oscillator

expression for the behavior of a particular device or system. However, in practice it may prove difficult to write such a mathematical expression. The nature of a given device may not be known, or if it is known, the necessary parameters may be unavailable. In such cases it is necessary to derive a transfer function using frequency-response curves based on experimental data.

To plot frequency response, the output and input sine waves are compared on a frequency-by-frequency basis. The magnitude ratio in decibels may be computed as

$$\text{Magnitude ratio} = 20 \log_{10} \left(\frac{\text{peak output value}}{\text{peak input value}} \right) \text{dB} \tag{8-2}$$

After the frequency-response data are plotted, the transfer function may be obtained by comparing the resulting curves with theoretical frequency plots such as those shown in Fig. 6-10 or 6-13. From the slope of the magnitude curve and/or the phase angle at high frequencies, one can determine whether a first- or second-order term should be used in deriving the transfer function. The various parameters such as corner frequency, natural frequency, and damping ratio can then be determined.

EXAMPLE 8-1

A frequency-response test on the electropneumatic servomechanism shown in block-diagram form in Fig. 8-5 produced the experimental data shown in Table 8-1. From these data, plot the frequency-response curves and determine a transfer function for the system.

Solution

Table 8-1 includes both the experimental data and the calculated values needed to plot the frequency-response curves. Since experi-

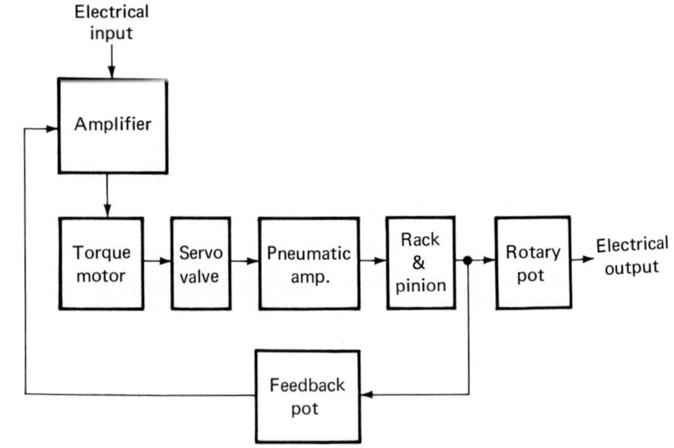

Figure 8-5 Block diagram of electropneumatic servomechanism for Example 8-1

TABLE 8-1 EXPERIMENTAL AND CALCULATED DATA FOR EXAMPLE 8-1

Experimental Data						*Calculated Magnitude Ratio*	
Frequency		*Input Sine Wave*		*Output Sine Wave*			
Hz	*rad/s*	*Peak Amplitude, V*	*Phase Shift, degrees*	*Peak Amplitude, V*	*Phase Shift, degrees*	$\frac{E_{out}}{E_{in}}, \frac{V}{V}$	$\frac{E_{out}}{E_{in}}$, dB
0.1	0.628	2.0	0	2.0	0	1	0
0.2	1.26	2.0	0	2.0	0	1	0
0.4	2.51	2.0	0	2.0	0	1	0
0.8	5.03	2.0	0	2.0	−2	1	0
1	6.28	2.0	0	2.0	−3	1	0
2	12.6	2.0	0	2.1	−5	1.05	+0.4
4	25.1	2.0	0	2.4	−13	1.20	+1.6
7	44.0	2.0	0	4.2	−34	2.10	+6.4
8	50.3	2.0	0	5.2	−55	2.60	+8.3
9	56.5	2.0	0	7.0	−90	3.50	+10.9
10	62.8	2.0	0	3.4	−115	1.70	+4.6
12	75.4	2.0	0	1.6	−148	0.80	−1.9
16	100	2.0	0	0.70	−163	0.35	−9.1
20	126	2.0	0	0.45	−168	0.23	−12.8
30	188	2.0	0	0.20	−174	0.10	−20.0
40	251	2.0	0	0.10	−175	0.05	−26.0

mental data are usually taken as a function of frequency in hertz rather than radians per second, the frequency has been converted by multiplying by 2π.

The various magnitude ratios are determined by taking the ratio of output amplitude to input amplitude and converting this ratio to the corresponding decibel value using Eq. (8-2). The various phase-shift values are calculated by comparing the phase angle of the output sine wave to that of the input sine wave.

The calculated data of Table 8-1 are plotted in Fig. 8-6. The experimental frequency data are in hertz so the frequency axis in Fig. 8-6 is plotted in these units. However, corresponding radian per second values are also given.

An examination of Fig. 8-6 indicates that the system is second-order. The response shows positive decibel values over a band of frequencies from about 2 to 12 Hz. At higher frequencies the curve follows a -40 dB per decade slope. The phase diagram further verifies the second-order assumption since the phase angle approaches $-180°$ for high values of frequency.

A transfer function can be obtained by comparing Fig. 8-6 with the general response curves for a second-order system.

Since the data suggest a second-order system, we can write

$$G(j\omega) = \frac{1}{1 + j2\zeta\omega/\omega_n + (j\omega/\omega_n)^2}$$

The resonant frequency occurs at about 9 Hz giving an ω_n of $9 \times 2\pi = 56.5$ rad/s. Comparing the peak magnitude ratio of 10.9 dB with the second-order curves in Fig. 6-13 suggests a damping ratio of 0.2. Thus

$$\frac{2\zeta}{\omega_n} = \frac{2 \times 0.2}{56.5} = 7.08 \times 10^{-3}$$

and

$$\frac{1}{\omega_n^{\ 2}} = \frac{1}{(56.5)^2} = 3.13 \times 10^{-4}$$

Substituting these constants into the general equation gives the transfer function of the system as

$$G(j\omega) = \frac{1}{1 + j7.08 \times 10^{-3}\omega + 3.13 \times 10^{-4}(j\omega)^2}$$

8-6 TRANSFER-FUNCTION MEASUREMENT

The sinusoidal transfer function of a control system can be measured either open-loop or closed-loop and the performance of the system

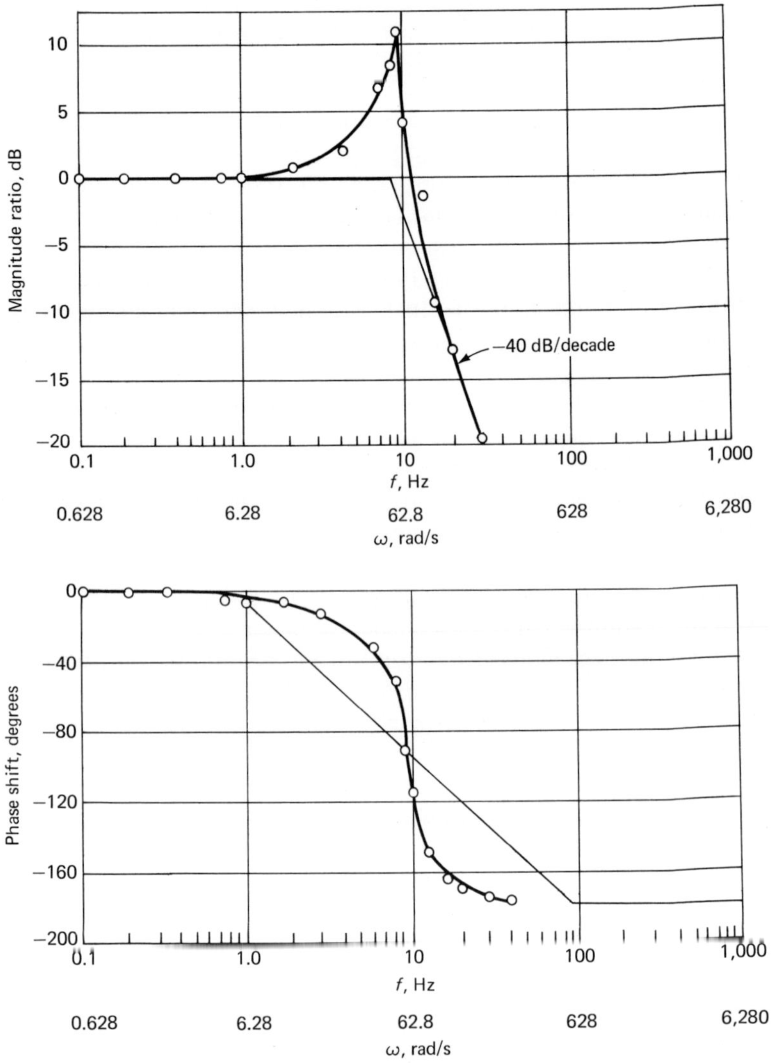

Figure 8-6 Frequency-response curves for Example 8-1

analyzed from these measurements. However, there are practical limitations on transfer-function measurement, and it is the purpose of this section to make the reader aware of these factors.

Transfer-function measurements are invariably performed with electrical signals, in which case a voltage input and a voltage from an output transducer are required. The transfer function of the transducer must

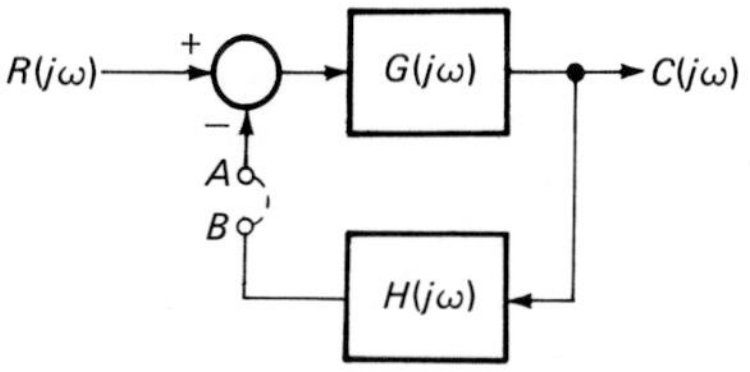

Figure 8-7 Measurement of open-loop transfer function

be either known or measured separately to determine the transfer function under consideration.

Figure 8-7 shows the block diagram of a closed-loop system in which the feedback loop has been broken from points *A* to *B*. A small sinusoidal signal is injected at point *A*. Assuming a linear system, the output at point *B* will be a sine wave whose magnitude and phase define the open-loop transfer function $GH(j\omega)$. The inherent 180° phase shift that is necessary for negative feedback must be considered in measuring the open-loop phase shift.

In measuring open-loop transfer functions which have large gains and/or complicated response characteristics, it is often advisable to measure the transfer function in parts. For example, if $G = G_1 \times G_2$, G_1 and G_2 can perhaps be better measured separately.

A control system usually consists of a number of interconnected components, each of which has one or more input and output signals. Two components are said to be *interacting* when the transfer function of one component is affected by a change in the characteristic of the other component. In analyzing the response of a control system, components which interact are best considered as a single unit.

A transfer function may include a pure integral term as illustrated in Fig. 8-8. This can pose difficulty in measuring the response because drifting of the dc level and large swings at low frequencies can cause excursions beyond the range of the output transducer. In such cases it is desirable to measure the derivative of the output variable in place of the variable itself. For example, rather than measuring the position of a motor shaft using a potentiometer as the transducer, the velocity may instead be measured using a tachometer. The integral that is known to be present can subsequently be added to the measured transfer function. An alternative approach to this problem is to place a small amount of negative feedback around the device under test. This feedback can reduce the dc drift and low-frequency swings to acceptable levels without appreciably altering the response over the frequency range of interest. This approach is also indicated in Fig. 8-8.

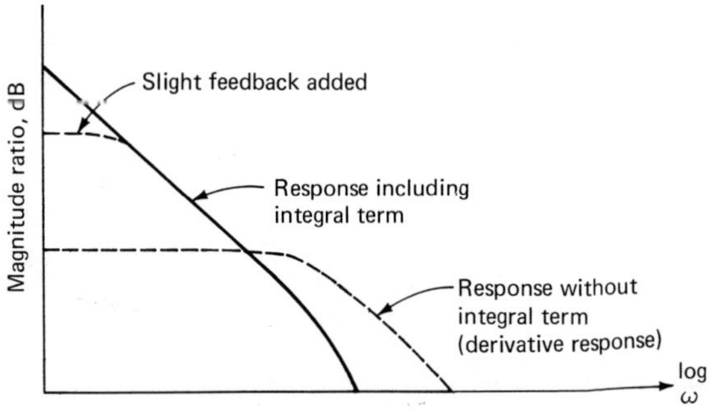

Figure 8-8 Transfer function containing integration

In transfer-function measurement, care should be taken to maintain signals at a level at which nonlinear operation is avoided, or is at least negligible. Saturation of the response variable is evidenced by clipping of the peaks of the sinusoidal response. This common form of limiting is readily detected with a cathode-ray oscilloscope.

Two other forms of limiting in physical systems are not as easily discovered. *Velocity limiting* is the limit in the rate at which the response variable can change. As the velocity limit is reached, the sinusoidal response tends to become triangular as indicated in Fig. 8-9. For a sine wave the maximum rate of change occurs at the zero crossing and is equal to $2\pi Af$, where A is the peak amplitude of the sine wave and f is the frequency in hertz. For a given velocity limit, the maximum amplitude for undistorted sinusoidal response at frequency f is then

$$A_{\text{max}} = \frac{\text{velocity limit}}{2\pi f} \tag{8-3}$$

Acceleration limiting is the limit in the second derivative of the response variable. For a given acceleration limit the maximum ampli-

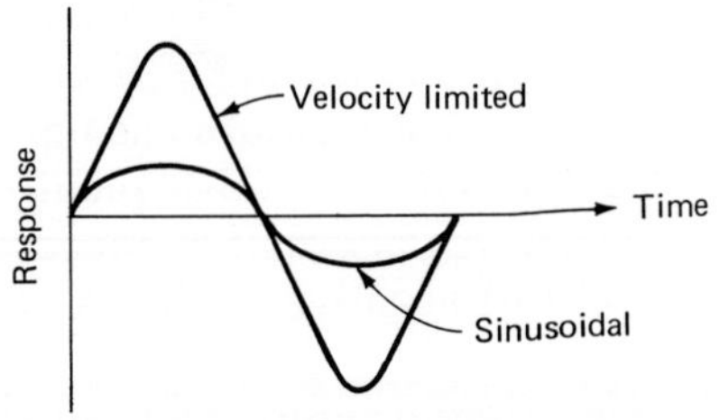

Figure 8-9 Effect of velocity limiting on sinusoidal response

tude for undistorted sine-wave response is

$$A_{max} = \frac{\text{acceleration limit}}{(2\pi f)^2} \tag{8-4}$$

Another potential problem in transfer-function measurement is random noise in the output. Such noise may obscure the desired output signal and make readings difficult, especially at higher frequencies where the amplitude of the response is small. Low-pass filters may be used to reject most noise; however, the filter cutoff should be well above the frequency range of interest so as not to affect the measured transfer function.

EXAMPLE 8-2

It is desired to measure the transfer function $\theta(j\omega)/V(j\omega)$ of a servomotor. The motor has a maximum speed of 600 rad/s and a maximum acceleration of 20,000 rad/s². Determine the maximum amplitude of the response so that the velocity and acceleration limits are not exceeded over the frequency range of 0.01 to 10 Hz.

Solution

Applying Eq. (8-3) for $f = 10$ Hz,

$$A_{max} = \frac{600}{2\pi \times 10} = 9.55 \text{ Hz}$$

Applying Eq. (8-4) at the same frequency,

$$A_{max} = \frac{20{,}000}{(2\pi \times 10)^2} = 5.07 \text{ Hz}$$

Thus at $f = 10$ Hz, the acceleration limit of the motor is the determining factor for the amplitude.

Figure 8-10 shows the velocity-limit and acceleration-limit curves of

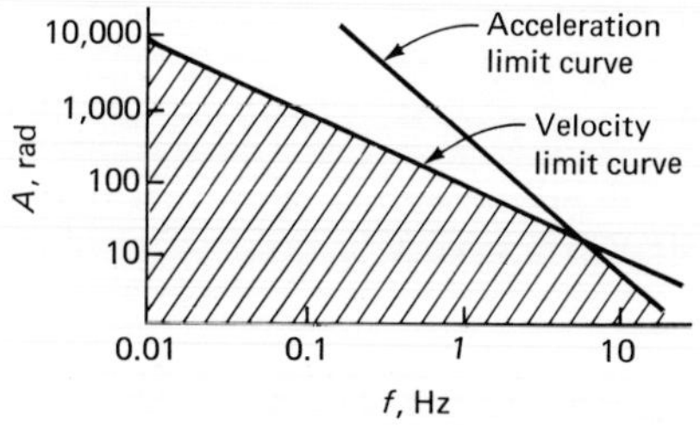

Figure 8-10 Velocity- and acceleration-limit curves for Example 8-2

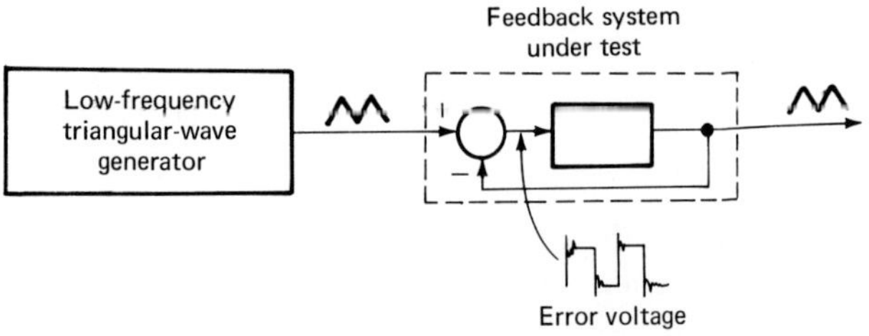

Figure 8-11 Velocity-response measurement

the motor over the designated frequency range. Note that the maximum velocity of the motor establishes the limit on A at low frequencies, while the acceleration limit dominates at high frequencies. The shaded area indicates the region of linear operation.

8-7 VELOCITY-RESPONSE TESTING

In certain control systems, especially positional-type servomechanisms, it is desirable to experimentally determine the value of the velocity constant K_v. This can be done by measuring the error response to a ramp input. A triangular input wave is frequently used for this purpose since it causes the output to respond in first one direction and then the other at a constant speed.

As shown in Fig. 8-11, the signal input is applied and adjusted to provide a response that is convenient for the purpose of determining the velocity error. The error voltage may be conveniently displayed using either a cathode-ray oscilloscope or pen recorder. This error has the form of a square wave with spikes. The spikes are associated with the transient that occurs when the direction of the controlled variable changes. The square portion of the wave corresponds to the steady-state velocity error. The peak-to-peak magnitude of this portion of the waveform represents twice the velocity error.

EXAMPLE 8-3

The scale factor of the output of a positional servomechanism is $0.5\ V/°$. It is desired to measure the velocity error corresponding to a tracking velocity of $40°/s$, which corresponds to a voltage change of $40 \times 0.5 = 20\ V/s$.

Solution

Using the test setup of Fig. 8-11, the steady-state error voltage is measured to be 0.25 V peak-to-peak. The velocity tracking error is one-half of this value, or 0.125 V, which corresponds to 0.25°.

8-8 CONTROL SYSTEM SPECIFICATIONS

Specifications are defined as a detailed description of requirements, materials, etc., for something to be built, installed, or manufactured. As applied to control equipment or systems, the specifications define the performance requirements. In defining performance specifications for feedback systems, the closed-loop characteristics are specified since this is the configuration in which the system operates.

The purpose of this section is to discuss the general nature of control-system specifications and to identify the terms ordinarily used in such specifications. A number of these terms have already been defined at appropriate points in earlier chapters. However, the various specifications are seldom independent of one another. Hence it is desirable that they be incorporated together as part of a discussion of testing.

Control-system specifications take on many different forms. The specifications for a relatively simple system such as a small servomechanism may not exceed a single page. On the other hand the specifications for a complete process control system may fill a small book. However, the specifications will usually include the following factors.

General Specifications

The general specifications define various constraints for the system as well as the environment and operating conditions in which the system is expected to perform. Typical constraints include such factors as weight, volume, power consumption, and cost. Typical operating and environmental specifications are loads on the output; ambient temperature, pressure, and humidity, mechanical loads of vibration and shock; and input signals and sources. Such general specifications are pertinent to testing because they specify the different conditions under which operating tests may need to be performed.

Static Specifications

Static performance is generally characterized by the single term *accuracy*. The accuracy of a system is its ability to have the controlled variable maintain a desired value with a minimum of error and is most often defined in terms of the steady-state error in the output. This is frequently expressed as a percentage of the full-scale range of the output. A number of factors may contribute to the inaccuracy of the output and may be included in the static specification of the system. These include the following:

1. *Linearity:* maximum deviation of the actual characteristic from the best straight line representing the actual curve. Linearity is frequently expressed as a percentage of the total range of the characteristic.

2. *Threshold:* maximum change in the input amplitude to the system which results in no change of output amplitude. The threshold value defines the sensitivity of the control system.

3. *Resolution:* smallest increment of output change which can be achieved by varying the input. This is also expressed as a percentage of the total range.

4. *Blacklash or hysteresis:* difference in output value obtained when approaching the same value of input from both directions.

The above terms are associated with variations in amplitude and all affect linearity in one way or another.

Dynamic Specifications

The two basic factors in specifying dynamic performance are speed of response and relative stability or damping. As we have seen, dynamic performance may be defined in either the time domain or the frequency domain. Most dynamic-performance terms are usually interpreted in terms of the definitions of an underdamped linear second-order system. These include the damped natural frequency ω_n and the damping ratio ζ.

The transient response to a step input is a common means of specifying dynamic performance in the time domain. Several of the terms that are used in specifying speed of response were defined in Sec. 4-10. These include the rise time t_r, which is the time for the output to rise from 10 to 90 percent of the final value, and the settling time t_s, which is the time required for the output to remain within a certain percentage (typically 5 percent) of the final value. The time constant τ, which is the time for the output to reach 63 percent of its final value, can also be used to define speed of response. Other terms that may be encountered are the time for first crossing t_c, which is the time required for the output to cross its final-value amplitude for the first time, and the time to first peak t_p, which is the time required for the output to reach

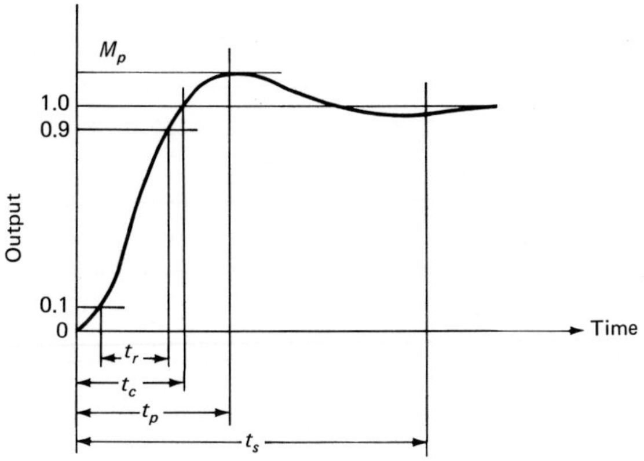

Figure 8-12 Dynamic-performance specification in time domain

the peak of its first overshoot. These terms are identified on the response curve shown in Fig. 8-12.

Relative stability is generally characterized by the damping ratio ζ. The term usually used to define relative stability in the time domain is the overshoot M_p, which is the maximum rise of the output with a step-function input. Figure 4-22 shows the relationship between this overshoot and the damping ratio.

Although the specifying of dynamic performance in the frequency domain was discussed in Sec. 6-11, several comments might be appropriate here. As was shown in Fig. 6-16, the magnitude frequency-response curve of the output of a closed-loop system generally peaks at some frequency and then attenuates as the frequency is increased further. The speed of response may be defined in terms of the system bandwidth ω_b, which is the frequency at which the response is 3 dB down from the low-frequency value. The frequency ω_m at which the peak value occurs may be also specified. The peak value M_m of the response provides an indication of the relative stability of the system. Figure 6-17 showed the relationship between this peak magnitude and the damping ratio.

For a system that is second-order, or may be approximated as second-order, the response in the time domain is related to the frequency response according to

$$t_p = \frac{\pi}{\omega_m} \tag{8-5}$$

where t_p is the time to reach the first peak in the step response and ω_m is the frequency at which M_m occurs.

8-9 CONTROL SYSTEM ADJUSTMENT: ZIEGLER–NICHOLS METHOD

One application of control system testing is to establish system operation by examining the response of the system. Based on the measured response, field adjustments may be made to the system to obtain satisfactory or possibly optimum operation. In this sense, the field adjustment of a control system may be viewed as a form of design. However, because field adjustment is closely related to test results and is often done on a trial-and-error basis, it is perhaps more appropriately included in a discussion of testing.

Field adjustments are most commonly made with process control systems where process characteristics may be difficult to measure or determine. It frequently proves less costly and more effective to adjust the system controller in the field rather than attempt to completely define the transfer functions and analyze the system beforehand. On the other hand, the transfer functions associated with electromechanical systems such as servomechanisms may be determined well enough to establish an accurate design through analysis. In this case field adjustment is frequently unnecessary or limited in scope.

A method for systematically determining controller settings for best performance was developed by J. B. Ziegler and N. B. Nichols in 1942 as a result of empirical studies of a wide variety of systems. The Ziegler-Nichols method has been widely applied on single-loop process control systems. It is primarily experimental in nature and uses the step response of the open-loop system (excluding the controller) to characterize the dynamics of the system. As shown in Fig. 8-13 the closed loop is opened immediately following the controller and a small step change Δm is applied to the final control element. The effect of this change is determined by recording the feedback signal. The step change must be small enough so that none of the variables attains a maximum or minimum value or is limited in any way.

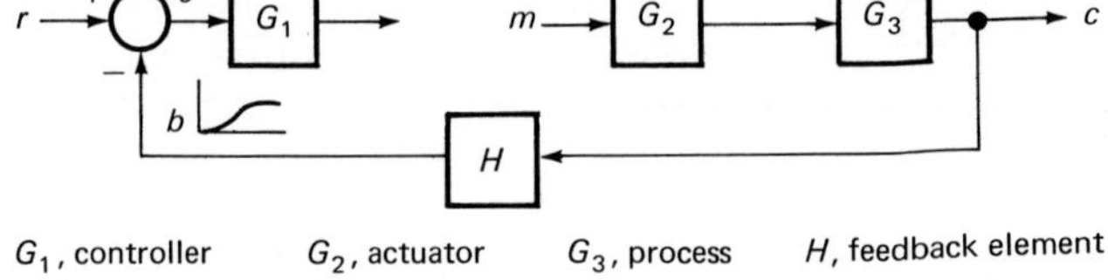

Figure 8-13 Open-loop test for Ziegler–Nichols method

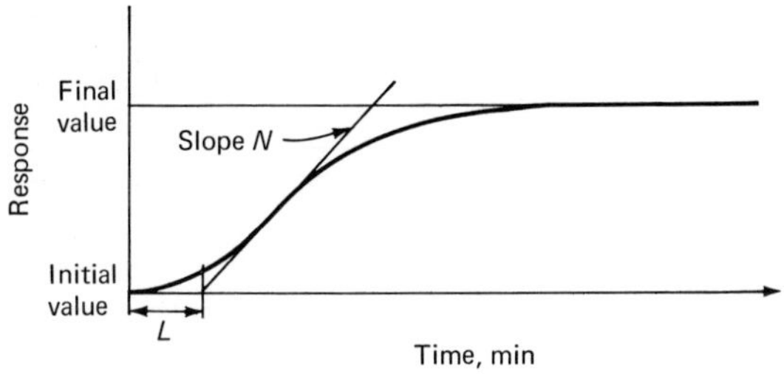

Figure 8-14 Reaction curve for Ziegler–Nichols method

The open-loop step response, which is often referred to as the *reaction curve,* is almost always an S-shaped curve as shown in Fig. 8-14. Note that the time axis for this curve is in units of minutes. The response is characterized by measuring the two parameters identified in Fig. 8-14. The maximum slope N is found by drawing a line tangent to the curve at the point of inflection. The apparent dead time L is fixed by the intercept of the maximum-slope line with the initial value of the response variable.

Using the Ziegler–Nichols method, the following equations may be used for calculating the controller constants.

For proportional control only

$$K_p = \frac{\Delta M}{NL} \tag{8-6}$$

where L = delay time, min
N = maximum slope of reaction curve
ΔM = magnitude of step change in input

For proportional plus integral control

$$K_p = \frac{0.9\,\Delta M}{NL} \tag{8-7}$$

$$T_i = 3.33L \text{ min} \tag{8-8}$$

For proportional plus integral plus derivative control

$$K_p = \frac{1.2\,\Delta M}{NL} \tag{8-9}$$

$$T_i = 2L \text{ min} \tag{8-10}$$

$$T_d = 0.5L \text{ min} \tag{8-11}$$

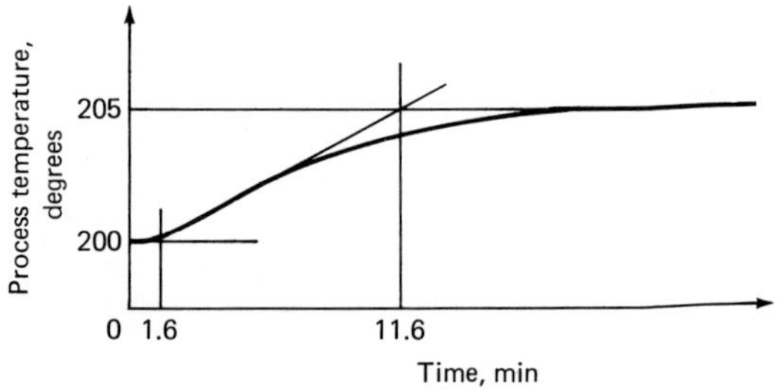

Figure 8-15 Reaction curve for Example 8-4

EXAMPLE 8-4

An open-loop test of a temperature control system yields the reaction curve shown in Fig. 8-15 for a 10 percent change in the setting of a control valve. Use the Ziegler–Nichols method to determine the proportional, integral, and derivative settings of a three-mode controller.

Solution

From Fig. 8-15,

$$L = 1.6 \text{ min}$$

$$N = \frac{205 - 200}{11.6 - 1.6} = 0.5^\circ/\text{min}$$

Applying Eqs. (8-9) to (8-11),

$$K_p = \frac{1.2 \times 10}{0.5 \times 1.6} = 15 \text{ percent}/^\circ$$

$$T_i = 2 \times 1.6 = 3.2 \text{ min}$$

$$T_d = 0.5 \times 1.6 = 0.8 \text{ min}$$

An alternative and somewhat simpler approach to the field adjustment of a controller is called the *ultimate-cycle method,* also developed by Ziegler and Nichols. Two measurements are required: the ultimate gain K_u and the ultimate period P_u. The ultimate gain is defined as the gain at which the closed-loop system just begins to oscillate, and the ultimate period is the period associated with the frequency of the oscillation.

Setting the integral and derivative modes of the controller at their least effective settings and starting with minimum proportional gain, the proportional gain is slowly increased until the system just begins to oscillate. (It may be necessary to apply a small step input to the system to initiate oscillation.) The gain setting is the ultimate gain, and the period of oscillation is the ultimate period.

In terms of the parameters K_u and P_u the controller settings are calculated from the following equations.

For proportional control alone

$$K_p = 0.5K_u \tag{8-12}$$

For proportional plus integral control

$$K_p = 0.45K_u \tag{8-13}$$

$$T_i = \frac{P_u}{1.2} \tag{8-14}$$

For proportional plus integral plus derivative control

$$K_p = 0.6K_u \tag{8-15}$$

$$T_i = \frac{P_u}{2} \tag{8-16}$$

$$T_d = \frac{P_u}{8} \tag{8-17}$$

The methods of Ziegler and Nichols are concerned with a specific aspect of control, namely, establishing the optimum control settings for an existing process control system. Only simple field tests are required to arrive at suitable controller settings using the procedures outlined above.

PROBLEMS

8-1. The test of a pressure gage yields frequency-response data as shown in the diagram below. Determine the transfer function $G(j\omega)$ for the gage.

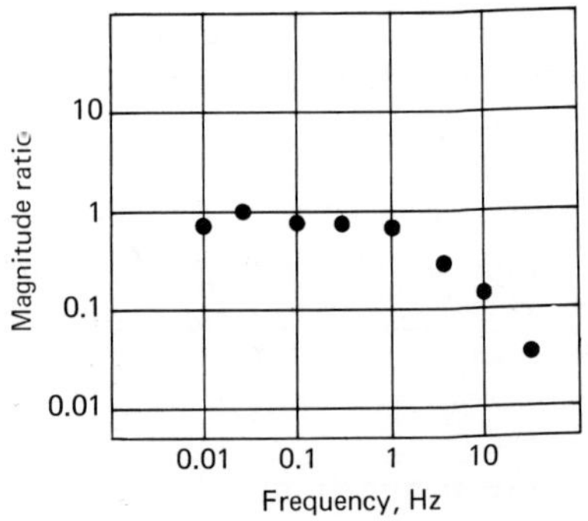

Prob. 8-1

8-2. The frequency-response test of a servomotor with armature control yields the data shown in the following table. From these data, plot the frequency-response curves and determine the motor transfer function $\dot{\theta}(j\omega)/V_a(j\omega)$.

Frequency, rad/s	$\dot{\theta}(j\omega)/V_a(j\omega)$	
	Magnitude, dB	*Phase Angle, °*
3	20	−10
5	21	−20
7	20	−25
10	19	−35
20	16	−50
30	12	60
50	9	−75
70	6	−80
100	2	−85

8-3. An ac voltage having a 10-V peak-to-peak amplitude is applied to the control field of a servomotor. The following data show the peak-to-peak variation in motor speed as a function of the frequency of the applied voltage. Plot the magnitude-response curve, and determine the transfer function $\dot{\theta}(s)/V_f(s)$.

Frequency, Hz	*Peak-to-Peak $\dot{\theta}$, rpm*
0.1	500
0.2	480
0.3	360
0.4	300
0.7	200
1	140
2	65
3	40
4	25
7	10
10	5
20	1

8-4. A control motor has a maximum speed of 4,000 rpm and a maximum stall torque of 10 lb-ft. The inertia of the motor and load is 10^{-2} lb-ft-s^2. Plot a curve showing the peak amplitude of the shaft-angle swing for undistorted frequency response over the range from 0.1 to 50 Hz.

8-5. A gas-fired furnace has its temperature controlled by a controller followed by a current-operated control valve. When the control loop is opened and a 10-mA step increment of current is applied to the valve, the temperature response approximates a time delay of 30 s followed by a linear rise of 30°F over a 2-min period, which then levels out to give a total increase of 75°F.

a. What is the reaction rate of the process?
b. Determine suitable controller settings for proportional action only.
c. Determine suitable settings when integral action is added to eliminate offset.

8-6. An open-loop transient test shows a pressure control system to have an apparent dead time of 0.04 min, a reaction rate of 2.5 psi/min, and a process change of 10 psi for a change of the control-valve setting of 5 percent. Determine suitable settings for a three-mode controller.

8-7. With the integral and derivative settings ineffective, the proportional gain of the controller in Prob. 8-6 is increased until the output starts to oscillate. The gain setting is 100 percent/psi, and the frequency of oscillation is 10 Hz. Use the ultimate-cycle method to determine the controller settings.

8-8. The transient-response characteristic of a closed-loop system is shown below.

a. Determine the values of t_r, t_c, t_p, t_s, and M_p.
b. What damping ratio is exhibited by the system?
c. If a frequency-response test of the system were performed, what value of ω_m would be expected?

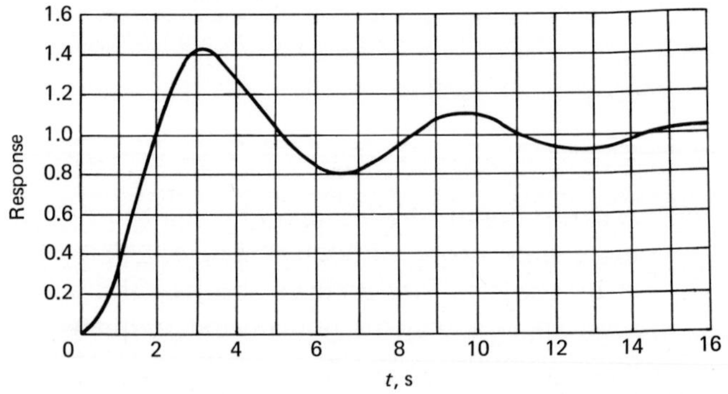

Prob. 8-8

Chapter Nine

nonlinearities in control systems

9-1 INTRODUCTION

In the preceding chapters we have limited our consideration to linear control systems. This has permitted the use of linear differential equations, which can be solved by such general methods as the Laplace transformation.

Unfortunately not all physical systems are linear. In fact, strictly speaking, all physical systems in nature are to some extent nonlinear. Although there are operating regions which may be regarded as being linear, it should be appreciated that under very small and very large input conditions the systems are indeed nonlinear.

The distinction between linear and nonlinear systems was first pointed out in Sec. 2-2, where it was noted that the superposition principle is not applicable to nonlinear systems. Since the Laplace transformation and other general methods of analysis depend on superposition, they cannot be directly applied to nonlinear systems. Indeed there are no general methods that are applicable to a large class of nonlinear differential equations. A few specialized analytical techniques are available to handle certain nonlinearities; however, these methods are beyond the scope of this text.

Many nonlinear control systems can be analyzed by linearizing the nonlinear element or elements and then applying linear methods to the resulting problem. One linearization technique that has proven useful in treating certain nonlinearities is based on an approximation known as a describing function. This method provides information in the frequency domain concerning system stability. If applied carefully, linearization by methods such as the describing function can yield useful results.

As mentioned in Chap. 2 analog simulation is extremely useful in the analysis and design of control systems. This is especially true in the case of nonlinear systems since nonlinear characteristics are readily introduced into the simulation.

As might be expected the field of nonlinear systems is important in automatic control and it is continually expanding as a result of research activity devoted to this subject. This chapter is intended to introduce some of the basic nonlinearities that may exist and to briefly present the describing-function method of analysis. The reader wishing to pursue this subject further is referred to texts of a more advanced nature.

However, it should be emphasized that a large number of control problems are suited to linear analysis, especially since linear approximations are possible for many nonlinearities.

9-2 SMALL-SIGNAL LINEARIZATION

When the coefficients of the terms in the system differential equation vary as a function of the dependent variable, the equation is nonlinear. If the coefficients are *continuous* functions of the dependent variable and its derivatives, the system is described as having a *continuous nonlinearity*. In general a continuous nonlinearity has a characteristic in which there are no discontinuities or sharp transitions.

Continuous nonlinearities may be eliminated from the differential equation by assuming a limited operating range for the variables. To illustrate this method of linearization, consider a control element with input signal X and response Y and with the nonlinear characteristic shown in Fig. 9-1. The value X_o is nominal input which establishes an operating level Y_o for the response. If we assume the input varies between X_{min} and X_{max}, the response will range from Y_{min} to Y_{max}. If

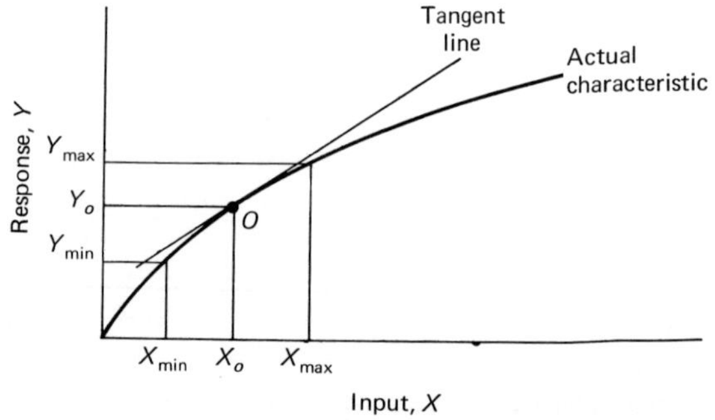

Figure 9-1 Linearization of nonlinear characteristic in specific operating range

a tangent to the curve is constructed at the operating point 0, the change in Y is approximately equal to the slope of the tangent line multiplied by the corresponding change in X. Thus we may use the equation

$$y = Kx \tag{9-1}$$

to linearize the operation of the element, where K is the slope of the characteristic at point 0 and the lowercase letters represent variations around the point. Obviously the value of K is different in different ranges of the variable.

It is apparent that the response will not be exactly as defined by Eq. (9-1). However, the difference should be small as long as the input signal is small. In addition using linear approximations of continuous nonlinearities gives the correct analysis of stability providing the disturbances are small.

Two instances where small-signal linearization has been used in this text are the linearization of turbulent flow in Sec. 2-7 and the speed-torque characteristic of an ac motor in Sec. 3-13.

EXAMPLE 9-1

A hemispherical tank whose radius is 10 ft contains liquid at a nominal height of 8 ft. Determine an approximate linear relationship between the volume of liquid and its height.

Solution

The volume v of a spherical segment of height h and radius r is

$$v = \tfrac{1}{3}\pi h^2(3r - h)$$

The equation may be linearized around the nominal height by finding the slope. Thus

$$\frac{dv}{dh} = \pi h(2r - h)$$

which for $h = 8$ becomes

$$\frac{dv}{dh} = 96\pi \text{ ft}^3/\text{ft}$$

9-3 DISCONTINUOUS NONLINEARITIES

Certain types of nonlinearities exhibit a discontinuous characteristic. In discontinuous cases there are abrupt changes in the characteristic,

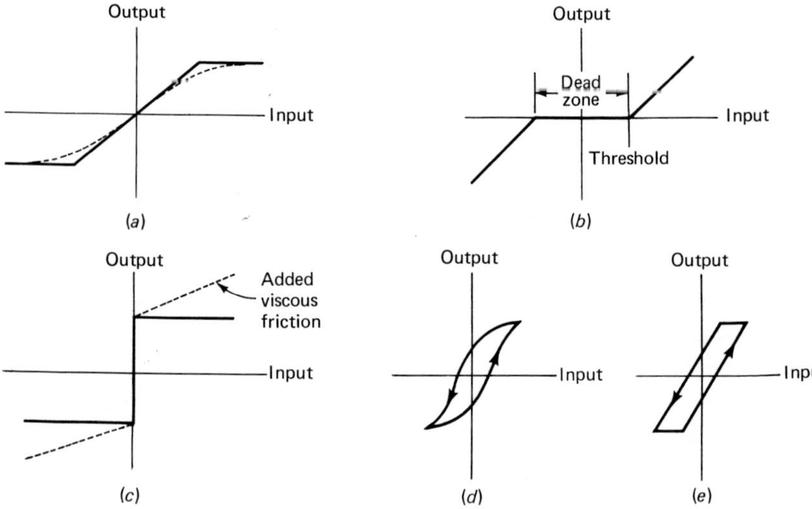

Figure 9-2 Common discontinuous nonlinearities: (a) saturation; (b) dead zone; (c) coulomb friction; (d) and (e) hysteresis

and so it is not readily approximated by a single straight line. Several forms of discontinuous nonlinearities occur fairly often in control systems and will be described in this section. These include saturation or limiting, dead zone, coulomb friction, and hysteresis.

We have seen that many components are linear up to a given input signal. For larger signals the output may no longer increase proportionally and can actually reach a maximum or limiting value as shown in Fig. 9-2*a*. This characteristic, called *saturation,* is common to many electrical and mechanical devices. For example, the output of an electronic amplifier has a maximum value, and a spring reaches a physical limit in either tension or compression. Although the transition between linear operation and saturation may be gradual, the characteristic is often represented by three straight-line segments. Such an approximation is known as *piecewise linearization*. The solid curve in Fig. 9-2*a* shows the ideal saturation characteristic, while the dashed curve shows the more practical case.

A *dead zone* is a form of nonlinearity in which there is no output for a range of inputs. This characteristic is shown in Fig. 9-2*b*. Dead zone exists in a motor because a certain voltage must be applied before the torque can overcome the static friction and produce motion. The minimum value of input to produce an output is termed the *threshold*. In a real sense, all physical components exhibit dead zone; the practical consideration is one of degree.

A phenomenon that exists in all mechanical systems involving sliding motion is *coulomb friction*. As shown in Fig. 9-2*c*, a friction force of constant magnitude opposes the motion. This differs from viscous friction, described in Sec. 2-4, in which the friction force is proportional to velocity. The presence of both coulomb and viscous friction would produce the characteristic shown by the dashed line in Fig. 9-2*c*.

Hysteresis is a nonlinear characteristic in which the output does not follow the same path depending upon whether the input is increasing or decreasing. One form of hysteresis shown in Fig. 9-2*d* is commonly associated with magnetic circuits. Backlash in gear trains, which was discussed in Sec. 3-14, also appears as hysteresis as shown in Fig. 9-2*e*.

9-4 EFFECTS OF NONLINEARITIES

Certain effects that occur in closed-loop control systems cannot be explained by linear control theory but can be related to system nonlinearities. Two of the more common effects that may be recognized as caused by nonlinear elements are hunting and jump resonance.

According to linear control theory, the output of a stable system will exponentially arrive at the desired value or, in the case of an unstable system, increase without limit. However, in most practical control systems, unstable behavior results in a fixed-frequency, constant-amplitude oscillation called *hunting*. In some cases, the amplitude of the oscillation is determined by the saturation characteristic or other nonlinearity of some element in the control system. The effect is then known as *limit cycling*.

An effect which sometimes occurs in measuring the frequency response of a system that includes a nonlinear element is known as *jump resonance*. This effect is indicated by Fig. 9-3. If the magnitude of the sinusoidal excitation is held constant and the frequency increased from

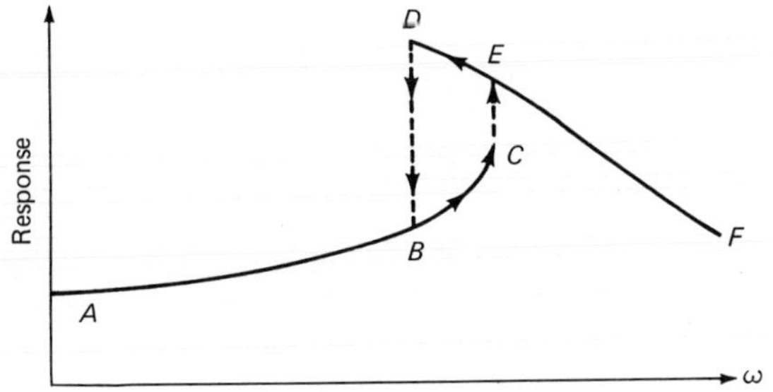

Figure 9-3 Jump-resonance effect

dc, the output response reaches point C and then suddenly jumps to point E from which it continues toward F. If the frequency is then reduced toward 0, the curve reaches point D and then jumps to point B and continues toward A. This phenomenon is usually caused by saturation in some system element. A change in the input magnitude will produce a different jump-resonance characteristic.

9-5 DESCRIBING FUNCTIONS

The material in Chaps. 6 and 7 has described the use of frequency-response methods in the analysis and design of linear control systems. The *describing-function* method is a technique that handles nonlinear systems in the frequency domain. As the name suggests, it is an approximate linear frequency-response function that represents or "describes" a nonlinear characteristic.

The describing-function method is based on a sinusoidal input being applied to a nonlinear element. Although the output of the element will be distorted by the nonlinearity, it will be of the same fundamental frequency as the input sine wave. As is shown in most basic circuits texts, a nonsinusoidal waveform can be represented mathematically by a series of terms called a *Fourier series*. This series is composed of a number of sinusoidal functions called harmonics whose frequencies are integer multiples of the fundamental frequency and whose amplitudes and phase angles vary according to the particular waveform.

The describing function assumes that a nonsinusoidal waveform can be described by its fundamental component and that the harmonics can be neglected. Consider the basic control system shown in Fig. 9-4, in which G_n represents a nonlinear element and G_s is the linear part of the system. If E is a sinusoidal input, M will be nonsinusoidal. Most control systems act as a low-pass filter so that the harmonics contained in M are attenuated by G_s. If the harmonics are sufficiently small compared to the fundamental, they can be neglected and the describing-function method is applicable according to the preceding assumption.

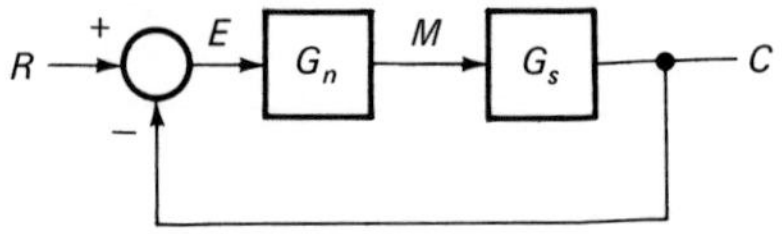

Figure 9-4 Control system containing nonlinear element

The describing function is defined as the ratio of the fundamental component of the output of a nonlinear element to the sinusoidal input. Thus

$$G_n(j\omega) = \frac{M_1(j\omega)}{E(j\omega)} \tag{9-2}$$

where $G_n(j\omega)$ = describing function
$M_1(j\omega)$ = fundamental component of the output
$E(j\omega)$ = sinusoidal input

$G_n(j\omega)$ may be a function of the frequency as well as the amplitude of the input to the nonlinear element. If no energy-storage elements are contained in the nonlinear element, $G_n(j\omega)$ is a function only of the amplitude of the input.

To calculate the describing function for a nonlinear device, the waveform of the output must be determined and analyzed in terms of a Fourier series. The resulting describing function may then be applied to the frequency-response analysis of the system. The describing-function method is especially useful in predicting the amplitude and frequency of a limit-cycle type of oscillation in the output.

Finally the method is applicable only to a single nonlinear element. If two or more nonlinearities are present, they must be combined into a single block since otherwise the input to the second nonlinear element will not be sinusoidal.

9-6 SATURATION AND DEAD ZONE

Two of the most common discontinuous nonlinearities in control systems are saturation and dead zone. The describing functions for these relatively simple cases will be discussed in this section. Other nonlinearities that involve energy storage have more complex describing functions and will not be considered. In addition the derivations of describing functions by means of Fourier analysis methods are beyond the scope of this chapter and are not included.

The phenomenon of saturation was described in Sec. 9-3. The ideal saturation characteristic from Fig. 9-2*a* is reproduced in Fig. 9-5 with a sinusoidal input $e = E_m \sin \omega t$ applied. The output M is proportional to the input until the output reaches the saturation value $\pm S$. For larger amplitudes the output remains at $\pm S$. Comparing the two curves in Fig. 9-5, we see that the shape of the output waveform (and consequently its harmonic content) depends on the ratio of the magnitude of the input to the saturation level. To find G_n it is necessary to determine M_1, the magnitude of the fundamental in the output.

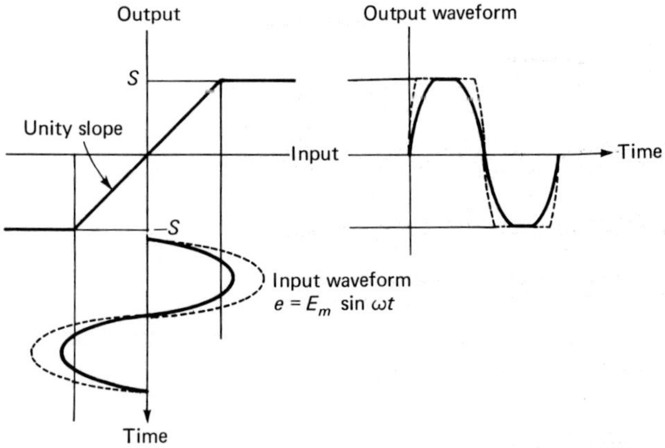

Figure 9-5 Saturation characteristic and waveforms

Figure 9-6 is a plot of the describing function of the saturation characteristic with the ratio E_m/S as the input. The function G_n is the ratio of the amplitude of the fundamental component of the output M_1 to the input E. As expected this function is unity for $E_m/S \leq 1$ inasmuch as the output does not reach saturation.

The fundamental of the output is in phase with the input, and so there is no phase shift associated with this saturation describing function. From Fig. 9-6 we may deduce that the effect of saturation is to reduce the apparent gain of the element being saturated, since G_n decreases as E becomes large. If the slope of the saturation charac-

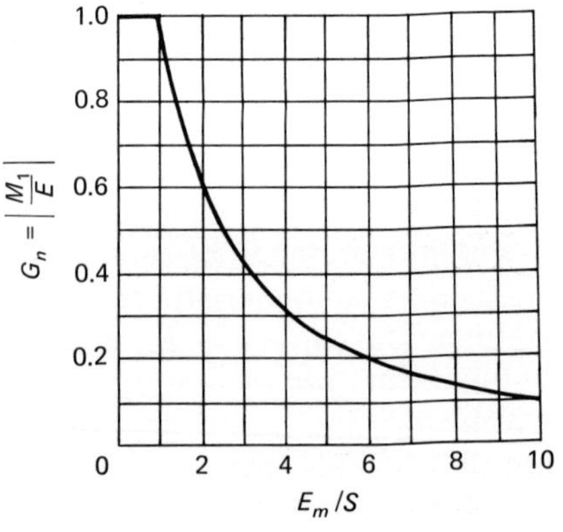

Figure 9-6 Plot of describing function for saturation

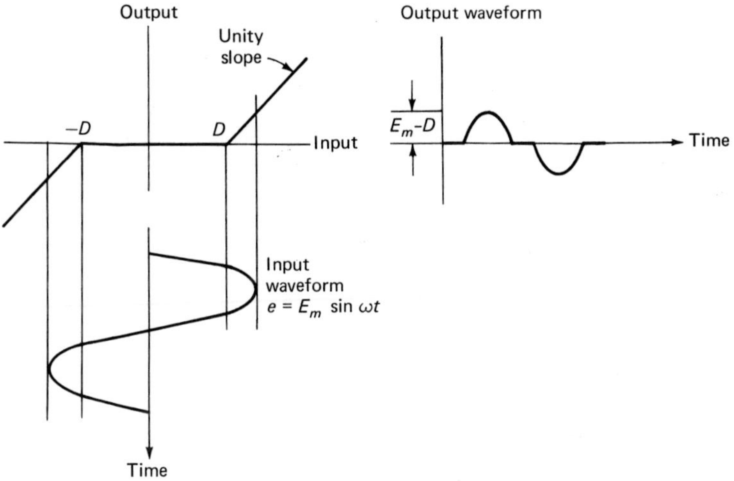

Figure 9-7 Dead zone characteristic and waveforms

teristic is other than unity, the describing function in Fig. 9-6 may be preceded by a gain block of magnitude equal to the slope.

The coulomb-friction characteristic shown in Fig. 9-2*c* is a special case of saturation in which the output is constant for all values of the input. The magnitude of the describing function may be calculated as

$$G_n = \frac{4S}{\pi E_m} \tag{9-3}$$

where S is the constant output and E_m is the peak value of the input.

The nonlinear characteristic identified as dead zone was shown in Fig. 9-2*b*. This characteristic is repeated in Fig. 9-7 with a sinusoidal input $e = E_m \sin \omega t$. The overall width of the dead zone in Fig. 9-7 is $2D$. The output changes abruptly at $\pm D$ even though an actual dead zone is more likely to exhibit a gradual change in the vicinity of $\pm D$. However, the idealized curve is simpler to analyze.

Figure 9-8 shows the dead zone describing function with the ratio E_m/D as the input. If E_m is smaller than $D(E_m/D < 1)$, there is no output and the describing is 0. If E_m is greater than D, the output increase is proportional to the value of $E_m - D$. As with the saturation describing function, the fundamental component of the output is in phase with the input so there is no phase shift.

If both saturation and dead zone are present, the resulting nonlinear characteristic is shown by Fig. 9-9. The describing function in this case depends upon the relative values of S and D. A family of curves of

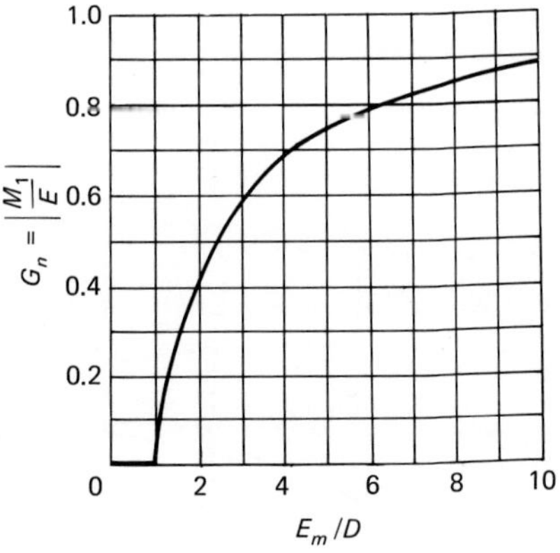

Figure 9-8 Plot of describing function for dead zone

this function is plotted in Fig. 9-10, with E_m/D as the input for several values of S/D. For $S = D$, the value of the describing function is 0 for all inputs.

EXAMPLE 9-2

A control valve has an overall dead zone of 2 percent of the total

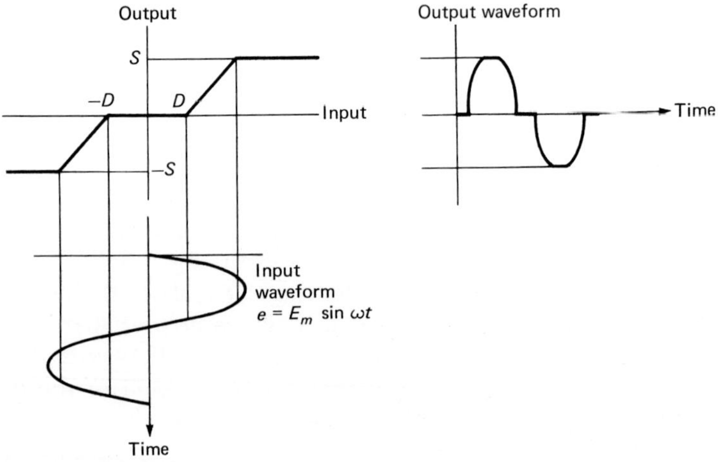

Figure 9-9 Combined saturation and dead zone characteristic and waveforms

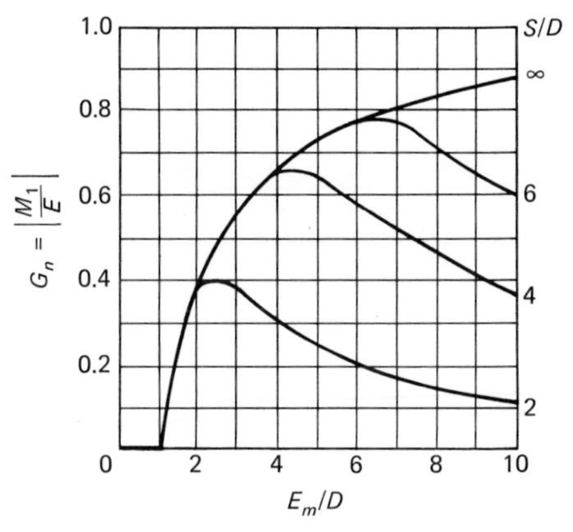

Figure 9-10 Plot of describing function for combined saturation and dead zone

range. Estimate the relative gain of the valve when the peak input signal is 5 percent of the range.

Solution

The ratio $E_m/D = 5$. From Fig. 9-8 the gain is determined to be 0.74.

9-7 HYSTERESIS

Hysteresis is a nonlinearity that occurs most often because of backlash and subsequent friction in gearing, electromagnetic effects, etc. As shown by the characteristic in Fig. 9-11, hysteresis causes the output to be double-valued for a cyclic input; that is, there can be two values of output for one value of input depending upon whether the input is increasing or decreasing. For example, in the case of gear backlash the output gear remains in contact with the input gear until the velocity reaches 0. Upon reversal of the input, the output remains stationary until the backlash is taken up, at which time it is assumed that the output instantaneously takes on the velocity of the input.

Hysteresis not only causes a variable gain but also a phase lag. This is shown by the describing-function plot in Fig. 9-12. Note that the phase shift is a function of amplitude and not of frequency as is the case for a linear element. In automatic control system applications the gain and phase effects associated with hysteresis tend to cancel each other. However, the overall effect of hysteresis is to cause less effective control action.

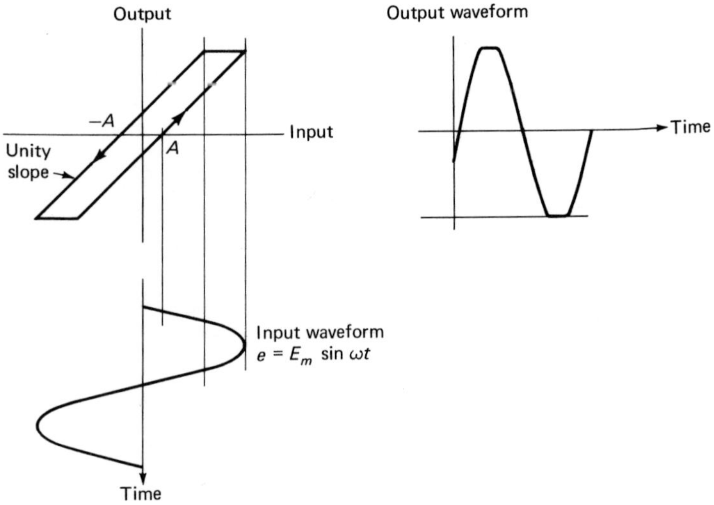

Figure 9-11 Hysteresis characteristic and waveforms

9-8 STABILITY ANALYSIS

We have seen that the stability of linear control systems is dependent only upon the characteristics of the system itself. Use of the Routh or Nyquist criterion will indicate whether the output will settle down in the case of a stable system or increase or oscillate with an increasing amplitude for an unstable system.

For nonlinear systems stability depends not only on the system parameters but also on the amplitude of the signal. In this section we

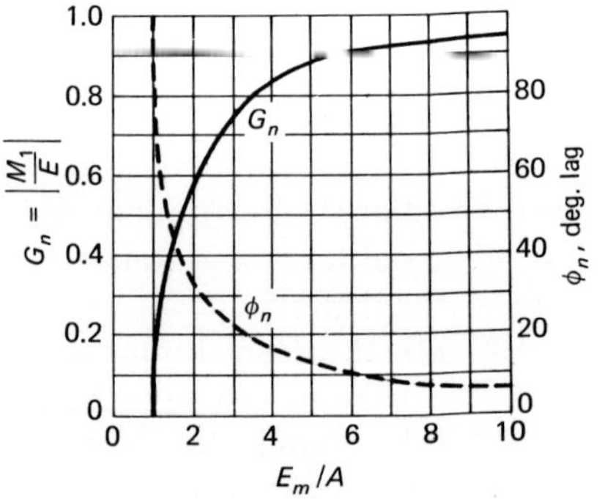

Figure 9-12 Plot of describing function for hysteresis

will briefly examine how the describing function of a nonlinear system may be used to obtain some information about the system's stability. Since the describing function is essentially a frequency-response technique, stability may be analyzed using the Nyquist criterion first presented in Sec. 6-12. Considering the system given in the block diagram of Fig. 9-4, the closed-loop transfer function is

$$\frac{C(j\omega)}{R(j\omega)} = \frac{G_n G_s(j\omega)}{1 + G_n G_s(j\omega)} \tag{9-4}$$

where G_n is a describing function that is dependent on signal amplitude but is independent of frequency.

The stability criterion in the nonlinear case may be determined in a manner similar to the reasoning employed in Sec. 6-12. There the result was that encirclement of the $-1 + j0$ point on the polar plane established the criterion for instability. This criterion may be extended to include the describing function G_n by setting the characteristic equation equal to 0. Instability will result when

$$1 + G_n G_s(j\omega) = 0$$

or

$$G_s(j\omega) = -\frac{1}{G_n} \tag{9-5}$$

Equation (9-5) suggests that two loci can be drawn on the polar plane: one of $G_s(j\omega)$ as for a linear system and the other for $-1/G_n$ for all possible values of the signal amplitude. The locus of $-1/G_n$ establishes the critical point which was simply $-1 + j0$ for the linear system. This concept is illustrated in Fig. 9-13. The frequency variant portion of the

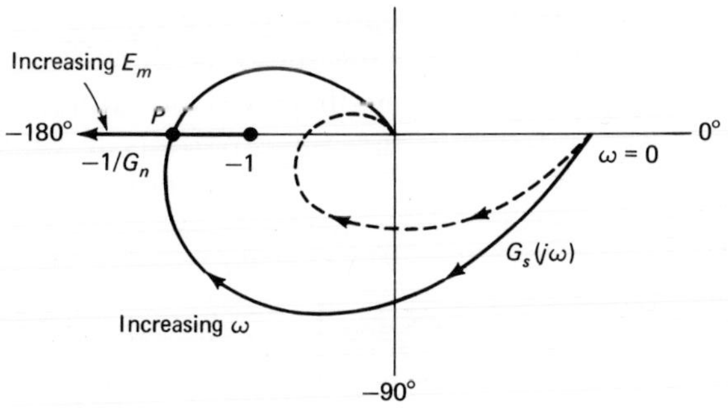

Figure 9-13 Polar diagram showing Nyquist criterion applied to nonlinear system

system is plotted on the polar plane together with the magnitude variant portion $-1/G_n$ (which in this case is a saturation characteristic). The relative positions of the G_s and $1/G_n$ loci give an indication of the stability condition of the system. The locus of the $-1/G_n$ plot must lie to the left of the G_s curve for the nonlinear system to be stable. Unstable operation can occur if the G_s curve touches or encloses any point on the $-1/G_n$ locus.

The transfer function represented by the dashed G_s curve in Fig. 9-13 lies to the right of the $-1/G_n$ plot, and this system would therefore be stable. However, the transfer function given by the solid G_s curve intersects the $-1/G_n$ plot at point P. For signal amplitudes for which the $-1/G_n$ plot lies to the right of point P, the system will be unstable. Conversely, signal amplitudes for which $-1/G_n$ falls to the left of point P represent stable operating conditions.

Stable or unstable operation of a nonlinear system may lead to oscillations that either decrease or increase in the same manner as for a linear system. However, a change in the amplitude of the oscillation in the nonlinear system affects the value of G_n. The resulting relationship may be such that the instability is constrained to occur only at the point of intersection of the G_s and $-1/G_n$ curves. Under this condition sustained oscillations of a constant or stable magnitude will result, and the system is then said to be operating in a *stable limit cycle*. If the oscillation is sufficiently small, this form of stability may be considered acceptable.

EXAMPLE 9-3

To illustrate the use of the describing function in nonlinear analysis, we will consider a positional servomechanism in which backlash is present in the gear train. Figure 9-14 shows the block diagram of a system of this type. The linear part of the system $G_s(j\omega)$ is made up of the blocks between the error detector and the motor shaft. The transfer function of the linear portion of the system is

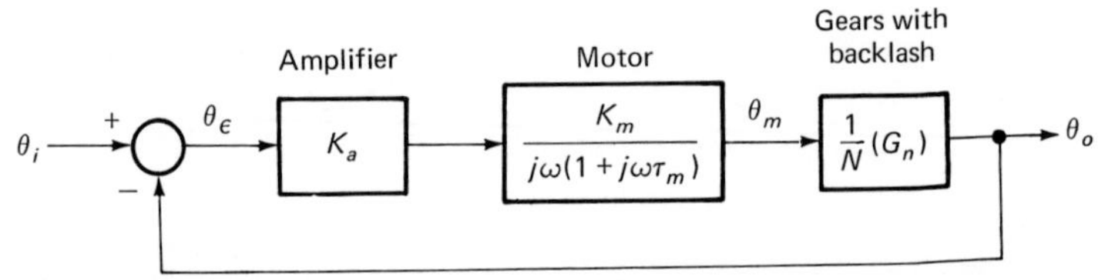

Figure 9-14 Block diagram of servomechanism for Example 9-3

$$\frac{\theta_m(j\omega)}{\theta_\varepsilon(j\omega)} = G_s(j\omega) = \frac{K_a K_m}{j\omega(1 + j\omega\tau_m)}$$

where K_a = amplifier gain
K_m = velocity gain constant of motor
τ_m = mechanical time constant of motor

The describing function G_n for backlash is shown in Fig. 9-12. The proportionality constant associated with the describing function is the reciprocal of the gear ratio.

For this example, the following values are assumed: $K_a = 10$, $K_m = 0.1$, $\tau_m = 10$, $N = 1.0$, and the overall backlash (defined as $2A$ in Fig. 9-11) is 0.05 rad.

Solution

The stability of the servomechanism is determined using Eq. (9-5). Thus

$$G_s(j\omega) = \frac{1}{j\omega(1 + j\omega 10)} = -\frac{1}{G_n}$$

The $G_s(j\omega)$ and $-1/G_n$ characteristics are plotted in the polar diagram in Fig. 9-15. The two loci intersect at approximately $\omega = 0.28$ rad/s and $E_m = 0.09$ rad. This intersection indicates that the system will oscillate at a frequency and amplitude determined by the point of intersection.

The stability of oscillation can be evaluated by assuming that the amplitude of the oscillation increases beyond that determined by the point of intersection. The operating point shifts to the outer or stable side of the $G_s(j\omega)$ curve. The amplitude of the oscillation now tends to decrease, moving the operating point back to the original inter-

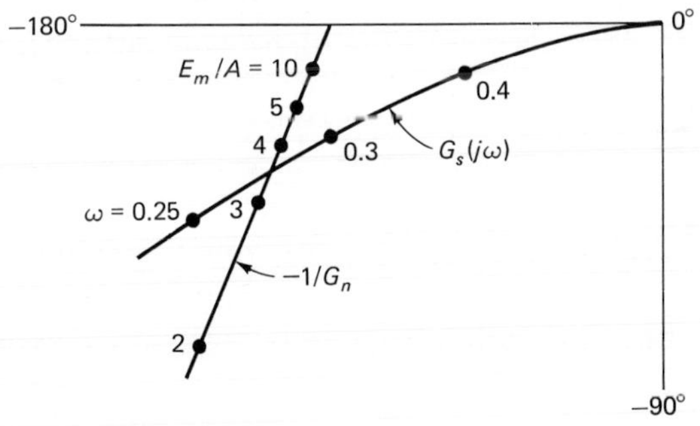

Figure 9-15 Polar diagram for describing-function analysis for Example 9-3

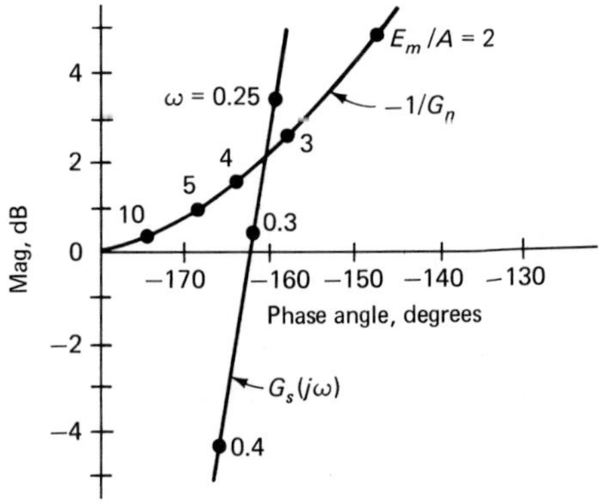

Figure 9-16 Nichols diagram for Example 9-3

section. In a similar manner a slight decrease in amplitude shifts the operating point to the unstable side of $G_s(j\omega)$. The amplitude tends to build up, which moves the operating point back to the original intersection. The sustained oscillation is therefore stable, and the steady-state operating condition is a limit cycle of amplitude and frequency defined by the point of intersection.

From the above example we may note that systems have stable limit cycles if the describing function locus with increasing amplitude crosses the $G_s(j\omega)$ locus from a region of instability to a region of stability. If the reverse is true, that is, an increasing amplitude takes the describing function locus into a region of instability, the oscillation is unstable and will either die out or increase in amplitude.

The method of plotting is not limited to the polar diagrams that have been used in explaining the theory of stability. It is often more convenient to use a Nichols chart plot that shows the magnitude of $G_s(j\omega)$ in decibels versus the phase angle of $G_s(j\omega)$ in degrees and the magnitude and phase angle of $-1/G_n$. This plot may be interpreted in the same way as the polar plot in analyzing stability. Figure 9-16 shows the $G_s(j\omega)$ and $-1/G_n$ characteristics of Example 9-3 plotted on a Nichols diagram.

Since the describing function is an approximate method of nonlinear analysis, the calculated amplitude and frequency of oscillation may differ somewhat from those determined experimentally. The describing functions we have considered are not functions of frequency. If a system has a describing function which is dependent on amplitude and frequency, a series of describing-function loci must be plotted for a range of frequency. This case is generally more difficult to analyze.

PROBLEMS

9-1. For turbulent flow in a pipe the pressure drop p varies as the square of the flow q; that is,

$$p = Kq^2$$

Linearize the pressure equation for a small variation in flow about a nominal value q_o.

9-2. A spring characteristic is defined by the relationship

$$F_s = 10y^{1.1} \text{ lb}$$

where y is the deflection in inches. The spring is to operate with a nominal deflection of 0.1 in. Derive a linear approximation for the incremental spring force ΔF_s, and calculate the maximum percent error in the total spring force if the linear approximation is used over the range $0.09 \leq x \leq 0.11$ in.

9-3. A system element saturates at a signal of 10 V amplitude. Calculate the attenuation of the element if a sinusoidal signal of 30 V peak amplitude is transmitted to the element.

9-4. A motor has a dead zone corresponding to 25 percent of its rated voltage and saturates at 150 percent of its rated voltage. Determine the relative gain of the motor when the peak input voltage is twice the rated voltage.

9-5. A unity-feedback positional servomechanism has a motor whose transfer function is

$$G(j\omega) = \frac{0.1}{j\omega(1 + j0.05\omega)(1 + j0.01\omega)}$$

The motor is driven by an amplifier with an unsaturated gain of 100. The amplifier saturates when its output is at 80 percent of the rated voltage of the motor. Draw the block diagram of the system, define the describing function, and investigate the stability on a Nichols diagram.

9-6. A control system has the linear transfer function shown below. (a) What will be the stability condition if the system contains a saturation characteristic as given in Fig. 9-5? (b) What will be the stability condition if the system contains a dead zone as given in Fig. 9-7?

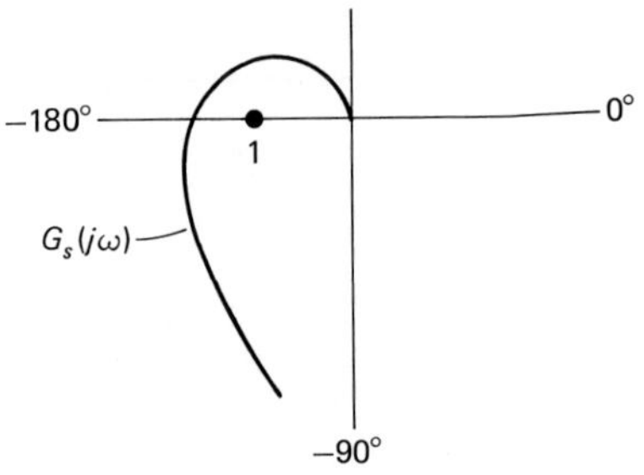

Prob. 9-6

9-7. Discuss the stability of the nonlinear system shown in the polar diagram below.

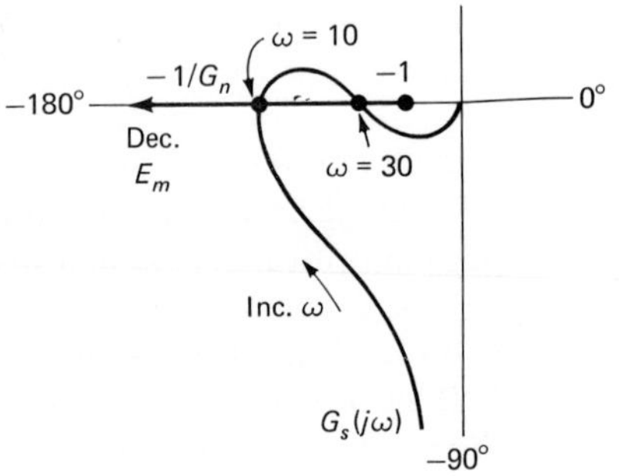

Prob. 9-7

Appendix A

table of Laplace-transform pairs[1]

$F(s)$	$f(t) \quad 0 \le t$
1. 1	$u_1(t)$ unit impulse at $t = 0$
2. $\dfrac{1}{s}$	1 or $u(t)$ unit step at $t = 0$
3. $\dfrac{1}{s^2}$	$tu(t)$ ramp function
4. $\dfrac{1}{s^n}$	$\dfrac{1}{(n-1)!}t^{n-1}$ n is a positive integer
5. $\dfrac{1}{s}e^{-as}$	$u(t-a)$ unit step starting at $t = a$
6. $\dfrac{1}{s}(1 - e^{-as})$	$u(t) - u(t-a)$ rectangular pulse
7. $\dfrac{1}{s+a}$	e^{-at} exponential decay
8. $\dfrac{1}{(s+a)^n}$	$\dfrac{1}{(n-1)!}t^{n-1}e^{-at}$ n is a positive integer
9. $\dfrac{1}{s(s+a)}$	$\dfrac{1}{a}(1 - e^{-at})$
10. $\dfrac{1}{s(s+a)(s+b)}$	$\dfrac{1}{ab}\left(1 - \dfrac{b}{b-a}e^{-at} + \dfrac{a}{b-a}e^{-bt}\right)$
11. $\dfrac{s+\alpha}{s(s+a)(s+b)}$	$\dfrac{1}{ab}\left[\alpha - \dfrac{b(\alpha - a)}{b-a}e^{-at} + \dfrac{a(\alpha - b)}{b-a}e^{-bt}\right]$

[1]From John J. D'Azzo and Constantine H. Houpis, *Feedback Control System Analysis and Synthesis*, 2d ed., McGraw-Hill, New York, 1966.

	$F(s)$	$f(t) \quad 0 \le t$
12.	$\dfrac{1}{(s+a)(s+b)}$	$\dfrac{1}{b-a}(e^{-at} - e^{-bt})$
13.	$\dfrac{s}{(s+a)(s+b)}$	$\dfrac{1}{a-b}(ae^{-at} - be^{-bt})$
14.	$\dfrac{s+\alpha}{(s+a)(s+b)}$	$\dfrac{1}{b-a}[(\alpha - a)e^{-at} - (\alpha - b)e^{-bt}]$
15.	$\dfrac{1}{(s+a)(s+b)(s+c)}$	$\dfrac{e^{-at}}{(b-a)(c-a)} + \dfrac{e^{-bt}}{(c-b)(a-b)} + \dfrac{e^{-ct}}{(a-c)(b-c)}$
16.	$\dfrac{s+\alpha}{(s+a)(s+b)(s+c)}$	$\dfrac{(\alpha-a)e^{-at}}{(b-a)(c-a)} + \dfrac{(\alpha-b)e^{-bt}}{(c-b)(a-b)} + \dfrac{(\alpha-c)e^{-ct}}{(a-c)(b-c)}$
17.	$\dfrac{\omega}{s^2+\omega^2}$	$\sin \omega t$
18.	$\dfrac{s}{s^2+\omega^2}$	$\cos \omega t$
19.	$\dfrac{s+\alpha}{s^2+\omega^2}$	$\dfrac{\sqrt{\alpha^2+\omega^2}}{\omega} \sin(\omega t + \phi) \qquad \phi = \tan^{-1}\dfrac{\omega}{\alpha}$
20.	$\dfrac{s \sin\theta + \omega\cos\theta}{s^2+\omega^2}$	$\sin(\omega t + \theta)$
21.	$\dfrac{1}{s(s^2+\omega^2)}$	$\dfrac{1}{\omega^2}(1 - \cos\omega t)$
22.	$\dfrac{s+\alpha}{s(s^2+\omega^2)}$	$\dfrac{\alpha}{\omega^2} - \dfrac{\sqrt{\alpha^2+\omega^2}}{\omega^2}\cos(\omega t + \phi) \qquad \phi = \tan^{-1}\dfrac{\omega}{\alpha}$
23.	$\dfrac{1}{(s+a)(s^2+\omega^2)}$	$\dfrac{e^{-at}}{a^2+\omega^2} + \dfrac{1}{\omega\sqrt{a^2+\omega^2}}\sin(\omega t - \phi)$ $\phi = \tan^{-1}\dfrac{\omega}{a}$
24.	$\dfrac{1}{(s+a)^2+b^2}$	$\dfrac{1}{b}e^{-at}\sin bt$
24a.	$\dfrac{1}{s^2+2\zeta\omega_n s+\omega_n^2}$	$\dfrac{1}{\omega_n\sqrt{1-\zeta^2}}e^{-\zeta\omega_n t}\sin\omega_n\sqrt{1-\zeta^2}\,t$

	$F(s)$	$f(t) \quad 0 \le t$
25.	$\dfrac{s+a}{(s+a)^2+b^2}$	$e^{-at}\cos bt$
26.	$\dfrac{s+\alpha}{(s+a)^2+b^2}$	$\dfrac{\sqrt{(\alpha-a)^2+b^2}}{b}e^{-at}\sin(bt+\phi)$ $\phi=\tan^{-1}\dfrac{b}{\alpha-a}$
27.	$\dfrac{1}{s[(s+a)^2+b^2]}$	$\dfrac{1}{a^2+b^2}+\dfrac{1}{b\sqrt{a^2+b^2}}e^{-at}\sin(bt-\phi)$ $\phi=\tan^{-1}\dfrac{b}{-a}$
27a.	$\dfrac{1}{s(s^2+2\zeta\omega_n s+\omega_n^2)}$	$\dfrac{1}{\omega_n^2}-\dfrac{1}{\omega_n^2\sqrt{1-\zeta^2}}e^{-\zeta\omega_n t}$ $\sin(\omega_n\sqrt{1-\zeta^2}\,t+\phi) \qquad \phi=\cos^{-1}\zeta$
28.	$\dfrac{s+\alpha}{s[(s+a)^2+b^2]}$	$\dfrac{\alpha}{a^2+b^2}+\dfrac{1}{b}\sqrt{\dfrac{(\alpha-a)^2+b^2}{a^2+b^2}}e^{-at}\sin(bt+\phi)$ $\phi=\tan^{-1}\dfrac{b}{\alpha-a}-\tan^{-1}\dfrac{b}{-a}$
29.	$\dfrac{1}{(s+c)[(s+a)^2+b^2]}$	$\dfrac{e^{-ct}}{(c-a)^2+b^2}+\dfrac{e^{-at}\sin(bt-\phi)}{b\sqrt{(c-a)^2+b^2}}$ $\phi=\tan^{-1}\dfrac{b}{c-a}$
30.	$\dfrac{1}{s(s+c)[(s+a)^2+b^2]}$	$\dfrac{1}{c(a^2+b^2)}-\dfrac{e^{-ct}}{c[(c-a)^2+b^2]}$ $+\dfrac{e^{-at}\sin(bt-\phi)}{b\sqrt{a^2+b^2}\sqrt{(c-a)^2+b^2}}$ $\phi=\tan^{-1}\dfrac{b}{-a}+\tan^{-1}\dfrac{b}{c-a}$
31.	$\dfrac{s+\alpha}{s(s+c)[(s+a)^2+b^2]}$	$\dfrac{\alpha}{c(a^2+b^2)}-\dfrac{(c-\alpha)e^{-ct}}{c[(c-a)^2+b^2]}$ $+\dfrac{\sqrt{(\alpha-a)^2+b^2}}{b\sqrt{a^2+b^2}\sqrt{(c-a)^2+b^2}}$ $e^{-at}\sin(bt+\phi)$ $\phi=\tan^{-1}\dfrac{b}{\alpha-a}-\tan^{-1}\dfrac{b}{-a}-\tan^{-1}\dfrac{b}{c-a}$

	$F(s)$	$f(t) \quad 0 \le t$
32.	$\dfrac{1}{s^2(s+a)}$	$\dfrac{1}{a^2}(at - 1 + e^{-at})$
33.	$\dfrac{1}{s(s+a)^2}$	$\dfrac{1}{a^2}(1 - e^{-at} - ate^{-at})$
34.	$\dfrac{s+\alpha}{s(s+a)^2}$	$\dfrac{1}{a^2}[\alpha - \alpha e^{-at} + a(a-\alpha)te^{-at}]$
35.	$\dfrac{s^2 + \alpha_1 s + \alpha_0}{s(s+a)(s+b)}$	$\dfrac{\alpha_0}{ab} + \dfrac{a^2 - \alpha_1 a + \alpha_0}{a(a-b)}e^{-at} - \dfrac{b^2 - \alpha_1 b + \alpha_0}{b(a-b)}e^{-bt}$
36.	$\dfrac{s^2 + \alpha_1 s + \alpha_0}{s[(s+a)^2 + b^2]}$	$\dfrac{\alpha_0}{c^2} + \dfrac{1}{bc}[(a^2 - b^2 - \alpha_1 a + \alpha_0)^2 + b^2(\alpha_1 - 2a)^2]^{1/2}e^{-at}\sin(bt + \phi)$ $\phi = \tan^{-1}\dfrac{b(\alpha_1 - 2a)}{a^2 - b^2 - \alpha_1 a + \alpha_0} - \tan^{-1}\dfrac{b}{-a}$ $c^2 = a^2 + b^2$
37.	$\dfrac{1}{(s^2+\omega^2)[(s+a)^2+b^2]}$	$\dfrac{(1/\omega)\sin(\omega t + \phi_1) + (1/b)e^{-at}\sin(bt + \phi_2)}{[4a^2\omega^2 + (a^2 + b^2 - \omega^2)^2]^{1/2}}$ $\phi_1 = \tan^{-1}\dfrac{-2a\omega}{a^2 + b^2 - \omega^2}$ $\phi_2 = \tan^{-1}\dfrac{2ab}{a^2 - b^2 + \omega^2}$
38.	$\dfrac{s+\alpha}{(s^2+\omega^2)[(s+a)^2+b^2]}$	$\dfrac{1}{\omega}\left[\dfrac{\alpha^2 + \omega^2}{c}\right]^{1/2}\sin(\omega t + \phi_1) + \dfrac{1}{b}\left[\dfrac{(\alpha - a)^2 + b^2}{c}\right]^{1/2}e^{-at}\sin(bt + \phi_2)$ $c = (2a\omega)^2 + (a^2 + b^2 - \omega^2)^2$ $\phi_1 = \tan^{-1}\dfrac{\omega}{\alpha} - \tan^{-1}\dfrac{2a\omega}{a^2 + b^2 + \omega^2}$ $\phi_2 = \tan^{-1}\dfrac{b}{\alpha - a} + \tan^{-1}\dfrac{2ab}{a^2 - b^2 + \omega^2}$

	$F(s)$	$f(t) \qquad 0 \le t$
39.	$\dfrac{s+\alpha}{s^2[(s+a)^2+b^2]}$	$\dfrac{1}{c}\left(\alpha t + 1 - \dfrac{2\alpha a}{c}\right) + \dfrac{[b^2+(\alpha-a)^2]^{1/2}}{bc} e^{-at}\sin(bt+\phi)$ $c = a^2 + b^2$ $\phi = 2\tan^{-1}\left(\dfrac{b}{a}\right) + \tan^{-1}\dfrac{b}{\alpha - a}$
40.	$\dfrac{s^2+\alpha_1 s+\alpha_0}{s^2(s+a)(s-b)}$	$\dfrac{\alpha_1+\alpha_0 t}{ab} - \dfrac{\alpha_0(a+b)}{(ab)^2}$ $-\dfrac{1}{a-b}\left(1-\dfrac{\alpha_1}{a}+\dfrac{\alpha_0}{a^2}\right)e^{-at}$ $-\dfrac{1}{a-b}\left(1-\dfrac{\alpha_1}{b}+\dfrac{\alpha_0}{b^2}\right)e^{-bt}$

Appendix B

partial-fraction expansion

The response function $F(s)$ for which an inverse transform is desired often appears as a rational fraction that can be represented as

$$F(s) = \frac{N(s)}{D(s)} = \frac{a_m s^m + a_{m-1}s^{m-1} + \ldots + a_1 s + a_0}{s^n + b_{n-1}s^{n-1} + \ldots + b_1 s + b_0} \tag{B-1}$$

The indices m and n are real positive integers, and the a's and b's are real constants. It is necessary that the coefficient of s^n be made equal to unity. In addition the highest power of s in the denominator is equal to or greater than the highest power of s in the numerator; that is, $n \geq m$. Because of the nature of control systems and the excitation functions commonly encountered, the requirement $n \geq m$ is invariably met.

The process for obtaining the inverse transform of Eq. (B-1) may not always be handled solely by direct reference to a table of transform pairs. In this case the equation may be rearranged to a form that may be more readily handled. This form is the sum of partial fractions with constant coefficients. The inverse transform of each of these simple partial fractions can be readily determined from a table of transform pairs. The total time solution is then the summation of the time solutions of the partial terms.

To express $F(s)$ as a sum of partial fractions, the denominator can be factored to give the form

$$F(s) = \frac{N(s)}{D(s)} = \frac{N(s)}{(s + r_1)(s + r_2)\ldots(s + r_n)} \tag{B-2}$$

where $-r_1, -r_2, \cdots, -r_n$ are the roots of $D(s)$ and may be real or complex numbers. Depending upon the form of $D(s)$, the expansion is carried out as follows:

1. $D(s)$ contains only simple roots, none of which are repeated.

The partial fraction expansion can be written as

$$\frac{N(s)}{D(s)} = \frac{C_1}{s + r_1} + \frac{C_2}{s + r_2} + \cdots + \frac{C_n}{s + r_n} \tag{B-3}$$

where the C's are the coefficients of the partial fractions. The typical constant C_k can be evaluated by multiplying both sides of Eq. (B-3) by the denominator of the Kth fraction, namely $(s + r_k)$. Since $D(s)$ contains the factor $(s + r_k)$, the left side of Eq. (B-3) contains $(s + r_k)$ in both numerator and denominator and these terms divide out.

When s is set equal to $-r_k$, the term on the left becomes a constant. All the terms on the right reduce to 0 except C_k. Hence each constant can be evaluated according to

$$C_k = \frac{N(s)}{D(s)}(s + r_k)\bigg|_{s=-r_k} \tag{B-4}$$

This procedure is unaltered if one of the roots is located at the origin.

EXAMPLE B-1

Determine the partial-fraction expansion and inverse transform of

$$F(s) = \frac{s + 2}{s^2 + 4s + 3} = \frac{s + 2}{(s + 3)(s + 1)}$$

Solution

Expanding $F(s)$ in partial fractions gives

$$F(s) = \frac{C_1}{s + 3} + \frac{C_2}{s + 1}$$

Applying Eq. (B-4) to evaluate C_1 and C_2,

$$C_1 = \frac{(s + 2)\cancel{(s + 3)}}{\cancel{(s + 3)}(s + 1)}\bigg|_{s=-3} = \frac{-1}{-2} = \frac{1}{2}$$

$$C_2 = \frac{(s + 2)\cancel{(s + 1)}}{(s + 3)\cancel{(s + 1)}}\bigg|_{s=-1} = \frac{1}{2}$$

Therefore

$$F(s) = \frac{1}{2(s + 3)} + \frac{1}{2(s + 1)}$$

Using Transform Pair 7 in Appendix A, the inverse transform of $F(s)$ is found to be

$$f(t) = \tfrac{1}{2}(e^{-3t} + e^{-t})$$

The solution in this example may be verified by means of Transform Pair 14.

When $D(s)$ contains roots that are complex conjugates, the procedure is no different than that for any other kind of root. However, the evaluation of the C's is somewhat more difficult since they are complex numbers.

For complex conjugate roots of the form $(s + \alpha)^2 + \beta^2$, the partial-fraction expansion can be expressed as

$$F(s) = \frac{N(s)}{(s+\alpha)^2 + \beta^2} = \frac{C_1}{s + \alpha + j\beta} + \frac{C_2}{s + \alpha - j\beta} \tag{B-5}$$

The coefficients are found in a manner similar to that used before.

$$\begin{aligned} C_1 &= \frac{N(s)}{D(s)}(s + \alpha + j\beta)\bigg|_{s=-\alpha-j\beta} \\ C_2 &= \frac{N(s)}{D(s)}(s + \alpha - j\beta)\bigg|_{s=-\alpha+j\beta} \end{aligned} \tag{B-6}$$

The two coefficients are complex conjugate numbers, since $-j$ appears in C_1 wherever $+j$ appears in C_2. The sum of the imaginary parts is 0, and the sum of the real parts is twice the real part of either constant. The two fractions in Eq. (B-5) can be combined into a single term. The inverse transform of this term is a damped sinusoid that may be expressed as

$$f(t) = 2|C_2|e^{-\alpha t} \sin(\beta t + \phi) \tag{B-7}$$

where ϕ = angle of $C_2 + 90°$

EXAMPLE B-2

Determine the partial-fraction expansion and inverse transform of

$$F(s) = \frac{1}{s(s^2 + 2s + 10)}$$

Solution

Expanding $F(s)$ in partial fractions gives

$$F(s) = \frac{1}{s(s + 1 + j3)(s + 1 - j3)} = \frac{C_1}{s} + \frac{C_2}{s + 1 + j3} + \frac{C_3}{s + 1 - j3}$$

Evaluating the coefficients, we have

$$C_1 = \frac{1}{s^2 + 2s + 10}\bigg|_{s=0} = 0.1$$

$$C_3 = \frac{1}{s(s + 1 + j3)}\bigg|_{s=-1+j3} = 0.053 \angle 161.6°$$

Taking the inverse transform of each term and applying Eq. (B-7), the solution is

$$f(t) = 0.1 + 0.106e^{-t} \sin (3t - 108.4°)$$

This solution may be verified by applying Transform Pair 27 in Appendix A.

2. $D(s)$ contains multiple-order (repeated) roots.

If the denominator $D(s)$ contains repeated roots, the procedure outlined above cannot be used directly but must be modified. Consider a function containing repeated roots of the form

$$F(s) = \frac{N(s)}{D(s)} = \frac{N(s)}{(s + r_1)^m} \tag{B-8}$$

The order of $D(s)$ is m, and there are thus m factors. Expanding Eq. (B-8) in partial fractions yields a series of m terms,

$$\frac{N(s)}{D(s)} = \frac{C_1}{(s + r_1)^m} + \frac{C_2}{(s + r_1)^{m-1}} + \cdots + \frac{C_m}{s + r_1} \tag{B-9}$$

The coefficients for the repeated roots are determined by differentiation with respect to s according to the following relationships:

$$\begin{aligned} C_1 &= \frac{N(s)}{D(s)}(s + r_1)^m \bigg|_{s=-r_1} \\ C_2 &= \frac{d}{ds}\left[\frac{N(s)}{D(s)}(s + r_1)^m\right]\bigg|_{s=-r_1} \\ C_m &= \frac{1}{(m-1)!}\frac{d}{ds}\left[\frac{N(s)}{D(s)}(s + r_1)^m\right]\bigg|_{s=-r_1} \end{aligned} \tag{B-10}$$

Where first-order roots are present in addition to repeated roots, their coefficients are evaluated using the same method that has been previously described.

EXAMPLE B-3

Determine the partial-fraction expansion and inverse transform of

$$F(s) = \frac{1}{s(s + 1)^2}$$

Solution

Expanding $F(s)$ in partial fractions gives

$$F(s) = \frac{C_1}{s} + \frac{C_2}{(s + 1)^2} + \frac{C_3}{s + 1}$$

Applying Eq. (B-4), the constant C_1 is found as

$$C_1 = \frac{1}{(s+1)^2}\bigg|_{s=0} = 1$$

Applying Eq. (B-10), the constants C_2 and C_3 are evaluated as

$$C_2 = \frac{1}{s}\bigg|_{s=-1} = -1$$

$$C_3 = \frac{d}{ds}\left(\frac{1}{s}\right)\bigg|_{s=-1} = -\frac{1}{s^2}\bigg|_{s=-1} = -1$$

Taking the inverse transform of each term yields the solution

$$f(t) = 1 - te^{-t} - e^{-t}$$

This solution may be verified by applying Transform Pair 33 in Appendix A to the given function.

List of Symbols

Symbol	Meaning
a	Acceleration
A	Amplifier gain
A	Amplitude
A	Area
B	Damping coefficient
B, b	Feedback signal
C	Capacitance
C, c	Controlled output
C_d	Coefficient of discharge
C_v	Valve size coefficient
dB	Decibel
e	Base of natural logarithms, 2.718
E, c	Error signal
E, e	Voltage
f	Frequency
$f(t)$	General function, time domain
F	Force
F	Friction coefficient
$F(j\omega)$	General function, frequency domain
$F(s)$	General function, Laplace transform
g	Acceleration due to gravity, 32 ft/s^2
G	Strain gage factor
G	Forward transfer function
G_1	Control elements or controller
G_2	System elements or process

Symbol	Meaning
G_c	Compensating elements
G_n	Describing function
G_s	System elements
G_v	Reference input elements
H	Feedback transfer function
H, h	Head
H, h	Height
I	Input function
I, i	Current
J	Mass moment of inertia
j	$\sqrt{-1}$
K	Gain constant
K_a	Acceleration error constant
K_a	Amplifier gain constant
K_b	Motor voltage constant
K_d	Derivative gain constant
K_g	Tachometer constant
K_i	Integral gain constant
K_i	Motor torque constant
K_m	Motor velocity gain constant
K_p	Position error constant
K_p	Proportional gain constant
K_t	Motor torque constant
K_u	Ultimate gain
K_v	Velocity error constant
L	Delay time
L	Inductance

Symbol	Meaning
L	Length
$\mathfrak{L}$	Laplace operator
M	Mass
M, m	Manipulated variable
M, Mag	Magnitude
M_m	Maximum magnitude ratio
M_p	Transient response overshoot
N	Describing function
N	Gear ratio
N	Phase angle
N	Reaction curve slope
O	Output function
P, p	Power
P, p	Pressure
P_u	Ultimate period
Q, q	Electric charge
Q, q	Heat flow rate
Q, q	Volume flow rate
r	Root of equation
R	Radius
R	Resistance
R, r	Reference input
s	Laplace transform operator $= \sigma + j\omega$
S	Specific heat
t	Time
T	Temperature
T	Torque

Symbol	Meaning
U, u	Disturbance input
v	Velocity
V	Volume
V, v	Desired value
V, v	Voltage
W	Weight
X	Reactance
X, x	Displacement
Y, y	Displacement
Z	Impedance

Greek Letter Symbols

Symbol	Meaning
α	Angular acceleration
α	Magnitude ratio
α	Temperature coefficient
γ	Fluid density
Δ	An increment
ε	A small value
ε	System error
ζ	Damping ratio
θ	Angle of rotation
ρ	Resistivity
Σ	Summation
τ	Time constant
ϕ	Magnetic flux
ϕ	Phase angle
ω	Angular frequency
ω	Angular speed
ω_b	Bandwidth
ω_c	Cutoff or corner frequency
ω_d	Damped natural frequency
ω_n	Undamped natural frequency

Answers to Selected Problems

Chapter 1

9. (a) $\frac{C}{R} = \frac{G_1G_2G_3}{1 + G_2H_1}$ (c) $\frac{C}{R} = \frac{G_1G_2}{1 + G_2H_1 + G_1G_2H_2}$

11. $\frac{C}{R} = 1.98$

13. $H = 0.7$

15. $\frac{C}{R} = \frac{G_1G_2G_3}{1 + G_3H_3 + G_2G_3H_2 + G_1G_2G_3H_1}$

Chapter 2

2. (a) $M\frac{d^2y}{dt^2} + B\frac{dy}{dt} + (K_1 + K_2)y = 0$

4. $M_1\frac{d^2y_1}{dt^2} + B_1\left(\frac{dy_1}{dt} - \frac{dy_2}{dt}\right) + K_1(y_2 - y_1) = 0$

$M_2\frac{d^2y_2}{dt^2} + (B_1 + B_2)\frac{dy_2}{dt} + (K_1 + K_2)y_2 - B_1\frac{dy_1}{dt} - K_1y_1 - 0$

7. $\frac{dn}{dt} - Kn = 0$

9. $q_i = \frac{h_1 - h_2}{R_1} + A_1\frac{dh_1}{dt}$

$q_1 = \frac{h_2}{R_2} + A_2\frac{dh_2}{dt}$

11. $RC\frac{dT}{dt} + T = T_w$

13. $\frac{dn}{dt} = -0.4$

15. $\left(1 + \frac{R_2}{R_1}\right)\frac{de_o}{dt} + \frac{e_o}{CR_1} = \frac{R_2}{R_1}\frac{de_i}{dt} + \frac{e_i}{CR_1}$

Chapter 3

1. $e_o = 0.763$ V
3. $\theta_e = 19.76°$
5. 3.28 V/rad
7. 0.5 percent
9. $E_{S_{31}} = 42$ V; $E_{S_{12}} = -42$ V; $E_{S_{23}} = 0$ V
11. $e_o = 0.5$ V
13. $\theta = 120°$
15. $E_{R_{24}} = 11.18$ V; $E_{R_{13}} = 0.31$ V
17. $\theta_e = 3.96$ minutes
19. $\Delta e_o = 0.498$ percent
21. 0.284 in.
23. $\Delta R = 0.24\ \Omega$
25. $e = 10.62$ mV
27. $K_t = 0.4$ oz-in./V; $B = 1.91 \times 10^{-2}$ oz-in./(rad/s)
29. $K_i = 225$ oz-in./A; $B = 3.18 \times 10^{-3}$ oz-in./(rad/s)
31. $C = 0.955\ \mu$farad; $Z = 312 \angle 0\ \Omega$
33. $N = 8$; $\dot{\theta}_L = 1{,}250$ rpm
35. $J = 2.048$ in.-oz-s^2; $B = 12.29$ oz-in./(rad/s)
37. $N = 44$; $T_s = 0.154$ oz-in.
39. $A_p = 551$
41. $e_o = -0.1t$
43. $e_o = \dfrac{Ae_i}{1 + A} \simeq e_i$ for $A \gg 1$
45. $K_a = 625$

Chapter 4

1. (a) $32\dfrac{dv}{dt} + 0.1\,v = 1$ (b) $v_{ss} = 10$

   (c) $v = 10(1 - e^{-t/320})$

3. $\dot{\theta} = 205(1 - e^{-t/0.047})$ rad/s

5. $\dot{\theta} = 19.5(1 - e^{-t/0.00114})$ rad/s

7. $\tau_e = 2.5 \times 10^{-4}$ s; $\tau_m = 2.5 \times 10^{-2}$ s

9. $\dot{\theta} = \dot{\theta}_o e^{-t/\tau_m}$

11. $q_i = 0.2h + 5\dfrac{dh}{dt}$; $h_{ss} = 0.5$ ft; $5\tau = 125$ s

13. $\omega_n = 11.3$ rad/s; $\zeta = 0.071$

15. $\theta_o = \theta_i\left[1 - \dfrac{e^{-\zeta\omega_n t}}{\sqrt{1-\zeta^2}}\sin\left(\omega_n\sqrt{1-\zeta^2}t + \cos^{-1}\zeta\right)\right]$;
$\omega_n = \sqrt{K/J}$; $\zeta = B/2\sqrt{KJ}$

17. (a) $\omega_n = 92.4$ rad/s; $\zeta = 0.115$
(b) $\theta_o = 10[1 - 1.007e^{-10.6t}\sin(91.8t + 1.457)]°$

20. $K_d = 248$

Chapter 5

1. $F(s) = \displaystyle\int_0^{\infty} e^{-(s+1)t}\,dt = \frac{1}{s+1}$

2. (a) $\dfrac{24}{s^5}$ (c) $\dfrac{8}{s^2+4}$

3. (a) e^{-3t} (c) $e^t \cos t$
(e) $1 - 1.202e^{-2t}(\sin 3t + 56.3°)$

5. (a) $\dfrac{1}{s+5}$ (c) $\dfrac{1-2s^2}{s^2(s^2+3s+1)}$

7. (a) $y = 0.2(1 - e^{-10t})$ (c) $q = 3 - 0.5e^{-100t}$

9. $i = \dfrac{E - V_o}{R}e^{-t/RC}$

10. (a) $\dfrac{s}{s + R/L}$ (c) $\dfrac{1 + R_2Cs}{1 + (R_1 + R_2)Cs}$

11. $X(s) = \dfrac{(s+1)X_o}{Ms^2 + Bs + K}$

13. $\dfrac{1}{s + 1/RC}$

15. $\dfrac{\dot{\theta}(s)}{V_f(s)} = \dfrac{2{,}000}{(s + 10)(s + 3.33)}$

17. $\dfrac{K_a K_m}{Js/B + 1 + K_a K_m K_g}$

19. (a) $K_p = \infty$; $K_v = 10$; $K_a = 0$
Step error = 0; ramp error = 0.1; parabolic error = ∞

20. (a) $s + 10 = 0$ (c) $s^3 + 2s^2 + 5s + 5 = 0$

21. (a) Stable (c) Marginally stable (f) Unstable

22. (a) Stable (b) Unstable

23. $0 < K < 8$

26. $K = 31$; $\omega_d = 5.20$ rad/s

Chapter 6

1. (a) $5 \angle 53.1°$ (c) $8.07 \angle 132°$

2. (a) $5 \angle -90$; $0.5 \angle -90$
(d) $0.49 \angle -101.4°$; $0.022 \angle -153.4°$

5. (a) 2 octaves (b) 4 decades

6. (a) 6.02 (c) 18.06 (f) 32.0

11. (a) $\dfrac{4}{1 + 0.01s}$ (c) $\dfrac{10(1 + 10s)}{(1 + 20s)(1 + 2s)}$

13. $\dfrac{1}{(1 + j10\omega)(1 + j100\omega)}$

15. $\dfrac{\dot{\theta}(j\omega)}{V_f(j\omega)} = \dfrac{200}{(1 + j0.1\omega)(1 + j0.008\omega)}$

17. $\dfrac{H(s)}{\theta(s)} = \dfrac{2}{(1 + 0.75s)}$

21. $M_m = 1.24$; $\omega_m = 1.73$; $\omega_b = 3.07$

24. $\dfrac{C(s)}{R(s)} = \dfrac{1}{1 + s/K + 0.1s^2/K}$

$K = 10$ for $\zeta = 0.5$

Chapter 7

1. $K_v = 10$; 6 dB; 18°

3. $K_v = 10$; 7 dB; 34°

7. (a) $K_v = 7$; 1 dB; 4°
(b) $M_m > 12$ dB; $\omega_m = 3.5$

8. $$G_c(j\omega) = \frac{1 + j10\omega}{1 + j60\omega}$$

9. $$G_c(j\omega) = \frac{5(1 + j10\omega)}{1 + j50\omega}$$

11. $$G_c(j\omega) = \frac{1 + j3.74\omega}{1 + j0.27\omega}$$

15. $$\frac{R_2 + R_3}{R_1}\left[\frac{1 + j\omega CR_2R_3/(R_2 + R_3)}{1 + j\omega CR_3}\right]$$

17. 2.27°F

18. $\Delta T_o = -0.32e^{-t/31}$°F

20. (a) $$\frac{s + 5}{s^2 + 6s + 25}$$ (b) 2°F

22. $G_c = 1/G_2$

Chapter 8

1. $$G(j\omega) = \frac{1}{1 + j\omega/6.28}$$

3. $$\frac{\dot{\theta}(s)}{V_f(s)} = \frac{500}{(1 + 3.3s)(1 + 0.29s)}$$

5. (a) N = 15°F/min (b) $K_p = 1.33$ mA/°F
(c) $K_p = 1.20$ mA/°F; $T_i = 1.67$ min

7. $K_p = 60$ percent/psi; $T_i = 0.05$ min; $T_d = 0.0125$ min

Chapter 9

1. $\Delta p = 2Kq_o\,\Delta q$

3. 57 percent

4. $G_n = 0.72$

6. (a) Stable oscillation (b) Unstable oscillation

7. Stable at $\omega = 30$; unstable at $\omega = 10$

Index

D

J

K

L

M

N

O

P

T

U

V

Z